THE
FIRE
WITHIN
THE
EYE

THE FIRE WITHIN THE EYE

A HISTORICAL ESSAY ON THE NATURE AND MEANING OF LIGHT

David Park

PRINCETON UNIVERSITY PRESS

PRINCETON, NEW JERSEY

LIBRARY OF CONGRESS CATALOGING-IN-PUBLICATION DATA

PARK, DAVID ALLEN, 1919–

THE FIRE WITHIN THE EYE : A HISTORICAL ESSAY ON

THE NATURE AND MEANING OF LIGHT / DAVID PARK.

p.   cm.

INCLUDES BIBLIOGRAPHIC REFERENCES AND INDEX.

ISBN 0-691-04332-9 (cl : alk. paper)

1. Light—History. 2. Optics—History. I. Title.

QC352.P34 1997   96-45573

535—DC21

THIS BOOK HAS BEEN COMPOSED IN GALLIARD TYPEFACE

PRINCETON UNIVERSITY PRESS BOOKS ARE PRINTED ON ACID-FREE PAPER,

AND MEET THE GUIDELINES FOR PERMANENCE AND DURABILITY OF THE

COMMITTEE ON PRODUCTION GUIDELINES FOR BOOK LONGEVITY OF THE

COUNCIL ON LIBRARY RESOURCES

PRINTED IN THE UNITED STATES OF AMERICA

BY PRINCETON ACADEMIC PRESS

1 3 5 7 9 10 8 6 4 2

For Jessica, Katharine, Paul, Rachel,

THE GOOD COMPANIONS

CONTENTS

WE PERCEIVE light with our eyes; we measure it with our instruments. Still, if we say that light dawned, we may mean the sun has come up, but we may also mean that someone has finally been penetrated by an idea. "I see" may refer to my eyes, or may mean that I understand. The word "vision" suggests eyes again, or some kind of apparition, or else a mental project for a better world. Light is life; darkness is death. Can we toss these meanings aside as nothing but metaphorical ways of speaking? Of course, but they were not always metaphorical, and they throw some light on the questions we are coming to. When the Bible says the world began with a great flash we may imagine, as some have done, that this was so God's heavenly host could see and understand what was going on, but there is also a suggestion that light itself is an instrument of God's power. When it says that "God is light and in him is no darkness at all," we may see a metaphor, but I shall argue later that the author of these words may have intended them as a plain statement of fact. There were times when understanding and vision, or goodness and light, were so close that you could not slip a straw between them, but the history of thought resembles the astronomers' expanding universe. As our mental cosmos widens, ideas once closely related spread out and become isolated and lose all connection except, perhaps, what is remembered in a few words. Medicine once went hand in hand with astronomy; now little remains of their relation except the term "influenza," which refers to heavenly influences.

If we start the history of light where we must start, with the experience of it, and ask what thoughts have arisen out of this experience, the subject becomes vast and impossible to focus. I will therefore deal mostly with questions concerning the nature of light and vision, understanding that in the beginning, and for a long time afterward, it is impossible to consider these questions apart from other strands of meaning. The search for light's nature leads to questions about vision: How does it happen that when we open our eyes our minds are filled with images of things close and far away? Why must there be light before we can see? For a thousand years people debated what the visual message is: What is it that carries the message, and does it move from the object to the eye or issue from the eye to play on the object? Either way, how do we manage to fit the image of a mountain through the little hole? What is color, and how do we distinguish one color from another? These and other questions have histories that are all related and are parts of this book.

What, finally, is what we call light? It brings life to the world around us, and it is the means by which most of us see and act. Yes, but what actually *is* it? Even this is a historical question, and if I were to jump the history and present a lecture on modern physics I would not be answering it. You will get a little of the lecture at the end, but light is enfolded in our words, our habits, our mental image of the world, and as that image changes, light changes with it. We shall watch as sensory knowledge, slowly accumulated, leads to abstract speculations and finally to a scientific knowledge of light that transcends experience. Does that mean we understand it better than we did at the beginning? You judge.

This is supposed to be a nontechnical book. It is written for readers who have no special interest in knowing how a laser works, as well as for those who have. (All the same, you will briefly be told.) There is no mathematics, and I hope that none of the illustrations need be puzzled over. It is a history of thought about light, rather than the story of a gradual climb toward truth, in which I have tried to present ideas as they presented themselves.

Leading ideas start growing, often taken for granted and hardly noticed. They develop slowly; after a while they age and die. How they start depends on soil and climate; I shall digress to describe them. Creating digressions has been part of the pleasure of writing this book, and I hope you will not be impatient with them. Some are in the form of potted history, since I find that I cannot enter the thought of a long-dead thinker if I know nothing of the time and society in which he lived. The "he" is intentional here. Throughout the writing I have been uncomfortable with the insistent repetition of this pronoun. It accuses me: are you not following a tradition that erases the contributions of women from the history of thought? Yes, but not blindly. I can work only with the written record, alert to the possibility of erasures.

The written record (and my ability to read it) also determines the boundaries of the space and time surveyed. I start with Presocratic Greece because the record of speculations, faulty though it is, starts there. Roughly, the trajectory is from Greece to the Near East and Africa to Western Europe and America. In order not to load you with a long list of names to remember I have written about some influential thinkers and omitted many, such as Averroes and Bishop Berkeley, whom you might have expected to find here.

The decision to focus on thinking more than on results has led to a book whose shape at first appeared strange, for the twentieth century and its immense accumulation of knowledge occupy little more than a chapter. In the first place, much of that knowledge is at a level of abstraction and technical complexity that would be out of place, but more fundamentally, ideas ripen slowly, and my impression is that the twentieth century has been not much richer in basic scientific insights than the three that preceded it. Also, of course, we who live in this century do not need to have its leading intellectual assumptions explained to us.

A note on sources. I have been guided by two authors who have worked long and hard on the history of optics: Vasco Ronchi, in whose *The Nature of Light* [1970][1] I first found a general map of the territory I must traverse.

And David Lindberg's works on ancient, medieval, and Renaissance ideas of vision and light have been a model which, at a respectful distance, I have tried to follow.

A note on translations. I have been pedantic about including the Greek of important philosophical terms, for different translators have their own English versions of them. In general, I have compared existing translations with the original and made my own where needed. Old writers used a lot of what we would call extra words, and faithful translators tend to keep them. Hoping that I nowhere changed the sense and apologizing if I did, I have cut some of these words from my quotations, but all such changes are mentioned in the "References and Further Reading" section at the end of the book. At the end is also a glossary of technical terms in optics and philosophy that may be helpful.

I have profited from many conversations with Clara Park, and from criticism by Karen Kwitter, Katharine Park, and Kay Yandell, to whom my most devoted thanks. And thanks also to Ilse Browner for helping with the proofs.

---

[1] The date in square brackets means that the work is listed under Ronchi's name in the Bibliography.

THE
FIRE
WITHIN
THE
EYE

Light in Antiquity

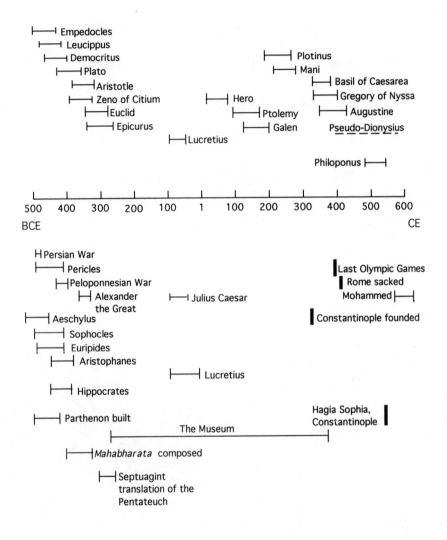

1

THE POWER OF LIGHT

At the very beginning the King made engravings in
the supernal purity. A spark of blackness emerged in the
sealed within the sealed, from the mystery of *En-Sof*, a
mist within matter, implanted in a ring, no white,
no black, no red, no yellow, no color at all.
(*Zohar* 15ª)

1. Prospectus

I CAME OUT of the darkness into my dark house and switched on the light. Instantly, I saw in front of me the fireplace with a tall old chair beside it. I saw the chair; that is to say, if someone wants to argue about it, I saw a familiar pattern of lines, colors, and textures that I recognized as the chair. Light came into my eyes, the chair came into my mind. But why the chair? Why not an aardvark or a bicycle? Try again. Not just light but an image carried by light came into my eyes; then something happened, and not just a chair came into my mind but a particular chair, located about ten feet in front of me. Perhaps it is not obvious how all this happened, or how it could happen so quickly.

We can study the mystery of vision most efficiently by breaking it down into small questions: What is light? Where does it come from? How does it carry an image into my eye? What is an image, anyhow? What happens inside an eye? Is an eye something like a camera, or is it more of a cerebral mechanism? What happens in my brain when my eyes see something? And finally, when it has happened, how does that chair get into my consciousness, my mind? But this is the modern style; by the time I have asked these questions I have already done much of the thinking that took twenty-five hundred years. Thought is integral. It grows in open air and is packaged later. Experience comes first: switch, light, see chair, all at once. Then, in an inquiring mind, a sense of curiosity: how did this happen?

The pages that follow will tell in a leisurely way how we arrived at our modern questions about light and vision and will suggest some answers. They will show, for example, how some people argued cogently that what passes through the pupil of the eye travels outward, not inward, while others maintained for centuries that nothing travels at all. Those ideas are no longer current, but I will have failed in my purpose if they seem to you absurd. And even if they seem a bit naive, so will our best thoughts, no doubt, after we have gone.

But why ask such questions at all? Well, for one thing, you can't hope to fit people with glasses or help their ophthalmia or make a telescope if you don't know what is going on, but behind that there stands another reason, the one Aristotle puts at the start of his *Metaphysics*: "All men naturally desire knowledge." We specially value knowledge of things that are important to us, as means to wise action and inner satisfaction. Then Aristotle continues, "Above all the other senses, sight helps us to know things and reveals many distinctions." Light enables vision and vision enables knowledge and action; it represents creative power. For minds touched by the Bible, the command "Let there be light" was the beginning of everything. It was not just that one could now see; light somehow energized the act of creation that transformed a dark, wet, chaotic half-existence into sea, land, a sunlit garden, a man and a woman.

The history of light is a long story which I intend to make longer still with many digressions. The first few chapters will be concerned with light as a means of seeing and knowing and as active power. We begin with some historical background, together with basic philosophical notions that for many centuries guided both questions and answers. We will see how light came to represent in people's minds the truth of an ideal world of which ours is only a moving model, in which the words "God is light" are a simple statement of fact. We will also have to discuss the physics of light. This starts early but will not become a dominant theme until the second half of the book.

Most of the ideas we shall first encounter originated, as far as we know, among the thinkers of the ancient Greek world. At least, they have come down to us with Greek names attached to them. The next few sections introduce some of the personalities and the ideas they put forth to explain both vision (their primary concern) and the nature of light. Out of this it will begin to become clear what the ancients thought was important in their philosophy of the created world.

2. The Roots of Matter

The name of Athens shines in history. Twice she led other nations in the defense of Greece when the Persians tried to conquer it. Later Athens became very rich and attracted writers and artists. She did not, however, produce many of them, or many of the thinkers who made the name of Greece immortal. They tended to come from the periphery, the shores of the Eastern Mediterranean, the Aegean, and the Black Sea.

From about 600 B.C.E. certain men around this rocky perimeter were identified by their reputation for wisdom. They were not professors like Plato and Aristotle, who lived two hundred years later. Many were civic leaders who had proved they were smart by solving political or military problems or by getting rich, or all three, so that people paid attention to what they said. They discoursed about the world, about Nature and the great questions of human life. Some wrote books, all of which disappeared after a while in the upheavals that destroyed the libraries of Athens, Pergamon, Alexandria, and smaller centers. We have only fragments of those books, chosen by later writers to illustrate some point of vocabulary or grammar or, more often, an opinion. We should never read one of these fragments in the belief that it is what the author actually wrote. But such as they are, by the beginning of the twentieth century scholars had collected them, arranged them, translated them, and begun to theorize about the intellectual culture from which they arose. It is dangerous work. From the fifth century on, we know much about the Greek world. For earlier times, the history that has been put together could be significantly changed by the discovery of only a few new fragments.

As far as we know, the first place where wise men achieved a special status was the city of Miletus, on the west coast of Asia Minor south of Ephesus. To the east was the kingdom of Lydia, which at the beginning of the sixth century was absorbed by the Persian empire. Nothing is left of Miletus now but a huge Roman theater and the remains of a public square paved in marble, but in its prime it controlled most of the trade along the northern coast and into the Black Sea, and in the sixth century it was the richest city of the Greek world. It was perhaps here that natural philosophy first caught the interest of the educated public. This is the philosophy that looks upward and asks questions about stars and clouds and winds and weather, and downward at things on earth, and wonders what they are made of and what principles explain the ways in which they change.

We live in a world of huge variety. Look around at the different things you can see, various solid materials, perhaps some liquids, and of course there is the air, which you feel but cannot see. Materials change. Living things grow and die, wood rots, iron rusts, liquids evaporate. Is there anything we can say about all these materials and processes in general? Are the various substances fundamentally different and continually changing their nature, or are they manifestations of some principle that does not change? A man named Thales, one of the first Milesian philosophers, suggested that water is the fundamental principle, though we do not know how he went on from there. Aristotle reports the suggestion and explains that for Thales Nature is life, and life needs water. What Thales proposed can be judged a bad guess if you wish, but the question he asks is not trivial, and physicists have pursued it ever since. At present they find themselves in a cloudy world inhabited by particles with names like quarks and gluons. Questions like "What is light?" and "How do I see the world around me?" must extend back into the night of time, but the first record of them comes from Miletus.

Other thinkers continued the Milesian game. I shall mention only the three who contributed theories of light and vision: Empedocles, Leucippus, and Democritus.

Empedocles (c. 495–435 B.C.E.) lived in Akragas, a large city on the south coast of Sicily that is now known as Agrigento. Pindar, who also lived there, called it the fairest city of men, and its massive temples of yellow limestone still stand among almond groves along the shore. Empedocles was well known, a leader of the colony's democratic forces, an orator, and tradition says he was also a doctor. He taught that matter is eternal and changes only its forms: "Before now I was born a boy and a maid, a bush and a bird, and a dumb fish leaping out of the sea." He wore fine clothes: "a purple robe and over it a golden girdle, . . . slippers of bronze and a Delphic laurel wreath." He glittered as he addressed his followers or walked the streets, and when he wrote down his philosophy it was in stately hexameters. Thales had one fundamental material principle; Empedocles has four, which eternally unite and separate and move about as the world goes its way. They are not substances, not things; he calls them *roots* and names them after figures from mythology: Zeus, Hera, Aidoneus (i.e., Hades), and a weeping nymph, Nestis. In a sense these are the elements of matter, and Aristotle, writing a century later, calls them Fire, Earth, Air, and Water. Nobody knows for certain which was which, but Nestis doubtless represented Water. The third-century (C.E.) biographer Diogenes Laertius says that Zeus is Fire and Hera is Earth, but whatever they were, the old writers understood that they are modes of existence, not substances, and that by themselves they have no properties. Because of their abstract nature I write them with capital letters. Earth is not dug with a spade, Fire is not hot, Water is not wet. We never encounter them in pure form because they represent extreme tendencies—Fire tries to move upward, Water and Earth try to move downward, Air tries to move sidewise, and if there is any hope of explaining the immense variety of natural substances it must be in terms of balances between contrasting principles.

Things made from these roots do not just sit there. Empedocles sees that change is everywhere and nothing lasts forever, so he introduces two further principles, Love and Strife. Love draws the roots together, and Strife tears them apart. In Nature they produce fixity and change; what we know as birth and death are only rearrangements. Concerning light, Aristotle says Empedocles taught that it travels at immense speed "between the Earth and its surrounding space,"[1] but he mentions him only to contradict him, and this is all we know. We shall discuss Empedocles's theory of vision in Chapter 2.

---

[1] Literally, the text says "the envelope," not surrounding space, but because Aristotle had no way of talking about empty space, or perhaps even imagining it, he always referred to space in terms of its bounding surface. See Park 1988, p. 226.

3. Atoms and Emptiness

Miletus was conquered and destroyed by King Darius of Persia in 494 B.C.E. It lost its rich hinterlands but started to rebuild itself as destroyed cities usually do, and a child named Leucippus was born there about twenty years later. By the time he grew up the city had recovered enough so that people had time to listen to a very original thinker. We know little about him, but Aristotle, though he did not believe in atoms at all, credits Leucippus with inventing them. Atoms represent another attempt to identify a permanent reality that underlies and explains the changing world. Leucippus's atoms are too small to see, infinite in number and in the variety of their shapes and sizes, and they move in empty space. We shall see in a moment why atoms have no place in Aristotle's philosophy of substance and change, but the particular point on which he concentrates his objection is empty space. The trouble really is that there was no way of talking about it that made sense. The atomists used the word *kenon*, meaning empty, but this is an adjective. They could of course say "the empty," and they did, but the phrase acquired the sense of "nonbeing," and then it was hard to answer questions like "Exactly what is nonbeing?" or "Would you please point to an example of it?" Today we would say they were talking about space, but they had no word for that. (The situation was similar in Greek arithmetic, which used a clumsy system akin to Roman numerals, apparently because a positional notation like ours or the Babylonians' requires a zero, a symbol that represents nothing at all.) Leucippus did not try to define the empty, causing critics to judge that he was a very poor philosopher, but imagined atoms moving in it, combining and separating so as to create the substances and the processes that surround us. They follow no plan, and whatever is combined today will some day separate. Atoms do not go where they want to go or where the gods want them to go. Instead, they follow their *logos*, which can be thought of as a necessity arising from the order and fitness of things. The only fragment we have from Leucippus says "Nothing happens at random; everything happens out of reason and by necessity," and this is the first suggestion that at some level of its structure the behavior of matter is governed by laws that are not violated. I used the word "law," but that metaphorical comparison with a legal statute came later, as did the notion of inviolability.

The second atomist was Democritus, a few years younger than Leucippus. He lived in Abdera, on the north shore of the Aegean near the border of Macedonia and is said to have traveled widely, perhaps to the Persian empire to learn astronomy from the Chaldeans, perhaps as far as India. One may suppose he visited Leucippus in Miletus.

Democritus saw some of the deeper implications of atomism, in particular how the theory deals with questions concerning what is really true. Suppose I sit in a room, reading. You come in from outside and remark how dark it is inside. It doesn't seem dark to me. We may both understand perfectly well why we don't agree, but how to explain the situation in exact language? —Even at this early time people wanted to use words exactly. What is the room, really: light, dark, or neither? Democritus answers by saying that questions like this are not questions of fact. He does not mention light, but "sweet exists by convention, bitter by convention, color by convention; in reality there are only atoms and emptiness" (fragment 9). Because this is an important point, let me make clear by an example what he is saying. The letters C-A-T do not look like a cat. The word "cat" does not sound like the noise a cat makes, but when we see the letters or hear the word we know the conventions that tell us what is meant. Long afterward, the German physicist and physiologist Hermann von Helmholtz wrote, "Our sensations are for us only *symbols* of the objects of the external world, and correspond to them only in some such way as written characters or articulate words correspond to the things they denote."

We cannot know atoms or emptiness, and if what we think we sense is only a convention, what can we know? Democritus continues, "We know nothing in reality, for truth lies in the abyss" (fragment 117). By now, twenty-five hundred years later, we know something about how to explain our sensations, but as to the truth about how things actually are, Democritus speaks for us all.

The atom was a stupendous invention, but it was only one small item in the Mediterranean supermarket of ideas. Plato bought it, but in *Timaeus*, the book where he discusses it, he reinterprets it in his own comprehensive system until it is almost unrecognizable. In Aristotle's metaphysics, on the other hand, a thing's properties belong to it as a whole, so that he cannot use the idea at all. Atoms mostly sat on the shelf until there was use for them in someone's philosophy. To tell that story I will jump a century, over Plato and Aristotle, who have little to do with it, and return to them shortly.

Epicurus (c. 342–270) was born to a prosperous Athenian family on the island of Samos, not far from Miletus. Like other young men of his class he studied and traveled, and in 307 he came to Athens, where he founded an institution that was something between a school and a social club. It met in his house and garden, where his followers, men and women and even slaves, sought a revivifying philosophy as Athenian greatness in art and political power faded into twilight. Epicurus tried to make his pupils see that fate and gods are not to be feared, that the gods are content and not vindictive, that enlightened people should ignore the supernatural and concentrate on living a life that is honest and frugal, serene and happy. Much of his argument rested on the teachings of Democritus.

How does Democritus enter this peaceful garden? His atoms and things made of them are impersonal, ethically neutral, and unaffected by supernatural forces. If there is a drought it is because atoms of wind have blown away atoms of cloud, and not because a god is angry. If something goes wrong you should work to make it better instead of killing a sheep. The simple machinery of atoms explains why real life has nothing to fear from malice on Olympus. It is not that gods don't exist, only that their existence is of a higher order. They are not concerned about us and will not bother us. "It is impossible," he writes, "to dispel our anxieties and enjoy our pleasures if we do not understand the nature of the universe and if we allow things learned from myths to frighten us."

The world runs by the operation of impersonal forces, but do we? That would be going too far, for there is no use in teaching a rule of life to someone who cannot freely choose to follow or reject it. And besides, we all know how it feels to make a free choice between alternatives. Epicurus uses rather strange terms to explain how this happens. He says that atoms have two kinds of motion: free fall, and motion imparted by a collision. Both, presumably, are rule-governed, but from time to time there is a third kind, a "swerve" that allows atoms to click together forming associations that would not otherwise have happened. These random events break the causal chain, but how he thought they allow free will we do not know, for his book *On Choice and Avoidance* is lost. All we have is a mysterious explanation given by the Roman poet Lucretius in his epic-length poem *De rerum natura*, or "Concerning the Nature of Things," written about 50 B.C.E. This is an account of Epicurean philosophy in more than seven thousand hexameter lines, clearly designed for immortality, and its fundamental purpose is to teach that Nature is impersonal and gods are not to be feared. We are lucky to have a very fine verse translation by Rolfe Humphries [1968].

Lucretius writes that particles go where they are sent, but "now and then they move a little from their path, barely enough so that one could say they are deflected." These are the words that promise each of us a certain freedom from causal constraints: "How does our will escape the chains of fate—this will through which we move forward as we please, swerving in our motion not at determined times or places but where our mind carries us?" Though our bodies are made of atoms we are not mechanical dolls. We make our own choices.

There is no occasion here to say more about the poem, and there can be no substitute for reading it, but I will pull out of it a few physical insights that have some bearing on light. First, the reality of atoms. Everything we have from Epicurus is in general terms, but a poem is best when it is specific. Lucretius gives examples of phenomena that atoms explain: the force of wind, the diffusion of a scent, and the drying of clothes on the line. Either the water departs a wet shirt invisibly, particle by particle, or else it simply ceases to exist, but Lucretius will not allow the latter. Nothing, he says, comes from nothing or vanishes into nothing.

Lucretius also gives some explanations that have not endured: he has atoms of heat, cold, and sound, but these sensations are now explained as the result of atomic motions and are not themselves atoms. From the properties of matter he deduces that the atoms of hard substances are firmly hooked together whereas those of a liquid are round and roll against one another. The atoms inside a lemon have sharp points. We can hear what is happening in the next room; therefore the particles of sound are small enough to go through a wall. Intelligible speech consists of a sequence of sounds, and so we must imagine a stream of these little particles, all different. The particles of light can pass through transparent substances. But now a difficulty arises. When we look at something we do not see it in temporal sequence as when we hear words; rather, we see it all at once, and so the thing that comes to our eyes must have some special character that makes it different from sound and allows it to create a picture in our mind that is different at each moment. But what is this thing? How Epicurus and Lucretius answered that difficult question will be told in Chapter 2 when we discuss theories of vision.

Outside a small circle, people greeted the philosophy of Epicurus with the special rage and scorn reserved for those who question prevailing beliefs. References in later philosophy are almost all contemptuous. We rarely use the term "epicure" to praise someone. The Stoic philosopher Epictetus, writing shortly after Lucretius, summarizes what most people thought of the Epicurean principle in a single sentence: "Eat, drink, and satisfy your passion for women; relieve yourself, and snore." Perhaps he is right; nowhere in Epicurus is there any mention of the tears and promises with which we try to change our bad ways and make ourselves better than we are.

Lucretius does not speak of gods or suggest any general rule for living, but particles swerve and we are not bound by fate, and the last two books of *De rerum natura* are devoted to simple, causal explanations of events that were then, and are still by many people, regarded as signs of wrath in Heaven: eclipses, thunder, lightning, waterspouts, earthquakes, and volcanoes. Not all the explanations still fly, but the principle does: exhaust every natural explanation before turning elsewhere.

Empedocles, Leucippus, and Democritus were all alive during the Periclean age, the middle of the fifth century. At the same time, in Athens, the four playwrights Aeschylus, Sophocles, Euripides, and Aristophanes were at work, the Parthenon was going up, and Socrates walked the marketplace. And as that shining age drew to a close, Plato was born.

4. Plato's Philosophy of Ideas

Every town laid out before the modern age has an open place, *agora* in Greek, *forum* in Latin, where markets were held and public business was transacted. In Athens, most of the time, the *agora* must have looked like a county fair in an old Currier and Ives print. The ground was dusty or muddy as the case might be, and most of it was covered with wooden booths where shopkeepers sold the things city-dwellers need. That is how it looked. Ordinarily it was also full of sound, for it was where men met to argue about people and news and politics and everything else in their world. Before printing, there was the voice. In the *agora* teachers of eloquence and argument, called sophists, attracted students by offering to uphold either side of a vexed question. Politicians made their noise, and in the crowd an ill-favored man with snub nose and bulging eyes, shabbily dressed, barefoot in most weather and generally regarded as a quibbling bore, joined group after group, listening for a while and then upsetting the easy flow of platitudes by asking exactly what they meant.

Sometimes Socrates was the center of his own circle, which Plato (427–347) joined in his twentieth year to discuss questions that are simple to ask and hard to answer. What is virtue? What is justice? It is easy enough to give examples of each, or of their contraries, but the questions remain. Each example of justice is unlike the rest; the more examples you give the harder it becomes to say what they have in common. It was the same question, in the moral order, that Thales had asked about matter. Examples are not enough, said Socrates, and the question is important, for only he who knows what justice is can be just. Then what is it? First of all, it is a word, needed because we need to capture thoughts in language so they can be understood and discussed. But a word ought to mean the same thing every time; it must refer to something. What does it refer to?

Before people awoke to questions like this, the specific was all. Aeschylus and Sophocles had put questions of justice and duty at the center of dramas depicting the world of myth, but myth is always specific; it describes the individual case and leaves us to deduce the generality. Myth begins the language of metaphor.

Philosophers before Plato, trying to express their ideas about nature, faced similar problems, and they tended to use terms such as Empedocles's four roots with mythological names, controlled by Love and Strife. Plato was of the last generation of Athenians who were at home in myth and familiar with its language. He understood that leaders of a modern state must find some other way to talk, but still, the old one was so natural to him that when he tried to explore the true meaning of words he tended to start with a myth.

The dialogues Plato writes usually have Socrates in the middle of them, arguing, insisting, changing tactics, laughing, walking away, but the upper regions of his talk are probably all Plato's. Imagine a realm of existence, Plato sometimes calls it the Heaven above Heaven, which never changes. It is populated by what he calls *eide*, translated variously as Forms or Ideas, one *eidos* called Virtue, another called Justice, another called simply the Good, and so on. Perhaps every word which in our world denotes a definite concept has its counterpart in that Heaven of Heavens. The world we live in is an imperfect version of that Heaven, imperfect because things here change all the time, and the examples of virtue and justice we encounter are never the same, never quite pure, never unmixed with their opposites. But still it is possible for a prepared and properly educated person, from time to time, to pierce through the barrier between the two worlds and discern the reality that shines above our cloudy air. In his effort to achieve consistency, never successful, Plato at one point tries to furnish his Heaven of Heavens with the Ideas of beds and tables, adding, as Aristotle says, that of course the objects around us are different from the ideal ones from which they get their names, but "it is because they participate in an Idea that objects of the same kind have the same name." A critic might say that Plato was having problems with words, that instead of solving them the Ideas raised more and worse problems; later he retreats and is not sure that he can have any material objects in Heaven without, as he says, "tumbling into a bottomless pit of nonsense."

A person about to recite a myth does not announce "I am going to tell you a myth." Instead, a story begins. Presently it reaches its end and stops, but myth never exhausts its meanings; that is why it endures. Plato did not present his mythical world of Ideas as a consistent doctrine that hearers were supposed to accept or reject; it was a way of talking about problems important to him and his friends and followers. Plato's myth is not important for the theory of light and vision to be described in the next chapter, but its influence on later thought through figures like Plotinus and Saint Augustine will become obvious as we go along. Myths in Plato's *Phaedo* and *Republic* speak to his readers through their imaginations and remain in their minds as pictures. One in *The Republic* is essential to our story. It is known as the Myth of the Cave, and I will sketch it because it provides a picture that has helped many generations to think about the world of Ideas.

Imagine, says Socrates, that there are men imprisoned in a cave with a low opening to the outer world along its side, but the prisoners are chained so that though daylight is at their shoulder they cannot see it. A fire is kept lighted behind them and in front of it people pass back and forth carrying various objects, so that all the prisoners can see is moving and flickering shadows cast on the wall in front. These shadows, being all they can see, define the world of their words and thoughts, so that when they believe they are naming a thing they are naming only its shadow. But suppose one of them were suddenly freed and made to look out at the real world beside him. His eyes would be dazzled and hurt by the light, he would see nothing familiar, nothing he saw would make any sense, and he would want to sit down again and face the familiar shadows. Only gradually, taken from the cave, would he learn how the world really looks and what is really in it.

The soul, says Socrates, is like an eye. When it looks toward the Heaven of truth and beauty it knows them and understands them, but when it looks into our world of becoming and passing away it has no knowledge but only opinions that change and change again. The Good is to the soul as the sun is to the eye, and Plato carries the analogy further: just as the sun sustains all of Nature's life and growth, so the Good, through its action in the ideal world, is the cause of all things in our world; "it causes their state of being, though the Good itself is stronger and more venerable than being."

Let us try to read these words not just as a metaphorical statement of Plato's view of the Good but, as many read them afterward, an inspired utterance that explains why the world exists and opens the Good to human understanding. We are not to take it literally but rather to look for the deeper truth it expresses. If we do this, we may be led to ponder what it implies about the Nature and powers of light. If the Good is the highest element of the Ideal world, fructifying every other form of existence, does it not seem to follow that the sun occupies an analogous position in our own world? If so, its powers extend as far beyond its effects on the eye as the powers of the Good extend beyond ordinary goodness. This was how it seemed to many of Plato's followers, and even centuries later we shall find them attributing powers to light far exceeding those suggested by experience and common sense.

Alfred North Whitehead says somewhere that all later Western philosophy can be read as footnotes to Plato. Though this need not be seen as a desperately serious remark, it at least describes the historical weight and persistence of Plato's words. His successors molded his ideas and adopted them to their own purposes and in doing so kept them alive. Plato isolated the Good in a domain inaccessible to ordinary experience, but Platonists were followed by Neoplatonists. In Egypt, six hundred years later, one of them named Plotinus saw Good pervading the universe, emanating from the world's unitary principle, the One, in a sort of radiation like the sun's. The One is the origin of all being, but it cannot be analyzed or described, because to analyze is to divide into parts, and even a description would tend to focus on one or another aspect of something that does not have different aspects. It brings warmth and life, and at the same time it manifests the world's goodness and intelligible order. Its radiation spreads perfection over every level of existence; its power is both physical and ideal, and the light by which the sun pours its blessings onto the Earth shares in that power.

Dante imagines this light in Book XXVIII of the *Paradiso*. He stands at the edge of the Empyrean sphere, from which

> I saw a point that shone with light so keen,
> the eye that sees it cannot bear its blazing;
> the star that is for us the smallest one
> would seem a moon if placed beside this point.

The light is spiritual light, and the point is a spiritual point; but it is a point, for before Dante, Saint Thomas wrote that God is perfectly simple, he is not mixed with anything else, he has no parts, and a point, as Euclid defined it long ago, is that which has no parts.

Neoplatonic doctrines flowing from Plotinus's luminous vision spread over the Greek world. Later on in this book we shall find them in early Christian philosophy exemplified by the Gospel of Saint John and in the early Church Fathers' explanations of the command "Let there be light." In the ninth century, Arabic translations of Plotinus and his successors and commentators opened a channel through which Neoplatonic thought flowed into Mesopotamia and North Africa. We shall find it in the writings of Alkindi of Kufa, whose work, appearing in Latin in the twelfth century, inspired the Englishmen Robert Grosseteste and Roger Bacon and their followers to think of light as a physical power. Finally, in sixteenth-century Italy there was a resurgence of Neoplatonic philosophy, almost unrecognizable under a veil of magic charms and delightful ritual, that started when Marsilio Ficino translated Plotinus. We shall touch some of these branches of Plato's great tree but only touch them; to go farther would need many chapters and someone who has read more than I have.

5. Aristotle's Philosophy of Desire

In the western suburbs of Athens there was a grove of trees adorned by fountains and walks; near it an old gymnasium. The place was called *Akademe*, and for many years men had taken study and exercise there. About 387, Plato established a school which endured on this spot, with some hiatuses and many changes of direction, for some 925 years, until Justinian closed all the pagan schools. This was the world's first institution of higher learning; I would call it the first university. It is said that Plato put a sign over the gate, "Let no one ignorant of mathematics enter here." In fact, very distinguished mathematicians worked and taught there, but I do not think that the sign was supposed to be taken literally. Plato was announcing that everything would be discussed here in the cool and analytical style of mathematics. Athenians and foreigners, some of them future commanders and statesmen and a few of them women, came here to study how to think and speak and act.

Aristotle (384–322 B.C.E.) was never an Athenian. He was born in Stagira, on the border of Macedonia, came to Athens at nineteen to study with Plato at the Academy, and stayed there till he was almost forty. He found Plato teaching mathematics and expounding his version of what the leaders of a state should know, much of it expressed in the poetic language of Ideas. Aristotle and the other young scholars loved Plato and his talk of an ideal world but could find no way to express its meaning in coherent terms, and for them it remained a myth. Their task was to find some other explanatory framework with which to express the same exalted thoughts. For philosophic discussion they wanted a language that conveyed its message without symbol or metaphor, using words that meant the same every time they were used. This was something new in the Greek world, and it separated Aristotle's followers from scholars who preferred to think like Plato.

Aristotle's knowledge and intelligence were legendary, and people expected that when Plato died he would succeed him as director of the Academy. For some reason, perhaps his nationality, he was not chosen and went back north. A few years later he was tutoring the young son of Philip of Macedon while Philip conquered Greece, but when Philip died and Alexander became king, Aristotle moved back to Athens and established his own school, the Lyceum, on the grounds of another old gymnasium. It too had a grove of trees, and in time Aristotle's followers became known as peripatetics, people who walk around. The Lyceum must have had lecture halls also, where students scratched on wax tablets while a teacher read from a bundle of notes. What we have of Aristotle's writings seems to be mostly those notes, uneven in quality and sometimes no more than aids to memory, collected and edited and finally published after many adventures and changes of hands. They cover a staggering range of subjects: medicine, physiology, comparative anatomy, metaphysics, psychology, politics, logic, meteorology, physics, cosmology, ethics, literary criticism, and many topics in biology and natural history.

For years after Greece surrendered its political power to the Roman Empire people still went to Athens to study, bringing books and knowledge home with them. But as the Empire declined and the great libraries disappeared, few but Greeks remembered their language, and outside a few centers of learning most of the old books were lost. Of Aristotle, the logical works had been translated into Latin and every educated Roman had to study them. Several of Plato's dialogues seem to have been translated, but the only one that survived into the middle ages is the first part of one called *Timaeus* which describes the structure of the world in mythic terms. It was not until the twelfth century that scholars from northern Europe traveling in Moorish Spain discovered complete Arabic translations of works they had heard of but thought they would never see. The braver ones learned Greek and Arabic, and in the next two hundred years a great flood of translated Greek and Arabic philosophical and scientific writing, accompanied by brilliant Arabic commentaries, arrived in Europe; it is hard to imagine how we would be thinking and living today if this had not happened. Now began the age of Aristotle's influence in Europe.[2] Until the scientific renaissance of the seventeenth century, and in some places later than that, the language of science was largely Aristotelian, and we must learn some of it.

---

[2] There is a historical mystery here that I do not understand. In 1204 the Fourth Crusade strayed from its mission and conquered Constantinople, a city in which Greek texts would have been available by the cartload, but during more than sixty years of occupation no European scholar seems to have thought of going there to look for them and bring them back.

First, what does the physical world consist of? It consists of substances, specific and individual. This pencil is a *substance*.[3] Do I see the substance? No, I see only some of its *qualities*—size, shape, color—and it has others, such as weight, that I perceive in other ways. The pencil itself has an identity independent of all the qualities by which we know it, and that is its substance. Substance is Aristotle's answer to the question "What is it that persists as the world changes?" As the pencil gets shorter it is still the same pencil, the same substance. The qualities that enable us to identify this substance as a pencil—including what it does—are called its *properties*, and the sum of these properties, what we can know and understand about the pencil, is called its *form*. But a pencil has other qualities. It may be long or short, hot or cold, sharp or blunt, and it might be on this table or in the other room. Qualities such as these, which are not essential to its being a pencil, are called *accidents*.

The pencil is made of wood and other kinds of matter, but each kind consists of a mixture of elements. Aristotle proposes three elements in his *Physics* and four in *On Generation and Corruption*; the second view prevailed, and Empedocles's four roots became idealized substances, like the earth, air, fire, and water we know, but purer. They consist of *prime matter* together with the properties that make it become one element or another. There is only one kind of prime matter and, by itself, it has no qualities at all and hence no independent existence. It is not made of atoms, for atomicity would be a kind of property.

Finally comes *essence*, the inner nature of a thing that is responsible for all its properties; we use it today in the Aristotelian sense if we say that some act is the essence of futility.

The formulas get messy when one tries to apply them in specific cases. For example, because one can identify and talk about a particular form, one can even call certain forms substances. This extra complexity will be necessary later when we think about perception, for Aristotle taught that the world is perceived through the soul, and that the soul is such a form.

But Nature consists of more than forms and substances; it consists of processes as well. Nature doesn't just sit there, it also happens, and it is here that Aristotle's explanation of the world departs sharply from our own. Things change for many reasons—because they are hammered or heated or pushed around, but also because they are urged toward perfection as plants grow upward, by their own inner natures. Matter, says Aristotle, desires form "as the female desires the male or the ugly the beautiful."[4] A peach ripens because its material yearns toward a certain perfect form of ripeness. In Aristotelian jargon the ripe peach has a *material cause* and a *formal*

---

[3] This is not the modern meaning of "substance," and the same is true of words to come: "accident," "form," etc. The trouble is that the Greek words have no equivalents in English. I will use these terms, confusing though they are, in order to avoid barbarisms like "thinghood" for substance, and so that you can read the standard translations.

[4] In Aristotelian biology the female of any species is considered to be an imperfect male.

cause. But if you allow it, the peach continues past perfect ripeness because the larger plan of the universe requires that fruit should fall and rot and propagate the species. That is why a fruit exists in the first place, its *final cause* as Aristotle says; and there is a fourth, trivial kind of cause called *efficient*: if the branch is shaken the peach falls.

But of course not all changes are possible. A green peach never develops into a ham sandwich. In its green state it holds a potentiality that strives to actualize its own kind of ripeness. Most changes can be described in terms of transitions of this kind. Aristotle's arguments and distinctions are delicate and exact and can be found in Book IX of the *Metaphysics*, but these simple remarks will suffice for what comes later.

We have seen why things happen, but how do they happen? In Aristotle's system this was an important question because one of his basic principles is that everything that moves (he counts any kind of change as motion) is being moved by something else that is itself in motion. Book VIII of the *Physics* contains a long and uncomfortable discussion of this claim. Many instances are obvious, but when a peach drops to the ground, where is the mover? It is the process of natural growth that produced the peach at the end of its twig. But obviously there must be an exception to Aristotle's principle. Motion must originate somewhere in an Unmoved Mover, or Prime Mover, that causes change without itself changing. It lies outside the sphere that carries the fixed stars; "it thinks that which is most divine and precious; and it does not change, for change would be change for the worse, and this would already be a motion." It causes change by being loved; in Dante's words, this is "the love that moves the sun and the other stars." That is, it causes motion in the sphere that carries the fixed stars, and this motion cascades down through the spheres below it that carry the planets, until it reaches the domain of the four elements that make the Earth and everything on it.

The definitions and ideas briefly sketched here are worked out at length in Aristotle's *Physics* and *Metaphysics*; their purpose was to enable natural philosophy to do its job of explaining Nature, starting with substances and properties and potentialities. One says: the peach tends to fall because it is largely composed of Earth, and that is what Earth does. This is plausible, and it applies to heavy things we have handled, but does it apply to *everything* that is heavy? What makes us think that the element Earth always has this property? It is time to look for a moment at what we know about the world, and how we know it.

6. What Is Science, and How Do You Know It Is Right?

When Aristotle distinguishes between art and science he is only putting into words a distinction that had long existed in people's minds. Science (*episteme* *) is certain knowledge, and he starts with a definition of the

---

* Pronounced epistémē

word "know": "We all suppose that what we know is not capable of being otherwise. . . . Therefore an object of knowledge exists necessarily and is eternal, for things that are necessary in the unqualified sense are all eternal; and things that are eternal are ungenerated and imperishable." The tentative, the provisional, the approximate have no place in Aristotle's *episteme*, only unchanging first principles or knowledge gained from them by logical deduction. In this sense a science of accidents is impossible: sunlight falls on a stone and it becomes warm, but warmth is an accidental property, and there is no use trying to explain it from first principles because warmth comes and goes, while first principles do not come and go. They refer only to what remains, to substances and the properties that define them.

Scientific knowledge, in Aristotle's view, is knowledge of causes. It is universal and invariable. On the other hand, experience is neither, and the evidence of our senses is not always reliable. No two cases of epilepsy are exactly alike; thus there can be no science of epilepsy. In fact, our world contains very little knowledge that would qualify as *episteme*. We need other categories of knowledge. One is opinion, *doxa*, which "deals with that which can be otherwise than it is." Another is art, *techne*. It is "a capacity to make, involving a true course of reasoning. All art is concerned with coming into being." Farming and building are arts, and so is medicine. Art is learned by experience, whereas *episteme*, if it does not follow logically from other *episteme*, must be intuited or guessed. An example is the proposition, "The shape of the heavens is of necessity spherical: for that shape is most appropriate to its substance and is also primary by nature." Such a judgment does not follow from observation; it is, to put it bluntly, a guess, but one of which Aristotle was absolutely certain, and on it his whole cosmology is founded. Science, which traded in this kind of knowledge, was loftier than any art, and its contemplation, known as *theoria*, was considered the highest form of activity.

Understanding this way of thinking, we can see why Greek explanations of Nature are so abundant in unsupported hypotheses. Aristotle was one of the few who admitted that this is what they were, and he admits it quite often when he refers to "the science we are seeking." In modern science hypotheses are usually suggested by experiment and, if possible, tested by experiment. But what is an experiment? To experiment on something we change it in some way—perhaps heat it—and see what happens. Aristotle would say we play with accidents, not substance, and only skate on the surface of *episteme*. For thinkers of antiquity and the middle ages, hypothesis belonged to science; experiment and opinion belonged to art, and they existed in separate worlds. During the seventeenth century, the period known as the scientific revolution, they drew closer together, but as we shall see, it was not until two centuries later that something like the modern scientific viewpoint developed.

What are the first principles, and where do they come from? When Aristotle proposes to explain the appearance of the heavens and the motions of stars, he declares that the universe is contained in a sphere, but he does not just say this and then go on to deduce the consequences, for the rest of the chapter lists his reasons for thinking so: the geometrically simple nature of a sphere, the fact that the universe revolves around us, the fact that the Earth at its center is also a sphere, and so on. None of these proves the principle, but they add up to a conviction, and for the parts of his system that might conceivably be otherwise Aristotle never claims any more. His *Posterior Analytics* is a book on making error-free deductions from first principles. In its final chapter he has to face the problem of where first principles come from. Ultimately, he says, intuition extracts them from our experience of many particulars, out of which one universal truth "come[s] to rest in the soul." In a wonderful figure he tells how it feels as the many become one, "as in a battle, when a rout occurs, if one man makes a stand another does and then another, until a position of strength is reached." Think of the great scientific principles, natural selection for example. The more general and inclusive they are, the more likely they are to have arisen in a single act of intuition. But scientists have become skeptical of any principle reached in this way. Examine it from every side, consider alternatives, think of observations that put it to the test; finally, try to devise an experiment by which it stands or falls. And ten years later, reopen the question in the light of new knowledge. But this kind of thinking came much later.

I should add that in the middle ages another question was raised: is a "science" of physics even possible? The problem is that whereas *episteme* is based on necessary connections, it is not obvious that anything in Nature happens necessarily. Thunderstorms may have a cause, but no two are alike. There is no use producing a watertight argument to prove that tides must be highest at the time of full and new moons if it turns out that this is not always the case. Only very slowly have people become convinced that there is necessity in Nature, and that one can usually find some explanation for an apparent exception.

These paragraphs have introduced a few bits of Aristotelian lore that will be useful as we go along. Aristotle himself never regarded his theory as complete. His works show that he saw most of the difficulties very clearly, and of course his efforts to deal with them tended to make things more complicated. He spoke of "the science we are seeking," never of "the science we have found." For him, though, substances and forms and the rest had a real existence outside the mind. I have tended to present them more as devices that enable language to be used in a consistent way. During the middle ages people who held the first interpretation were called *realists*, while the others, who said no, substance and form are only names of concepts, were *nominalists*. The nominalists finally won, but the language remains, and so do various questions that are asked in it. People asked, for example, "Is light a substance or an accident?" —*Substantia aut accidens?* Some of the controversies that will occupy this book turn on that question. Later we shall find several reasoned answers, but people went on arguing. In Chapter 10 we shall see how modern physicists create a paradox when they treat light as *accidens* in their theory but speak of it as if it were *substantia*.

Why does anything happen? The world around us is in continual motion (remember that motion, for Aristotle, includes any kind of change—the rising moon, a burning fire, a blossoming rose). What drives potentiality to become actuality? I have mentioned the principle with which Aristotle begins Book VII of his *Physics*, which seems so natural that it is hard to see how any reasonable person could doubt it:

> Everything that is in motion must be moved by something.

But this, as we have seen, leads to an infinite regress unless one postulates an Unmoved Mover at the beginning of it all, for whose sake the heavenly spheres turn, bringing motion and change to the Earth below. Aristotle says little about how motion is transmitted from the outermost sphere—which, being largest, moves the fastest—down through the successive planetary spheres to the Earth—which does not move as a whole but suffers various other changes. That the spheres affect what happens on Earth was obvious. Anyone can see that the sun controls the seasons and the moon controls the tides, but how is it done? It was a long time before anyone attempted a serious answer to this great scientific question, and the first answer, when it came (§3.6), was more concerned to explain the stars' effects on humanity than on seasons and tides. But still, no one doubted the effects.

7. Planets Power the World, Light Stops a Planet

The stars called Mercury, Venus, Mars, Jupiter, and Saturn attracted attention very early in history because they move from night to night, some quickly, some slowly, against the background of the other stars. They were called planets. Greek *planetes* means wanderer, and since the sun and moon also change their positions they too were planets, bringing the total to seven, the number which shows that a list is complete. (Days of the week, stars of the Dipper, wonders of the world, gates of Thebes, openings of the head, sleepers of Ephesus, liberal arts, sacraments, deadly sins). The planets make a pattern in the sky that varies continually, but why, what's the point of them, what are they there for? The conviction that these movements must have some relation to events on Earth led ancient peoples toward systems of astrology. For these to be of any practical use it was necessary to be able to calculate where planets would be in the future, and so mathematical astronomy was born. The moon is the easiest to study, for it moves around the Earth at the fairly uniform rate of about

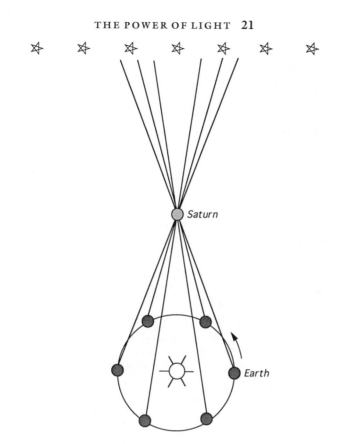

Figure 1.1. Saturn moves slowly, and suppose for the moment that it stands still. As the Earth goes around the sun, Saturn will seem to move back and forth against the background of the fixed stars.

29½ days from new moon to new moon. The sun is harder to observe, because one cannot see it against the starry background, but one can estimate where it is by looking at the stars before dawn and after sunset, and it too moves in a circle through the constellations in a period of about 365¼ days.

The other planets have more complex motions. Consider Saturn from the modern point of view. It moves slowly, taking about thirty years to go around the sun; to make the picture simpler, let us pretend for a moment that it does not move at all during a year, while the Earth goes around the stationary sun (Figure 1.1). If this were so, Saturn would seem to move back and forth against the starry background with a period of one year. Because Saturn does move, each of these loops should be a little ahead of the previous one, and this is what is observed. Until the Renaissance, when the Copernican theory took hold, everyone put the Earth at the center of the universe and required Saturn to move back and forth against the stars to produce the same visual effect. Should the sun be held responsible for Saturn's zigzag?

By the beginning of the Common Era, Greek intellectual influence was fading in the Roman world. Most writings of Eudoxus and Apollonius were lost, and most of the professional philosophers of Rome, even if they knew Greek and had heard of Aristotle, had never seen his works and knew little about them. At their best, the Romans were great poets and historians and statesmen, but there were no significant Roman students of Nature unless you count Lucretius. *Episteme* was a Greek specialty, but people in the Mediterranean world were also acutely alive to what we would call magic, and that was the Romans' specialty. They felt the unity of Nature very strongly and sought to forecast and influence the course of events by daily routines of spells and offerings. In the city, and in all the empire, astrology found eager buyers. Chaldean diviners appeared in Rome, and Jews and Christians in the eastern provinces quickly became experts. The Wise Men from the East, who followed a star, were probably Chaldeans.

There were at this time two views of astrology, with quite different implications. Suppose for example that the stars say there will be danger for you at a certain time and place. You can take this in two ways: either you had better not be there, or your fate is already decided so it doesn't matter what you do. The first is the Epicurean view; the second comes from the Stoics and was more prevalent in Rome. One can read about this side of Roman astrology, at paralyzing length, in the more than four thousand lines of Marcus Manilius's *Astronomica*, written in the first years of the Common Era.

The astrological interests of the Romans made them aware of planetary motions, and someone found a way to explain the planetary zigzags I have just mentioned. One finds the idea expressed, in the casual way that shows everybody knew it, by Vitruvius in his treatise on architecture, by Pliny the Elder in his gigantic *Natural History*, and by Lucan in his bitter epic, *Pharsalia*, about the Civil War between Caesar and Pompey. Lucan tells how Caesar, amazed at the wealth and luxury of Cleopatra's court, begins to think of conquering Egypt. He opens a campaign of flattery to win over its priesthood by asking an ancient priest about the secrets of the heavens. In the course of his answer come the words, "The sun divides time into parts, and changes day into night; / Sometimes the strength of his rays forbids the planets to move." Roman authors didn't think about the matter very hard and repeated Lucan's couplet as a bit of general knowledge, and so the idea stuck. Planets, after all, are little dots in the night sky; why shouldn't sunlight push them around? Aristotle and the other Greeks were not read in Europe until the thirteenth century; until then, people quoted Pliny and Lucan.

I can think of only one historical episode that suggests the power of the sun's light. It was in the year 212 B.C.E. when the Roman general Marcellus was besieging the Greek city of Syracuse. Among the fearsome devices that Archimedes, then seventy-five years old, invented to harass and destroy the attacking fleet was an arrangement of mirrors. "This contrivance he set to catch the full rays of the sun at noon, both summer and winter [the siege went on for three years], and eventually by the reflection of the sun's rays a fearsome fiery heat was kindled in the barges, and from the distance of an arrow's flight he reduced them to ashes."[5] True or not, this story would have been known to educated Romans, and perhaps they were persuaded that if light could stop a Roman fleet it could also stop a little white dot.

What would an educated Roman have answered if asked, *Substantia aut accidens*? I think he would have brushed it aside as one of those silly philosophical questions, but, if pushed a bit, he might have answered "*Substantia*, naturally. It takes a substance to burn a warship."

8. The Manichees: Light Defeated by Darkness

The question of evil opens an ancient theme. How is it that if the world and humanity was created by a benevolent power we have become such a flawed and twisted species, mixing cruelty with love and injustice with good deeds wherever we are found? What went wrong? Greek mythology answered with the story of Pandora. Christian theologians seized on the biblical tale of Adam's and Eve's disaster, labeled it the Fall of Man, and concluded that though the first humans were created sinless, they managed, before they had been alive for a full day, to ruin everything. But more important than asking what went wrong is deciding what to do about ourselves, how to make human life better.

The literature of the East is dense with answers to the question of evil. The cult of Mithra, god of light, goes back to the night of time when Persians and Hindus worshiped some of the same gods, and Persian mythology pitted him against forces of darkness led by the evil Ahriman. But let us focus for a while on a later story. It was the work of a Persian teacher named Mani (c. 215–c. 75), who started a movement that lasted for more than a thousand years and at different times reached from Rome to China. In this vast expanse of space and time Mani's teachings underwent many changes. The myth in which he tells what happened has been put together by scholars out of inscriptions and a few hymns; it probably does not correspond exactly with the belief of any one Manichaean. I tell it because it is a wonderful tale and because it embodies the Neoplatonic vision that light carries a power transcending its physical strength. Mani and Plotinus were exact contemporaries. Did one influence the other? I don't know.

---

[5] That some such thing is possible was shown long ago by the Comte de Buffon. With 128 glass mirrors, 6′ x 8′, mounted on a large wooden frame, he "very quickly" set fire to a piece of tarred wood at 150′. More recently a Greek engineer named Ioannes Sakkis made fifty or sixty bronze-coated mirrors 3′ x 5′ with which sailors at the naval base near Athens focused the sun's rays on a rowboat 165′ out in the bay. In seconds the target was smoking, and before two minutes had passed it was ablaze. (*New York Times*, November 11, 1973.)

If God is good, purely good, nothing but good, then he cannot be entirely responsible for us or the world around us; there must be something else, or someone else. For Mani, God is light; this is a fact, not a metaphor, and the place he lives in is also light. Not far away there is darkness also, not merely absence of light but presence of darkness, a focus of evil. It is an opaque material mass, the "Earth of Darkness," and Satan lives in the middle of it. God commands five powers: Intelligence, Knowledge, Thought, Deliberation, and Resolution. Satan commands five also: Smoke, Fire, Wind, Water, and Darkness, but in the beginning, down there in the Earth of Darkness, Satan's five hardly knew one another. "Only when one of them screamed did they hear him and impetuously turn toward the sound." From time to time they fought, and at one moment their struggle brought them close enough to the surface of the dark mass they lived in so that they caught sight of the realm of light. It was new to them, they wanted it, and under Satan they united for once and strove to capture it. Now comes the strangest part of the story: God is no match for these forces. Look at his powers: they are contemplative, while Satan's are full of furious energy. God is so purely good that he cannot fight back; he is almost passive in the face of evil. He is too innocent, and evil rages unopposed.

In his predicament, God creates a being known as the Mother of Life, and she gives birth to another being known as First Man, who gathers together five Powers of Light: gentle breeze, wind, light, water, and fire. These are the traditional elements, minus earth, whose nature is evil, plus light and breeze, but their nature is spiritual, not material. Together they form his soul, and with a soul First Man is truly alive. God sends him to save the world of light, but he is defeated in battle and dragged down into the Earth of Darkness, and God must try to get him out. With the aid of his five powers he succeeds, but he has not really won, for some of the light in First Man has been taken away and replaced by dark matter. Now begins a long effort to recover the lost light.

To this end, God creates the universe we live in, containing the sun and moon as well as the Earth of Darkness with its small quantity of stolen light. Matter craves soul to make it alive, and the five Powers of Light provide it. This soul is not like that of First Man, since its elements are corrupted by their connection with matter, and the bonding of light into matter makes its recovery and purification more difficult. The sun and moon are made as reservoirs for the recovered light; and to guard them, First Man lives in the sun and the Mother of Life in the moon. The cosmos is set in motion around the Earth, and the constellations of the zodiac serve as spaceships that deliver light to the sun and moon as they travel past.

But now Satan makes his move. He creates Adam to serve as a reservoir to hold most of the light he has stolen, and he makes him a sexual being. Then he makes Eve, so that Adam can be controlled through his desires. The brilliance of Satan's scheme becomes clear: by concentrating light in humans and then commanding them "Go forth, increase, and multiply" he hopes to replicate thousands of living reservoirs to contain the light he has stolen, so that God must struggle to recover it from every one. But if God wins Adam and Eve and their children, Satan loses everything.

God's good powers cannot act physically, but he creates teachers—Noah, Abraham, Gautama Buddha, Zoroaster, "Jesus the luminous"—to protect and teach humankind. It is Jesus, in the Garden, who encourages Adam to eat from the Tree of Knowledge, because this is the only way he can be saved. Once Adam has eaten the apple and his mind has awakened, Jesus tells him the story that explains how he was created. "Then Adam looked up and wept, he lifted up his voice like a roaring lion, he tore his hair and beat his breast and said, 'Woe, woe to the creator of my body! Woe to him who has bound my soul to it, and to the rebels who have brought me to servitude.'" Jesus warns Adam against procreation, and for a long time he does not approach Eve. Her first two children, Cain and Abel, are men of the Earth and Satan is their father, but finally Adam and Eve come together and with Seth, a "child of light," begins the multiplication of humanity. The struggle continues.

Some versions of the story have a strange end, a last Great War in which Darkness is defeated and locked up forever, together with those fragments of Light that were never rescued. We have seen that God by himself is no match for Satan, and in the end his great enterprise, the best he can do, does not entirely succeed. Of all the abominable heresies that Christian theologians found in Manichaeism, this was the worst.

In most versions Jesus is composed almost entirely of light. He is recognized as a form of the First Man, like him except that his small contamination of Earth enables him to suffer like one of us. And finally, in some centers of Manichaeism, initiates after years of study received the great secret: the First Man is God, and therefore humankind, made in his image, is above all things.

Thus Mani proposed a solution of the problem of evil that did not require God's goodness to be compromised. It did not demand that one worship the fearful God of the Old Testament, who for Manichaeans was nothing but Satan. There was no text that had to be interpreted as meaning something other than what it seems to say. And above all, he taught that what is good in us, what is really us, is perfectly good, whereas the part of us that sins is something for which we are not responsible. It is not we that sin; we are only vessels that seek to preserve the light that is in us while the war goes on toward its compromised ending, but it will not be our war or our compromise. Our duty is a hopeful passivity, and in his hymns the Manichaean, while taking pains to avoid sin, is only a spectator of the cosmic drama of evil and redemption.

Manichaeism was especially strong among the Christians of North Africa. Here Augustine of Hippo (354–430), bishop and saint, and one of the main creators of Christian doctrine, was captivated at first by the legend's dramatic and explanatory power but later became its enemy as he realized how much it conceded to evil and how little spiritual effort it expected of a believer. He preached and wrote against it ferociously, and in a hundred years its books and followers had vanished from the land. Farther east it continued to thrive, and in the eighth century it reached its summit among the Uigur Turks in the deserts of what is now Sinkiang Province in China. There it became a state religion; miniatures painted by Persian artists show the clergy in magnificent dress surrounded by servants and musicians, and an inscription near the capital city proclaims that under its gentle influence this "country of barbarous customs, full of the fumes of blood, was changed into a land where people lived on vegetables; from a land of killing to a land where good deeds are fostered." But in this nation that had established itself by force of arms the teachings of Mani had an effect that should have been predictable: the Uigur weakened, and after only eighty years they were swept away by less civilized Turks from the north.

As Manichaeism was dying in the empires of China and the Uigur Turks it entered Europe from the Near East and quickly became a Christian sect. Its first home was in Bulgaria, where its adherents were called Bogumils. Thence it spread to northern Italy and Provence, where they were called Cathari (*katharos* means pure). By the eleventh century Cathari were in Languedoc, where they became known as Albigensians from their settlement in the city of Albi.

The Manichaean view of the world presented itself as a vivid legend and easily answered the question that gave People of the Book so much trouble: why, under the domain of a just and loving God, did the lives of ordinary people know so much danger, pain, and grief and so little happiness? And there was another reason why so many from all classes of society became Cathari. Especially there in the south the open corruption of the clergy at all levels encouraged self-respecting people to separate themselves from the Catholic Church and join a sect that called itself pure. The nobility, disgusted by the Church, threatened by its rapacity, and greedy for its land, supported the heretics. By the beginning of the thirteenth century bishops, archbishops, and common clergy had joined the movement, and it seemed that in the Languedoc, roughly the region between Toulouse and Béziers, the Church was dead or dying. In 1207 Pope Innocent III persuaded the king in Paris and his northern barons to launch a crusade against the heretics under the direction of Count Simon de Montfort. Entire populations, Catholics and Cathari alike, were put to the sword. As the numbers of Cathari diminished they fell into doctrinal confusion, and except for recognizing evil as an active force opposed to God they largely forgot the teachings of Mani. By the mid-fourteenth century, under the ceaseless pounding of French armies, their lamp had flickered and gone out. They left no texts and few stones.

We shall see in Chapter 4 how light became identified with God and with the spiritual content of Christianity, but no Catholic writer testifies that the Manichees had anything to do with it. After all, light was not the invention of Doctor Mani, for the universe began in light, and every Jew and Christian knew it.

9. And There Was Light

In the beginning God created the Heaven and the Earth. And the Earth was without form and void;[6] and darkness was upon the face of the deep: and the Spirit of God moved upon the face of the waters. And God said: Let there be light: and there was light. And God saw the light, that it was good: and God divided the light from the darkness. And God called the light Day, and the darkness he called Night. And there was evening and there was morning, one day.

The Creation continued for five more days. It is described in the first chapter of Genesis and the first three verses of the second; then starts another and probably older version of the story that gives the events in a different order and is not divided into days but includes the Garden of Eden. Nothing more to explain what happened, and so you must believe the whole story is there, if only you read carefully enough. The scholars of the Talmud studied the text to search out its deeper meanings but did not try to fill the gaps in the narrative. The Koran says several times (for example in *Surah* 32) that God made the world in six days and on the seventh ascended the throne to govern the world he had created. It was Christians who wrestled with the details.

The monks and bishops who cultivated the seeds of Christianity that had taken root in Roman paganism felt the need to define their new faith in terms of a large number of propositions concerning humankind and the Holy Trinity, the history of the universe, and humanity's place in it. Faced with a hundred questions of detail that the Bible does not answer, they used reason and faith to fill in the gaps as well as they could.[7]

The western part of Cappadocia, once a Greek-speaking Roman province in the center of Asia Minor, is one of the strangest regions on Earth. An ancient volcanic eruption covered it deeply with white ash, which rain has

---

[6] These are the words of the King James translation. In the days of the Church Fathers there were several versions of the text to choose from. Those who knew Greek used the version known as the Septuagint, made in the second or third century B.C.E. by Hellenistic Jews. It says *aoratos kai akataskeuastos*, meaning invisible and without form. The earlier Latin Fathers used the version known as Old Latin, which says *invisibilis et incomposita*, exactly the same thing. They took "invisible" to mean that everything was covered with water. Jerome's translation, which later became canonical throughout the Latin Church, says *inanis et vacua*, meaning featureless and empty. The Hebrew, *tohu wa bohu*, is unspecific and means chaotic. These distinctions will be useful in the discussion to follow.

[7] In a famous essay, "The Scar of Odysseus," Erich Auerbach (1953) contrasts Homer's love of the details that make a scene live in a reader's mind with the reticence of the Old Testament, which only outlines events so that attention will rest on their deeper meaning. A thousand years after Homer, Greeks still needed the details.

sculpted away to produce a landscape of cones and valleys and cliffs soft enough so that people wishing to live in solitude could dig dwellings and even little churches out of the compact mass. The climate varies between extremes of heat and cold, but the soil, where it is exposed, is fertile. In the fourth century this fantastic place sheltered hermits, monks, clergy, and their congregations from the attacks of wandering tribes, and it became the intellectual center of the growing Christian movement.

Basil the Great (c. 330–79), was born in the town of Caesarea (now Kayseri), a few miles east of the cones. His family's wealth allowed him to study in Constantinople and Athens; later he returned to Caesarea and for a few years before his early death was bishop there. He strengthened his church and fought back its heretics, he founded hospitals and other works of mercy, and in a time of famine gave away his fortune; his authority was immense.

In his sermons on the Creation Basil saw no reason to tamper with what the Bible says. His *Hexaëmeron* (the word is from the Greek for Six Days) announces, "For me grass is grass; plant, fish, wild beast, domestic animal, I take all as they are said, 'for I am not ashamed of the Gospel.'" (He is quoting Saint Paul.)

He does not tamper, but he invents, and when his hearers want to know exactly what happened, he fills out the bare narrative of Scripture. At the time the story begins there was, he writes, an "older state," unformed, chaotic, and timeless; in the opening scene Spirit hovers above it. Then, "Let there be light: and there was light." The Gospel says nothing more about this light. What did it look like? What happened to it at the end of the day? Did it just go out, leaving the windy ocean in darkness as before? Saint Basil explains:

> The air was lighted up, or rather it made the light circulate mixed with its substance, and, distributing its splendor rapidly in every direction, dispersed itself to its extreme limits. . . . For the ether is such a subtle substance and so transparent that it needs not the space of a moment for light to pass through it. Just as it carries our sight instantaneously to the object of vision, so without the least interval, with a rapidity that thought cannot conceive, it receives these rays of light in its uttermost limits. . . . Since the birth of the Sun, the light it diffuses in the air when it shines on our hemisphere is day; and the shadow produced by its disappearance is night. But at that time it was not by the movement of the Sun, but by this primitive light spread abroad in the air, and withdrawn in a measure determined by God, that day came and was followed by night.

Neither the sun nor the round, solid Earth was yet created, so night could not have been caused in the usual way. Basil deduces that God simply caused the light first to grow and then to diminish, marking the period of one day, as the Bible says. In this way he established the daily rhythm of time that sun and moon would continue once he had made them. But when light was gone the primitive creation did not relapse into darkness. "Before the creation of light the world was not in night but in darkness," Basil writes. Now it was night.

Genesis says that on the fourth day "God made the two great lights: the greater light to rule the day, and the lesser light to rule the night; and he made the stars also . . . to divide the day from the night; and let them be for signs and for seasons, and for days and years." The sun measures days and the stars measure a year, for if you follow the sun's path as it moves against the starry background that is how long it takes to arrive back at its starting point. The Egyptians, and doubtless others, measured the year by noting the day on which the star Sirius could first be seen for a moment before sunrise obscured it. When the Bible says "for signs" he means that signs like this will tell farmers when to plant and when to reap. Basil is firm on that point. Some people were saying that "for signs" justifies astrology. That is impious and ridiculous.

Basil had a younger brother, Gregory (c. 332–c. 400). Seeing that he was brilliant but not a leader, Basil sent him to be bishop of a small town named Nyssa close by the cave-dwelling hermits. Gregory's sermons interpret the Creation according to Aristotelian tradition, and his conclusions are different from Basil's. Gregory's *Hexaëmeron* tells us that God is immaterial and not situated anywhere in space or time. Because any piece of matter is here now and later may be somewhere else, God cannot be connected directly with matter and thus cannot have created it. Instead, his Word created the qualities that belong to matter, and because there cannot be qualities without something for them to belong to, matter came into existence by itself. For the same reason God does not control what matter does, but in the beginning he installed in it a *logos*, that inherent tendency toward orderly and predictable development already named by Empedocles, which determined how everything happened afterward. Thus not only did God create a fruit tree; he also made it so that it would grow and bear fruit whose seeds would make new trees. Many of the early writers emphasized that the physical world consists not only of pieces of matter but also of the principles according to which they change and interact. If the stars moved at random they would be useless to us; if fire burned sometimes hot and sometimes cold it would not serve us. Now we speak of natural law, the term suggesting control by decree from outside, but a *logos* that acts from within is really more what we mean.

At first, writes Gregory, matter was chaotic and stirred together. "The brightness of fire, still hidden among the particles of matter, had not yet appeared." Then "Let there be Light," and something like a great ball of fire burst out and ascended according to its *logos*. "Leaving the other elements like an arrow from a bow," it rose to the highest heavenly sphere and then, because it must move and could rise no further, turned to travel in a circle inside the sphere, so that light and darkness alternated below. "God called the light Day, and the darkness he called Night." This must have happened three times, establishing the daily rhythm; then the sun was created in the second sphere, far below, and took it over.

During the next five days of creation the new substances assembled themselves into the material objects, the plants and animals and humankind, that make up the world. I am sorry that no painter ever gave us the scene of God, surrounded by his angels, at the hopeful launching of this immense enterprise that would finally cause him such care and trouble.

There are other versions of the story. Saint Augustine discussed it in several works, including Chapter 12 of his *Confessions* and a long and detailed book called *The Literal Interpretation of Genesis* from which most of the following account is taken.

The first creation was "invisible and formless," as in the Old Latin version of the text. Almost without form, that is, for if entirely without form, what could it be? Formless matter is an oxymoron. "Before [God] formed this unformed matter and fashioned it into kinds there was no separate being, no color, no shape, no body, no spirit. Yet there was not absolutely nothing: there was a certain formlessness devoid of any specific character." Later, Augustine explains that if before light there was darkness on the face of the deep, then there was a deep, even if it was unformed, and thus formless matter, Aristotle's prime matter, exists somehow outside of time and was already there. (Scripture mentions water, which implies a specific form, but Augustine says that Moses, the supposed author of Genesis, was only trying to make himself clear.)

Then, on command, light was created. The utterance "Let there be light" was spoken instantaneously, "with no division of syllables," Augustine says, so that the creation of light would not be the second thing that actually happened. The words "Heaven" and "Earth" refer to spiritual and bodily creation, and the nature of the new celestial light was spiritual.

Thus were created angels (Heaven) and formed matter (Earth), and the angels turned to behold God. The whole process was instantaneous, and it seems that it was repeated on each of the six days so that the angels could understand the logical sequence of what had been done.

Like Basil, Augustine sees the difficulty that arises if one says that everything was in darkness until the first moment of creation. It would imply, he says, that God had lived in darkness until then, and so he explains that darkness and light were both created at the same time. Darkness refers to all the unformed substance out of which substantial things would later be formed, but he cannot decide what happened to that first light "when evening came on, so that night could have its place."

Augustine is not sure of any of this, but he wants to help believers understand God's immense intellectual achievement in conceiving without error the interlocking complexity of the natural world, which could begin to function in all its detail the instant it was created, and of the moral order set in motion at the same time.

All the commentators stress that there are symbolic correspondences between these two orders, established in the first instant of time by the dual nature of light, physical and spiritual. Their explanations continued as every teacher needed to explain to his congregation the huge amount of truth contained in the few simple words of Genesis, but no authority at that time tried to assemble from them a definitive account of the Creation. They just weren't specific enough.

By the time of John Milton the impulse to interpret the scriptural text literally was largely spent. He wrote *Paradise Lost* with a larger purpose, "to justify the ways of God to men," and perhaps also to provide England with a national epic to match those of Homer and Vergil. He is of course faithful to the letter of the Bible, but whereas the Church Fathers expanded the Genesis story in order to make sense of it, Milton's additions reflect a wider purpose. Creation is seen first from Satan's point of view: he makes his way through chaotic darkness to a place from which he can see, far off, the wall of Heaven, shining in its own light. God allows him to come nearer, and close by the city he sees the new little world, hanging from a golden chain. From the brilliantly illuminated outside he is looking not at the Earth but at the outside of the sphere of the fixed stars, within which the machinery of orbs and epicycles is already moving the sun, moon, and planets around the tiny central Earth. The light inside comes from the sun. Then begins blind Milton's hymn to light,

> Hail holy Light, offspring of Heav'n first-born,
> Or of th'eternal Coeternal Beam
> May I express thee unblam'd? since God is light,
> And never but in unapproached Light
> Dwelt from Eternity, dwelt then in thee,
> Bright effluence of bright essence increate.
> Or hear'st thou rather pure Etherial stream,
> Whose fountain who shall tell? before the Sun,
> Before the Heavens thou wert, and at the voice
> Of God, as with a Mantle didst invest
> The rising world of waters dark and deep,
> Won from the void and formless infinite.

Later, in Book VII, Milton describes the Creation itself, when God's Son, his chariot followed by legions of angels, rides out of the city to the shore of the abyss and with a golden compass defines the sphere of the world that is to be created. Then God's Spirit (Milton does not say *logos*, but this is what the *logos* did) hovers over this formless waste and begins to bring order into it: the heavier elements fall and the lighter ones rise, "And Earth self-balanc't on her Centre hung."

> Let there be Light, said God, and forthwith Light
> Ethereal, first of things, quintessence pure
> Sprung from the Deep, and from her Native East
> To journey through the airy gloom began,
> Spher'd in a radiant Cloud, for yet the Sun
> Was not; shee in a cloudy tabernacle
> Sojourn'd the while. God saw the light was good.

All this is just the outer envelope of Milton's epic of fall and redemption, but we need not go further. It is here only to show how a blind poet dealt with the light of Creation.

This chapter has introduced some early ideas about the nature of light and its vast power to create and to perfect. It has also described some intellectual tools, especially those of Plato and Aristotle, which were in general use as late as the seventeenth century and occasionally afterward. Their importance is that they helped determine both the questions that were asked about the world and what could be accepted as answers. There is one immense fact, though, that I have scarcely mentioned: we can see. And how is it that light brings us such intricate information about the world in front of our eyes? How does so much detail pass so quickly from the outside world through those little holes and into our minds? Theories that answer this question have been with us since the beginning of philosophy, and the next chapter will tell about the earliest of them. Then Chapter 3 will show how a coherent theory of light and vision began to emerge during the following centuries.

2

THE IMAGE AND THE MIND

Zeus had smitten the swift ship with his bright thunderbolt,
and had shattered it in the midst of the wine-dark sea.
(Homer, *Odyssey* 5.132)

1. What Color Is Green?

IN THE GREEK of Plato and later writers the word *kyaneos* means
blue. Homer, writing in the archaic period, describes the sea as purple
(*Iliad* 16.391), as white (*Odyssey* 10.94), as winelike (*oinops*; *Odyssey*
2.421 and many other places), but never as blue. Cattle also can be *oinops*
(*Iliad* 13.703), and wine, sometimes, is black (*Odyssey* 9.196). The sky is
bronze (*Odyssey* 3.2) or iron (*Odyssey* 15.328), but not blue. Goethe seems
to have been the first to comment that Homer almost never says anything
is blue. One of the few times he does is when he describes the dead Hector
as Achilles's chariot drags him around the walls of Troy: his hair is *kyaneos*
(*Iliad* 22.402). In other places the word means dark or terrible or dangerous. Is it then just a mistake to say that *kyaneos* means blue? In *Iliad* 7.87
the house of Alcinous is decorated with a cornice made of *kyanos*, a glass
paste. Fragments of such a paste have been found in Tiryns, and it is blue.

Also, *chloros* as a color means yellow-green, the color of wheat just beginning to grow. (Chlorine is a greenish gas; chlorophyll is named from *chloros* plus *phyllon*, a leaf.) Homer uses "green wood," exactly as we do (*Odyssey*
9.320), and one also finds *chloros* referring to fresh meat or fish. When Odysseus tells of his encounter with the strengthless dead (*Odyssey* 11.43), he
says, "Green fear took hold of me." A few centuries later, in Euripides's
Medea (line 907), the leader of the chorus says that green tears have flowed
down her cheeks. And in his *Hecuba* (line 127) the chorus of Trojan women
tells Queen Hecuba that her daughter Polyxena is to be sacrificed so that her
blood will stream down over the tomb of Achilles. The blood will be *chloros*.

Everyone agrees that leaves were just as green in Euripides's time as they
are now, but there is another case that is not so clear. Sirius, also called the
Dog Star or Alpha Canis Majoris, is the brightest star in the sky. Find
Orion, find the belt, look to the left of it and a little below. The Dog Star
shines blue-white, but a Babylonian tablet dated at about 700 B.C.E. says it
shines like copper. Perhaps that is just to tell how bright it is, but it is not
the only indication that Sirius was regarded as red; and long ago, in En-

gland, one Thomas Barker [1760] pointed to a number of others. I will only mention three.

In about 50 C.E. the Roman playwright and polymath Lucius Anneus Seneca wrote a book called *Questions Regarding Nature*, and in the astronomical section his words seem clear: "In the heavens everything is not of one color, for the red [*rubor*] of the Dog Star is quite sharp, that of Mars less so, while Jupiter has no red, for he shows his splendor in pure white." *Rubor* is generally taken to mean redness, but Horace, for example, writes of *luna rubens*, the blushing moon (*Carmina* II.11.10). Color words do not always mean what they say. And other Roman writers such as Manilius (first century B.C.E.) in his *Astronomica* call Sirius blue.

In *Almagest*, the astronomical treatise of the second-century C.E. astronomer Claudius Ptolemy, there is a table of the magnitudes and positions of more than a thousand stars. It rarely mentions color, but six stars are identified with the term *hypokirros*, including Arcturus, Betelgeuse, and Sirius. The dictionary says that *hypokirros* means "somewhat yellow," and this describes Arcturus, but Betelgeuse looks red to us, not somewhat yellow at all. And in the constellation Canis Major (the Big Dog) Ptolemy again mentions "the star in the mouth, the brightest, which is called 'the Dog' and is *hypokirros*." Perhaps we do not know what color *hypokirros* denotes, but whatever it is, how can it apply to all three?

Explanations: many have been offered, both cultural and scientific. Could it be that for some mythological reason it became customary for people to call Sirius red, and they never really looked at it? This would be more credible if Sirius were not the most conspicuous fixed star in the sky. It is the only star bright enough to be seen when close to the horizon; then it is often reddish, and on certain nights when it twinkles strongly one can see flashes of red and blue; conceivably this is what Seneca was talking about. But why would Ptolemy the astronomer identify it with the color it has when it is near the horizon and he would not be observing it? Did a dust cloud pass between it and the solar system for a short time in the first two centuries C.E.? Maybe. Modern theories of stellar structure and evolution say that for a star to change color so quickly is impossible. Are they wrong? Perhaps. At any rate, it is one of history's little puzzles; apparently no one really knows.

These few examples show that as we discuss early speculations about the way people see, and what they see, we should remember that for whatever reason, the world may not have looked to them the way it looks to us.

2. The Arrow from the Eye

From earliest times everyone seems to have agreed that the most reliable of our senses is touch but the most useful is sight. Touch is easy to understand: body makes contact with some object, body is connected with mind, idea forms in mind. Today we might say something about nerves and brain, but the old explanation makes sense as it is. Scent and hearing can tell us of things too far away to touch, but what makes sight unique is that it brings us an image. One can make plausible guesses as to how the nose works or how we hear, but how does a visual image arrive in our mind? Sight has two limitations that give us hints toward an answer: we can see only that part of a thing that is reached by a straight line from the eye, and we cannot see in the dark.

The oldest theory that has come down to us, from about 450 B.C.E., was proposed by Empedocles. The language of the fragments is obscure, but the idea seems to be that an object gives off rays that carry information about its surface. To read this information, the eye projects forward a narrow *visual ray* that somehow feels the object's radiation and returns through the pupil into the sensitive part of the eye, where it creates an image in the mind. The visual ray is like a long finger projecting from the eye, and sight is a kind of touch.

Empedocles's fragments are expressed in language that is hard to decode. Aristotle objects that he does not even explain why we cannot see in the dark. This does no serious damage, because it was generally believed that some animals actually do see in the dark; the real question was why the rest can't, and probably somewhere in the missing parts of Empedocles's text was his answer.

Occasionally the visual ray itself can be seen. People who sit by campfires are used to seeing an animal's eyes gleam as it looks in from the darkness, and Homer uses this in a wonderful simile (*Iliad* 13.474) when he compares the hero Idomeneus awaiting an attack from Aeneas to a wild boar facing its pursuers with back bristles up and both eyes shining with fire. Of course we can read this as a literary trope, but that might be a mistake. There is another belief that supports the theory of the visual ray. The evil eye is feared all around the Mediterranean basin and in places far beyond. Obviously, that influence comes out of the eye.

Even though the theory of the visual ray seemed to fit so many of the facts, another hypothesis was soon being launched, and it sailed for a long time.

3. A World Full of Images

Leucippus of Miletus and his student Democritus were younger contemporaries of Empedocles. Leucippus not only invented the idea of an atom but also (they say) introduced two ideas concerning vision that are crucial for what follows. The first is that most sense experience is essentially passive. We actively explore the world only as far away as our fingertips will reach; all other sensations flow from the outside world into our senses, which somehow communicate them to the mind. This is easy enough for us to imagine when we smell a flower: it gives off tiny particles that sweeten the air; we detect them with a sense organ when we inhale.[1] It can even explain how we hear, if we believe, as later commentators certainly did, that there are atoms of sound. But sight presents a unique problem, for how is an image explained by any sort of emanation? Leucippus's second idea answers this question. He seems to have taught that under the influence of light the surface of any visible object continually produces thin veils of matter, perhaps only one atom thick, which peel off and retain their shape as they fly with immense speed in every direction. Though they may be quite far apart in space, they arrive close together in time so that our view of a changing scene is almost continuous, like what we see in the cinema. We need a word for such an image. The Greek is *eidolon*; later in Latin they were called *simulacra*, and later still *species*, meaning appearances, but by then the concept behind the word had changed. For the present I will use *eidolon*.

It is tempting to laugh at the idea of *eidola*, but if we start now we will have to laugh for a long time, and it may be better to put on a serious face and compare what has just been proposed with Empedocles's model. Both philosophers can explain why we cannot see an object put next to the pupil: Empedocles because it keeps the visual ray from going out, Leucippus because the *eidolon* is too big to get in. Leucippus may have explained why we cannot see in the dark, but Democritus, developing Leucippus's idea, claims that in fact we do see in the dark if our minds are sensitized by sleep, for there are *eidola* still floating around at night that fill our dreams. Plutarch, in one of his fifteen volumes of table talk called the *Symposium*, reports Democritus as saying that *eidola* escape from all of us, carrying ghostly copies of our mental impulses, designs, moral qualities, and emotions that speak to the dreamer as if they were alive. Empedocles, by invoking something analogous to the sense of touch, easily explains why we see only the surface of a thing that faces us but the matter is not so clear for Leucippus. If we receive the whole *eidolon*, why don't we see it all at once, front, back, and sides? If we receive only a part of it that moves toward us, then there really is no such thing as an *eidolon*, for it must fragment as soon as it forms, with different parts going off in different directions.

The theory of *eidola* contains another set of problems that dominated research in optics until the ninth-century Arab philosopher Alkindi began to show how it might be overcome. Suppose I look at a mountain. How does its *eidolon* get into the little hole in my eye? Connected with this, why does something far away look small? Is it perhaps that *eidola* get smaller as they travel? Also, how does an *eidolon* know where I am so as to attain exactly the right size and fly to that little hole? One might think that questions like these are enough to devastate the theory, but what was the alternative?[2]

---

[1] This sounds obvious, but you might like to compare it with Aristotle's ponderous explanations of smell in Book II of *On the Soul*.

[2] Toward the end of the long history of *eidola*, almost two thousand years later, Leonardo da Vinci has an answer: "The instant the atmosphere is illuminated it will be filled with innumerable *spetie* [his word for *eidola*], produced by the different bodies and colors located in it. And the eye is the target, the magnet of these *spetie*." Fewer *eidola* are needed if the eye attracts them toward it and draws them in.

Today it is easy enough to say that light from the perceived object comes into our eyes bringing the image we see. But how does this happen? Is it obvious, if one hasn't studied physics or perceptual psychology, even today?

"What is redness?" asks Leucippus's follower Democritus. Do we see red because a red thing gives off red atoms? No; atoms are simpler than that. They have only what are called primary qualities, in this case those appropriate to a particle: size, shape, position, and motion, perhaps weight, but nothing more. The atoms of the *eidolon* touch our eyes in a certain way that makes them say "red" to our mind. "Sweet exists by convention, sour by convention, color by convention; atoms and the void exist in reality." Sweet, sour, color—all the qualities that are not primary are called secondary. The idea that the particles which bring us the sights and sounds of the world have a very modest set of properties and what we really see is created by us as we receive them has prevailed since the Renaissance, but that is far in the future, and there was a long battle first.

The idea of *eidola*, attributed to the same person who gave us the idea of atoms, remained connected with it, and part of the reason is that both ideas were taken up by Epicurus. We have a careful discussion in a letter he wrote to someone named Herodotus giving a condensed outline of his theory of the material world and our perceptions of it. After judicious interpretation and editing, it says:

> There are images or patterns that have the same shape as the solid bodies we see but are very thin in texture. It is not impossible that emanations of this kind are formed near an object, and that as they flow away they retain its shape. We call them *eidola*. Furthermore, as long as nothing comes in the way to offer resistance, motion through the void covers any imaginable distance in an inconceivably short time.[3]

Epicurus was not given to abstract speculation, and he uses cautious phrases like "it is not impossible" and "nothing prevents us from believing" as he seeks to explore possibilities rather than announcing an unquestioned principle. His rules for living were simple and directly stated, and the atomic theory, from its beginning with Leucippus and Democritus, was a rock of specificity in the current of Greek thought that always flowed toward greater abstractness.

Epicurus's ideas about sensation endured in Lucretius's poem *Concerning the Nature of Things*. Here sensation is explained in terms of atoms and illustrated with analogies and examples: we can reasonably guess that we smell something by absorbing particles given off from it; and walking by the sea, if the wind is blowing our way, we can even taste the salt. Sound

---

[3] The source of this quotation is the late Greek biographer Diogenes Laertius [1925]. In making this version I have been guided by Siegel's careful analysis [1970]. A close reading of the original does not reveal whether Epicurus thinks that *eidola* are made of atoms or even whether (as Lucretius thinks) they are in the form of thin films.

also has its *simulacra*,[4] for when a speaker addresses a crowd, "the one voice splits at once into many voices," so that each ear hears one, and those unheard are wasted. And if there are *simulacra* of scent and taste, why not *eidola* of light? Lucretius specially emphasizes that *eidola* are not just a few random snapshots but represent whole sequences of actions, "so quickly they move, so numerous they are."

It was through Lucretius's poem that the theory of *eidola* became widely known. Since it was in Latin educated people could still read it even as knowledge of Greek declined, and it was so much admired as poetry that though pagan in inspiration and clearly inconsistent with Christian doctrine, it was always read and copied. There are manuscripts in Europe that go back to the ninth century, four hundred years before anyone knew more than a little of Plato and Aristotle, and the Arabs must have had it very early.

Light, for Lucretius, is something that is necessary before we can see an *eidolon*. It comes from the sun and the moon and from fires of various kinds, all of which produce not only light but also *eidola* that enable us to see them. He insists, though, that these *eidola* are of a different kind, for they do not get smaller as they approach us. He makes the peculiar statement that a fire burning on the ground seems almost the same size far away as close up, and deduces that sun, moon, and stars are "no larger than we see them to be," whatever that may mean. And what exactly is this sun? Consider what we actually see. It rises dim and red, at noon it shines brilliantly, and at the end of the day it is once more cool. A bonfire does the same thing: it starts small, grows as the wood catches fire, and dies in redness and ashes. Lucretius considers the possibility that the sun is a bonfire each day; on the other hand, perhaps it does not go out but continues its journey under the Earth. A bonfire derives its light from the wood it consumes, but where does the sun get its light? This is an important question. For him light is a substance, and he needs to know where it comes from. The sun can be imagined as a thing that creates its light, but he prefers the analogy of a spring of water. It does not create water; instead, water comes down as rain in the surrounding landscape and collects at the spring. Perhaps the sun gathers invisible heat from all around and emits it as light. Perhaps what the sun radiates is mostly heat, which produces light by igniting the surrounding air. Lucretius weaves the possibilities into his poem, leaves them open, and goes on. Having asked us to accept his two great hypotheses, atoms and the visible images made out of them, he does not insist that we accept the rest, for he is creating a tapestry out of images of Nature and reflections on life. No more than Plato does he pretend that what he writes is deductive reasoning.

---

[4] I use Lucretius's word *simulacra* here rather than *eidola* so as to reserve *eidola* for visual images.

4. Fire within the Eye

Plato's *Timaeus* gives his own version of the Creation, which like the versions described earlier is intended to make us see how everything fits together, and he does not expect us to take it any more literally than we take the myth of prisoners in a cave. Again and again he modestly labels it a likely story, but it is a good story. It enters the mind more easily than a philosophical argument, it does not assert its own truth, and it can say what he wants to say. It is rich and full of surprises, and there is no reason to summarize it all, but when Timaeus, the narrator, tells how a god designed humankind for his children to create, what he says about vision echoes what Empedocles had written seventy years earlier. It may go back beyond that to the early Pythagoreans, but Plato was the one who launched it onto the sea of ideas where we shall encounter it often as we voyage through the next two millennia.

> And of the organs they first contrived the eyes to give light, and the principle according to which they were inserted is as follows. So much of fire as would not burn, but gave a gentle light, they formed into a substance akin to the light of everyday life, and the pure fire which is within us and related to it they made to flow through the eyes in a stream smooth and dense, compressing the whole eye and especially the center part, so that it kept out everything of a coarser nature and allowed to pass only this pure element.

There are different kinds of fire, and only one of them is what we call a flame.

This visual ray joins with the ambient light to form a substance, perhaps something like a thread, that projects straight in front of the eye until it meets a surface. There it detects a radiation from the surface that reports its *chroma*, its entire condition, color, and texture, and conveys these qualities back to the observer's body (the whole body, not just the eyes), through which they reach the mind. At night the visual ray enters an alien environment. It is quenched, and we cannot see.

In Plato's *Meno*, Socrates mentions vision to illustrate another point; from this and from some comments of Theophrastus, who was Aristotle's successor in the Lyceum, we can get an idea of what happens according to the likely story. Things radiate sound and *chroma* in the form of particles of different shapes and sizes. Our sense organs are pierced by tiny pores into which these particles fit: sound fits the pores designed for sound and not those designed for light, and so on. Theophrastus writes of Plato, "He says that the interior of the eye consists of fire, while round about it is earth [i.e., solid tissue] and [watery] air. . . . The passages of fire and air are arranged alternately; through those of fire we perceive light objects, through those of the water, dark." The nature of fire is light. Because like responds to like, the nature of water must be dark, and so Anaxagoras, following this same line of thought, had been led to the conclusion, much wondered at in antiquity, that our senses deceive us and snow is really not white but perhaps even black.

Evidently the process of vision requires a very complicated action in the observer, whose body receives particles of different sizes, associates them with the direction of the eyebeam's gaze, decodes messages of shape and color, and delivers the information to the mind. No wonder our eyes so often deceive us. The rainbow is an illusion; it seems very large and very close, yet it is not there. Mirrors that are not flat give crazy images, and everyone has heard about mirages even if they have not seen one. The only source of clear truth, for Plato, is contemplation of the unchanging world of ideas, with the soul as an eye and the Good as the source of light. Much later, medieval treatises on optics recall this conviction as they spend a surprising amount of their space on illusions.

Timaeus was the only Platonic dialogue known in Europe before about 1150. All but the last part had been translated in the fourth century by Calcidius of Tyre, who provided a long commentary that quoted a few other dialogues. This version, which ends just after the discussion of color, had enormous influence in the new Christian world. First, it was the only Platonic text widely available in Latin, just as Aristotle was represented only by his works on logic. But more important, it happened that Platonic myths agreed with Christian doctrine in places where Aristotle didn't; for example, Aristotle held that the world is eternal while Plato taught creation by a loving and provident deity. It is therefore not surprising that we encounter echoes of Plato more often than Aristotle in early Christian and Muslim works.

In the middle of Augustine's discussion of light in the *Literal Interpretation of Genesis* is a layman's version of how we see, which shows that eight centuries after Plato his theory was still current:

> The shaft of rays from our eyes is a shaft of light. It can be pulled in when we focus on what is near our eyes and sent forth when we fix on objects that are at a distance, but when it is pulled in, it does not altogether stop seeing distant objects, although of course, it sees them more obscurely than when it fixes its gaze upon them. Nevertheless, the light which is in the eye, according to authoritative opinion, is so slight that without the help of light from outside we should be able to see nothing.

The Platonic influence in Augustine is very strong. For him "there are two classes of things that are known; one of them is of those the mind perceives through the bodily senses, the other is of those it perceives through itself." He distinguishes "the intelligible world, where truth itself dwells," from the sensible world, "which impinges on us in sight and touch." Of these, the first "is in itself true; the second resembles truth and is made in its image." But this is not Platonism. For Plato, the light of truth emanates from the Good, illuminating both the lower forms and the prepared mind. For Augustine the forms exist within the mind of God, and intellectual light is a divine illumination within the human mind.

Augustine's works belonged to the Church and have been copied and studied throughout its long history. He read Plotinus and parts of Plato in Latin translations, but knew little of Aristotle beyond some of the logical works. Aristotle's time of influence came long afterward, but now we must look at his theory of vision, the third of those that were born in antiquity.

5. The Activity of What Is Transparent

Aristotle was not convinced by either of the theories of vision that have been described. He did not like atoms, basically, I think, because they were incompatible with his ideas of quality and change. Suppose a peach ripens in a bowl. Aristotle would say that ripeness, a primary quality, was potential in the peach to start with and that in ripening it achieved actuality. For the atomists, on the other hand, qualities like color and flavor are secondary and change only because atoms change their positions and linkages. In this view, Aristotle's theory of qualities explains nothing.

But Aristotle, in turn, is unimpressed by the theory of vision that comes from Empedocles and Plato. "Why should the eye not be able to see in the dark? It is totally idle to say, as Timaeus does, that the visual ray coming forth in darkness is quenched. What is a quenching of light? . . . It is, to state the matter generally, an irrational notion that the eye should see in virtue of something issuing from it." And yet, in another place, he says that vision proceeds in straight lines, and he believes there is a visual ray that has something to do with it. In his *Meteorology*, to which we shall return for his theory of the rainbow, he describes an interesting medical case of a man with weak sight: "He always saw an image in front of him and facing him as he walked. This is because his sight was reflected back to him. Its morbid condition made it so weak and delicate that the air close by acted as a mirror . . . and his sight could not push it back."

Aristotle's theory of vision may use the ray, but he does not think it explains how the image of what we see is formed in the soul. For that, he discusses the purpose of vision. Earlier, we have mentioned that according to Democritus, *eidola* convey an object's visual qualities to the mind. That, says Aristotle, is not the point. "Democritus roundly identifies soul and mind, for he identifies what appears with what is true, . . . he does not employ mind as a special faculty dealing with truth." Vision sends its unedited message to the soul, but it is mind that decides what we have seen. To understand better what Aristotle says we turn to his marvelous book *On the Soul*.

Book I summarizes his predecessors and is the best place to learn what they said. Book II starts by telling what Aristotle means by soul. Recall, we have already said that a form can be a substance. He writhes in language trying to say it right: "The soul cannot be a body, since the body is the subject or matter, not what is attributed to it. Hence the soul must be a substance in the sense that it is the form of a natural body having life potentially within it. But substance is actuality, and thus soul is the actuality of a body of this kind." There are many kinds of actuality, and the one he is talking about is the one that distinguishes living creatures from everything else. Every living being has a soul, but soul is manifested in different ways. Aristotle distinguishes five powers that it can have: nutritive, appetitive, sensory, locomotive, and the power of thought.

> Every living creature must keep itself alive: this requires the nutritive power.
> Most plants stop there, but animals generally give some signs of preferring one thing to another: they avoid danger, move toward food, and so on. This is appetitive.
> Animals also show signs of sensory function.
> Most animals are able to go from one place to another, the locomotive power.
> Finally, "man, and possibly another order like man or superior to him," are able to think. (I suppose "superior" is a nod in the direction of the spirits that filled the Greek world.)

Now comes Aristotle's analysis of vision; I will use his own (translated) words where possible. When I look at a thing that is not transparent, my eyes bring me news of the *chroma* of its surface layer. *Chroma*, as he explains in his book *Sense and Sense Objects*, is what characterizes the surface of an object but may also exist inside it, even if not seen. He suggests that white and black are the fundamental visible properties of *chroma*, and that their mixture in a surface gives rise to the other colors.

Every *chroma* is able to set in movement—that is, to change—a medium that is transparent; that power constitutes its visual nature. Media like air and water contain a special substance called transparency (*diaphanēs*), and

light is the activity of what is transparent.

This means that air and water are not transparent in themselves but only when something activates the *diaphanēs* in them, and darkness is of course another kind of activity. We must imagine the transparent medium as full of images of the thing looked at, not traveling but just there.

As to why light is not a substance, we know that two rays of light can cross, but two streams of any material substance would collide. Further, Empedocles was wrong when he said that light travels at great speed. If the distance were always short we might believe him, but we instantly see things that are very far away, even stars, and it can't possibly move that fast.

So far we have only spoken of sensory impressions, but what are they, and how do they bring the mind to a knowledge of the truth? When something in the external world acts on a sense organ, says Aristotle, the organ receives the thing's form but not its material. As a piece of wax can receive different shapes and impressions without any change in its substance, so the sense organ becomes hot, green, and so on. To create a mental image of what the sensation represents, the soul uses a power called imagination. Suppose I see something green of a certain shape. Without reflection, imagination identifies it as a leaf, or even a particular kind of leaf. Any animal can do this, but often things are more complicated. Suppose I come near a crowd of people at night. My eyes and ears send their confused sensations to a faculty called common sense, which sorts them out and sets them together and tells imagination what is going on. Imagination is not a sense, since we can imagine a leaf even when it is not there; instead it is a kind of motion in the soul, and there are other motions such as judgment and thought, but we need not discuss them here.

The example of vision shows how Aristotle fits questions of perception into his scheme of soul, mind, substance, quality, and all the rest. The scheme is too complicated to be very satisfying, and he presents it only as a possible version of something that might be true. By talking in general terms he avoids the difficulties connected with *eidola* and visual rays, but of course he doesn't really explain anything about the process of vision or what the eye is doing. For him the pupil is not just the eye's entrance; it is the organ of sight, and it seems that the rest of the eye does not function in vision. He does not say why distant things look small, why we can look in a direction where there are several things and focus on them one at a time, why we see only the surface of a thing that faces us, why if we look at a coin lying on a table we don't always see its shape as a perfect circle, why we can't see something distinctly if it is too close; and to explain how we see a star requires the whole mechanism to work at distances that defy belief.

There is of course much more, but I hope this sketch will serve as a guide through some of the wilderness of Arabic and medieval European thought. Remember, though, that in Europe before the thirteenth century there is no Aristotelian tradition in philosophy, for the texts had been lost. Until then, these things were discussed in terms taken from summaries such as the one given by the sixth-century encyclopedist Isidore of Seville [1911; Book XI], who disposes of the entire Aristotelian theory in a single sentence. Aristotle began to be known among the Syrians and Arabs in the eighth century, but in the West only after 1150 when *On the Soul* was first translated. At about the same time arrived Arabic texts and commentaries bearing on both *eidola* and visual rays. It was the authority of the great teachers Albertus Magnus and Thomas Aquinas that imposed the Aristotelian view in the universities, but there was also another great authority who is important for our story but has not yet been mentioned: Galen the physician.

6. Galen and the Stoics

At almost the same time that Epicurus was finding a morality in Lucretian atomism, a contrary philosophy was being born. It started with the teachings of a Cypriot named Zeno (c. 384–c. 322 B.C.E.). He met his students in a *stoa*, a shaded space near the Athenian *agora*, and after a while they became known as Stoics. Like Epicurus he believed that the path to the good life starts with a knowledge of nature, but he imagined nature in a very different way. I will try to describe only as much of the Stoic philosophy as will be needed here.

The Stoic universe is full, and we move in it like fish in the sea. It is composed of Water and Earth, held together and given form by *pneuma*, and it floats somewhere surrounded by a limitless vacuum. *Pneuma* is a gas or vapor of Air and Fire; the word denotes breath, but the meaning is much broader, for it is nature, mind, rationality, vital spirit, even God. Diogenes Laertius wrote much later that for Stoics, "the universe is a living being, rational, animate, and intelligent," and that "Nature [as they use the term] is an artistically working fire, going on its way to create; it is a fiery, creative, or fashioning breath." The characteristic property of *pneuma* is tension; it is what holds things together and makes events happen, and it takes part in vision and the other sensory processes.

The early Stoics placed the soul in the heart, and believed that one part of the soul called the *hegemonikon* governs the rest of it. To direct vision toward some object the *hegemonikon* sends a stream of *pneuma* into the eye which sensitizes the air next to the pupil. Light from the sun or some other source activates this air, and then, as Diogenes Laertius says, "the thing seen is reported to us by the medium of the air stretching out toward it, as by a stick." From an optical point of view this is like a visual ray, but the machinery is different.

Galen of Pergamon (c. 130–200) was one of the two great doctors of antiquity; Hippocrates was the other. Pergamon was a city in Asia Minor dedicated to medical practice, and here, after study in Corinth and Alexandria, Galen sharpened his skills patching up the job-related injuries of gladiators in the city's arena. In time he moved to Rome, called by the Emperor Marcus Aurelius to be physician to his crazy son, Commodus. Eighty-three of his works have survived, and they became basic to medical study, first in Greece and Rome, then among the Arabs, and finally in Western Europe, where they were used by medical students as late as the seventeenth century.

Coming four centuries after Zeno, Galen is much wiser medically. Nerves have been discovered and plotted and identified as conduits for *pneuma*; the soul is located in the brain. *Pneuma* completes a circuit from brain out through optic nerve to eye to object, then back to eye, optic nerve, and brain. Inside the eye (see Plate 1, in the color section of this book) is the crystalline lens that is now known to be the structure that adjusts the eye's focus for near and distant objects. For Galen this is the part of the eye that actually sees: "The lens is the organ of vision, and all parts of the eye are created for its sake." By a stroke of genius he recognizes that the retina, including the optic nerve, is part of the brain, and writes, "Its chief function, that for which it was sent down by the brain, is to perceive the alterations that occur in the crystalline lens and to communicate them to the brain." He has nothing to say about how the image of a thing gets through the pupil and reaches the lens, how it is registered there, or in what form it travels through the spirit in the optic nerve to the brain. We shall see these problems attacked by the Arabs Alkindi and Alhazen in §3.6 and §3.7, by Johannes Kepler in §6.2, and in §9.8 by an army of modern workers who are starting to understand what nerves actually do.

Galen had a vast influence on the thinkers who came after him. He was thoroughly Aristotelian in his view of the world; that is, he saw it in terms of its qualities. The atomists went deeper, referring the qualities we perceive to a small number of primary properties concerned with the nature and arrangement of atoms. Their approach, when it was finally taken up again, developed into quantitative sciences that by now have had considerable success in explaining how the world works. Aristotle's long domination owes much to Galen. Aristotle opened the world to thousands of educated people, but fifty generations of doctors trained on Galen. In doing so they absorbed the Aristotelian worldview, and then they passed it to their students.

7. Aristotle Takes Aim at a Rainbow

The Great Flood ended, the waters dried from the Earth, and God allowed Noah to open his hatchway. Later, when he saw the little crowd of people and animals at the foot of the ladder looking out uncertainly over a ruined landscape, he spoke to Noah and promised that he would never do anything like that again: "And God said, This is the token of the covenant which I make between me and you and every living creature that is with you, for perpetual generations: I do set my bow in the cloud, and it shall be a token of a covenant between me and the Earth." After that, when rain flooded fields and turned roads into mud, the children of Israel looked into the sky, and from time to time, if the sun came out, they saw the sign that God remembers his promise.

In the Greek world also, rainbows had a meaning. Once when Menelaus, King of Sparta, was attending a funeral in Crete, his wife Helen ran off with a houseguest named Paris who was a son of the King of Troy. At once Hera, Zeus's wife, sent her messenger Iris racing along a rainbow to tell Menelaus what had happened. Menelaus came home, his brother Agamemnon raised an army, and so began the Trojan War. Just as the thunderbolt is the attribute of Zeus, Iris's is the rainbow. When it is seen, she is about her business, either carrying news or bringing rain, for she also uses it to draw water into the clouds.

In the sixth century B.C.E., at about the time of Pythagoras, Xenophanes of Colophon followed the style of philosophers in those days and tried to demythologize the rainbow. He wrote: "And she whom they call Iris, she too is actually a cloud, purple and flame-red and yellow to behold." But this does not explain very much. How does a cloud come to be stretched out into a perfect arc of a circle? Where does it get its color? And a cloud is in a definite place, but where is a rainbow? No matter where you are, if the rainbow is in front of you the sun is directly behind you; nobody ever saw a rainbow from the side. Also it may seem huge and far away, but perhaps you suddenly notice that its foot is between you and a nearby house. Suppose you decide to go and find it, for there is supposed to be a pot of gold . . .

Aristotle was the first philosopher to make a serious attempt to explain the rainbow; it occupies most of Book III of his *Meteorology*. After describing it and some related phenomena he starts his explanation by saying that the science of optics tells us "that sight is reflected from air and from any object with a smooth surface just as it is from water; also that in some mirrors the shapes of things are reflected, in others only their colors." Here he is speaking of the visual ray of Empedocles and Plato, for he can hardly do without some kind of ray. He goes on to explain that if a thing is small enough its surface cannot reflect an image, but it can still reflect color— think how a dewdrop sparkles in sunlight. A rainbow is never seen in a clear sky and must be produced by reflection from a cloud. Thus he concludes that a visual ray leaves the observer's eye, bounces off droplets of rain that are forming in a cloud, and then travels in the direction of the sun, behind the observer.

To explain the shape of the arc, Aristotle represents it with a geometrical diagram (Figure 2.1). Curve A represents the hemisphere of the sky, put it at any distance you like. The observer stands at the center, K, and the sun, for simplicity, is assumed to be just rising or setting, at G. The visual ray from K reflects from the cloud at some point M, and back to G. What determines how high the bow is? Here Aristotle makes a new hypothesis: that the ratio of the distances GM and KM is fixed at a certain value,[5] and if all points M must lie on the hemisphere A this forces them to lie on an arc, as shown. Now we see the reason why the distance from K to the sun is taken as the distance from K to the rainbow, in defiance of common sense: the construction does not work unless both G and M lie on the hemisphere A.

Aristotle has explained why the sun's reflection forms an arc and why the observer is always on the line between its center and the sun. By tilting the horizon line on the diagram one sees that if the sun is higher the fixed ratio

[5] By talking of the ratio he avoids having to specify the size of the hemisphere.

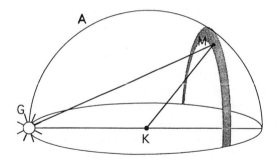

Figure 2.1. Aristotle's theory of the rainbow. A is the vault of the sky, the sun is on the horizon at G, K is the observer, and M is a point on the rainbow.

of GM to KM requires the bow to be lower; this shows why rainbows are usually seen early or late in the day and never at noon. But he has not yet explained the rainbow's color, and here the argument gets more difficult. Lacking the idea of a continuous gradation from one color to another he declares that a rainbow consists of three bands. The outer one is the largest and is red; the others, in order, are green and blue; yellow, coming between red and green, is explained as a subjective effect of contrast. Why is red at the outside of the bow? We know that light weakened by smoke or a cloud looks redder; further weakening gives green and blue. The outer band is largest and receives the most sunlight and is least weakened by cloud and distance; therefore it looks red. The other two colors result from dilution. Aristotle notes that there is sometimes a second rainbow, fainter than the first, and he conjectures that there may be more, but too faint to see. He explains why the secondary rainbow is colored oppositely from the first, with red on the inside of the bow, but the argument is tangled in obscurities, and I will not try to reproduce it.

This is a condensed version of the original explanation. The mathematics proves nothing, but the idea that the bow is formed by reflection from water droplets has endured; and in §7.8, when we have understood how that reflection actually occurs, we shall have the key to both primary and secondary rainbows and the exact heights at which they are seen. We shall understand why the colors of the secondary are reversed with respect to the primary, and glimpse with the mind other rainbows which, as Aristotle guessed, are too faint to be visible.

Finally, a novelty has been introduced, the idea that if you have expressed something in mathematical terms you can understand it better. When you think of what mathematics cannot touch—the true nature of things, for example, or their sensory properties—this seems a presumptuous claim. Perhaps it is, for Aristotle had no mathematical hypotheses based on experience to start with and so had to invent some ad hoc, but it sets the direction that physical science eventually took, and we shall watch the process as it goes on.

Aristotle's diagram represents a visual ray coming from the eye of the beholder at K by a straight line, and with that and a few assumptions he derives the curve of the rainbow. Geometry as known to Greeks of his time was largely a matter of straight lines and parts of circles, and in the following centuries it was brought to an astonishing degree of development, progressing from propositions that are easy to see to others that are hard to believe until you see a proof. The mathematicians who created this science reasoned geometrically much better than they calculated numerically, and generally the culture's mind-set assigned calculation to an inferior world of carpenters and shopkeepers.

The visual ray is a mathematical abstraction. When we look out at the world we see not in a straight line but in a sort of cone, in which things are seen clearly only in a small central part. The abstraction begins by replacing the cone by a bundle of diverging straight lines. Then finally, one draws just one of them, such as KMG, to stand for the rest. With this convention it becomes possible to treat certain parts of the theory of light by the exact and error-free techniques of geometry. The resulting science of rays was so successful that for a long time people were tempted to think that a cone of vision or a beam of light actually does consist of a bundle of rays, and we shall find people still using that mental picture in the nineteenth century.

8. Three Theories of Vision

When Thales declared that everything is made of water I do not think he expected to be taken literally. Probably the statement was part of a more general and abstract explanation of why the world is as it appears to us. Democritus has already told us that we see only the surface of things and that truth is hidden in the depths. In our time a scientific theory is judged by its agreement with experiment—a critic might call this a study of surfaces. What, then, has happened to truth? There is of course a truth about surfaces, but the deeper truth that Democritus conceived is the one that answers the question "Why?" and that question is not answered by experiment. There are people who claim that "Why?" is not even a scientific question. One must not argue about terms, but I do not think any scientist would agree with the claim. Historically the focus on experiment as the main criterion for judging a scientific idea dates from about the time of Isaac Newton, the late seventeenth century, and it has led to progress, from experiment to hypothesis to experiment to hypothesis, along many lines of inquiry. This did not happen to the theories of vision we have discussed.

Each was complete in itself; though suggested by experience they in turn suggested no experiments designed to test them, and nobody tried to develop them much further. But just for fun, let us examine them in the modern spirit.

Aristotle's theory really explains nothing; it serves only to provide a vocabulary for talking about the senses and to situate vision within the larger context of the operations of the soul. I have mentioned some of the questions not answered by the theory of a transparent medium filled with images. Perhaps the visual ray of the *Meteorology* is supposed to reach out into a transparent medium full of images and bring them back to the senses; perhaps these are two alternative theories of vision, but in no case is the theory satisfactory. The main advantage of the theory of images is that it was part of a wide-ranging and consistent theory of physical and mental function which explained how we know what we know in a way that satisfied readers for a long time.

Empedocles's theory and its progeny are a bit more plausible—if a visual ray reaches out of the eye and feels what is in front of it one can at least understand why we see only the side of a thing that faces us. It also explains why we can judge distances, for we may have some sense of how far our ray projects in front of us. But there are common experiences it cannot explain: for example, how we can lock onto a star so quickly. Of course, nobody knew in those days how far away the stars really are, but simple astronomical considerations already made it clear that their distance is much greater than the size of the Earth. There is another difficulty: the visual ray must be both wide and narrow, wide because when we look around we see many things at once, narrow because we can use it to focus on something very small. And exactly what happens to it when we look into a mirror? Finally, there is convincing evidence that the sun actually sends us something. We feel warmth in the light and can concentrate it with a concave mirror so as to ignite a piece of paper. Does the sun also emit rays, and if so what is their role in vision?

All these problems considered, there was room for another theory, but *eidola* also raise more questions than they answer. Why do the atoms of an *eidolon* stay together, even when forcing their way through water or glass? How does this ethereal thing fly straight to my eye and, stuffed through its pupil, create an image in my mind?

If one of these theories is right, the other two are wrong. Experiment, of course, would decide. There ought to be some other way of detecting flying clouds of atoms than just by receiving them in our eyes; we ought to be able to detect a beam of fire or other radiation coming out of an eye. Either theory could be blown away by an experiment. But suppose you claim that light is the activity of what is transparent and I maintain that it is not—how could we possibly decide? Experimental tests of anything were rare in the old days, and theories were argued again and again. Of the three hypotheses, the visual ray, which does not really conflict with Aristotle, was widely thought to make the most sense, and it was still being taught in the seventeenth century. As Aristotle's authority waned it became clear that terms like "activity" did not denote anything one could point to and, finally, denoted only concepts. They lasted longer inside some skulls than others. When they blew away it was through neglect and, finally, by laughter.

It may have struck you by now that though the ancients have been using words like "ray" and "image" and "soul" and "actuality" so as to give them meanings that are at least fairly definite, the word "light" has not done so well. It is there in the rhetoric, referring to something that is necessary if we are to see anything, but exactly how does it enter the discussion and what does it mean?

What do we actually see? Look at a leaf. You see a color, green, and a texture. These are its *chroma*. You see size and shape, and from the general visual context you see how far away it is. Have you seen light? No, you saw the leaf. Take a careful squint at the sun. Did you see light? No, you saw the sun. Then what is light, where does it enter the picture? Perhaps you think of a ray slanting through the air of a smoky room, but a moment's thought shows that what you are seeing is smoke. Is there really any such thing as light? Of course there is, for you can't see without it. Is light invisible? At night, in clean air, look across the headlight beams of a car. Light is there, but you look through it without seeing it.

Then what can be said about light? Though I am not aware that any early philosopher put the question this way, it is the puzzle that faced them, and that is why they had such trouble in saying what part light plays in vision. My question has answers, of course, but not within the scope of the three theories of vision mentioned in this chapter. Aristotle's theory comes closest to accommodating light, but when we have been told that it is the activity of what is transparent, we feel like someone who expected dinner and has been given a salted peanut. What kind of activity? It turned out that the fruitful question is not what light is but what it does, and the historical development of the subject breaks the question into two. First, what can be said about the behavior of light rays? How do they move (if they do move), how do they reflect from a shiny surface? How do they bring us news of the world outside? Second, what power resides in the rays? Of course, they act somehow on the eye so that we see, but that is not all. The sun sends heat as well as light; the moon sends light but also lifts the tides. Does this mean that there are radiations that we do not see, that act in other ways, to make something happen in the world?

3

RAYS

The Chaldeans, according to their own statements, believe
that a person's destiny is affected by the condition of the
moon at the time of his birth. . . . As a result, in forming their
judgments, they depend on the sense of sight, which is the
least trustworthy of the senses, whereas they should employ
reason and intelligence.
(Cicero, *On Divination* II.91)

1. What Are the Stars Saying?

THE LAST CHAPTER described some theories of vision. This one
is concerned with theories of light—not so much what light actually is as what can be said about where it comes from, what it does,
where it goes, what its powers are.

Until the end of the Renaissance there is no use trying to distinguish
between astronomers and astrologers, for no one doubted the influences
that astrology dealt with. There was little on Earth that escaped the action
of stars. As the sun (a star) rises and sets it brings warm days and cool
nights; by slowly moving north and south it brings summer and winter.
Crops must be planted or reaped when certain constellations begin to be
visible at dawn. The moon, because it is so close, produces big effects: it
controls the tides, it sets up a rhythm in a woman's body, and as the term
"lunatic" records, it affects the mind. Saint Basil, in his *Hexaëmeron*, tells
how the moon's influence changes living things: "When she wanes they
lose their density and become void. When she waxes and is approaching her
fullness they appear to fill themselves with her, thanks to an imperceptible
moisture that she emits mixed with heat, which penetrates everywhere."
He concludes, "The Moon must be, as Scripture says, of enormous size
and power to make all Nature thus participate in her changes," but he does
not think that analyzing the celestial influences enables one to predict anything at all about the tangled course of human events, for it is well known
that twins, conceived at the same moment and born minutes apart, often
lead very different lives and come to different ends. Augustine, a few years
later, mentions the question of the twins, but that is only a matter of detail.
There is a more serious question of principle: we must not try to shift the
responsibility for human choices and actions onto the stars or, through
them, onto God.

Still, no one doubted that the stars exert some recognizable influence on the course of a human life. Most of them move steadily overhead, and the sun and moon move among them, but the other planets travel in a more irregular way. Sometimes we can see several planets at once, sometimes none. What part do they play in the cosmic drama? What is the point of having the sun move through the constellations of the zodiac—what is the point of all the other planetary motions and configurations if not to give humankind something to ponder and interpret? If all Nature is one, created by a single hand and mind, these apparently aimless movements must have some relation with the world that touches us.

In the old days, for one reason or another, stars, planets, and constellations received names (the Dog, Mars, Venus, the Lion, the Maiden), and their presence, absence, or position in the night sky took on the color of these names as they helped determine the character and fate of a child at birth and the success or failure of an undertaking. The next step was obvious: if it were possible to know in advance where the planets were going to be, one could make provisions for the future.

Tradition says that it was specially in Chaldea, the southern part of Mesopotamia, that divination was studied from remote ages, and in time the Chaldeans became known for their skill. Almost nothing is known of their astrology. It came to the Greek world sometime around 400 B.C.E. but was slow to penetrate popular culture or religious practice. The omens a Greek tried to read were mostly closer to home. The flight of a bird as one stepped outside in the morning told of fortune or misfortune in the day ahead; in fact, the Greek word for omen was *ornis*, which means a bird. There were other omens. The chorus in Aristophanes's *The Birds* lists a few:

> An ox or an ass that may happen to pass,
> A voice in the street, or a slave that you meet,
> A name or a word by chance overheard,
> If you deem it an omen you call it a *Bird*.

(The galloping dactyls are from the translation of John Hookham Frere [1874]).

It was another two hundred years before Greek astronomical texts began to show much Babylonian influence, but then, as oriental wisdom spread about, those who looked for its intellectual roots found them without effort in Plato and Aristotle. *Timaeus* taught the unity of Nature and could have been designed as a first course in astrology. Plato's myth describes the design of the world by the Demiurge, a god who sought to make a material illustration, a working model, of the heavenly abstractions that are Ideas. The world he made has a soul, it is a living organism in which, as in any living being, every part has functional connections with every other. Gods and human souls are part of this great linkage. Most of the gods are stars, made of fire "so that they might be the brightest of all things and fairest to behold." Plato tells us that the stars belong to a cosmos that unifies the human and the divine. He says in so many words that God created the world, and later imagination was able to find other ways in which *Timaeus* was a divinely inspired anticipation of Christian belief. This gave authority to what it said about the astrological connections between man and the cosmos.

Aristotle adds his voice to the Platonic teaching:

> Our forefathers in the most remote ages have handed down to us their posterity a tradition, in the form of a myth, that the stars are gods. The rest of the tradition has been added later in mythical form with a view to the persuasion of the multitude and to its legal and utilitarian expediency. . . . But if we were to separate the first point from these additions and take it alone—that they thought the first substances to be gods—we must regard this as an inspired utterance, and reflect that, while probably each art and science has often been developed as far as possible and has again perished, these opinions have been preserved like relics until the present. Only thus far, then, is the opinion of our ancestors and our earliest predecessors clear to us.

I have mentioned Aristotle's doctrine of the Prime Mover, which sits beyond the sphere of the stars and for whose sake the sphere moves; that motion is the source of all change and motion on Earth. "We must," he says, "treat Fire and Earth and the elements like them as material causes of events in this world, . . . but must assign causality in the sense of the original principle of motion to the power of the eternally moving bodies." He does not say how this happens, but bequeaths to those who come after him a question that may be divided into two parts: how do the heavens exert their influence on Earth, and how does one movement on Earth act to cause another? He has already told us that everything that moves is moved by some other moving thing; in simple cases like a push or a pull this is obvious, but what can one say in general? We are going to find in the work of powerful and influential thinkers that the answer to both questions is light, and that is the reason why you have been reading a few pages on astrology.

2. Euclid Draws Pictures of Light

Sunlight makes a shaft of light in the air of a dusty room. We see in straight lines, meaning that I can see something only if an uninterrupted straight line can be drawn from it to my eye (with allowance for phenomena such as reflection from a polished surface). Clearly a rational account of the visual process starts with straight lines. Leucippus's *eidola* move in straight lines, and Empedocles's visual ray is straight. Straight lines belong to geometry, and Euclid the geometer wrote an *Optics* that contains some fragments of a mathematical theory of vision.

We know almost nothing about Euclid the man. He studied at the Academy, probably as a student of Plato's students, and sometime around 300 B.C.E. he was invited to the new Egyptian capital of Alexandria to add to its intellectual luster.

Alexander the Great admired Greek culture, even if his father had destroyed the Greek political system and reduced Greece to the status of a province. Alexander needed a city to serve as administrative center of Egypt, and in 331 he ordered one to be built in the Greek style near a mouth of the Nile. He never saw it, but eight years later while it was going up his body was brought there to be buried in a great tomb. A struggle for succession ensued from which Ptolemy, one of his generals, emerged as master of Egypt. He gave himself the name Soter (savior) and used Egypt's wealth to ornament his capital city. Like Alexander he had studied with Aristotle, and along with temples, monuments, and gardens he decided to establish a philosophical school, a sort of Institute for Advanced Study, to match, perhaps surpass the great schools of Athens. It was named *Mouseon* after the Muses; we call it the Museum. When complete its walls enclosed lecture halls, laboratories, observatories, living accommodations, and a dining hall. It may also have contained the great Alexandrian Library, which eventually housed half a million rolls and was one of the wonders of the world. As one would expect of a temple of the Muses, the Museum was directed by a priest. The king appointed its members, and they received room, board, lodging, and salary. As long as kings took an interest in it and were careful about the appointments, it remained the intellectual center of a rapidly changing Mediterranean culture. It acquired the remains of Aristotle's library and some manuscripts of the Greek dramatists, which its scholars edited for posterity.

We know almost nothing about the lives of any Alexandrian scholars, there being no Alexandrian Diogenes Laertius, and so we can only guess that most of them probably had something to do with the Museum and some may have been members. Euclid seems to have been among the first. He is known as the supreme architect of mathematics, for he collected the mathematical knowledge available in his time, provided rigorous proofs for theorems that lacked them, and arranged the whole into thirteen books so that the proof of each theorem depends only on the definitions, assumptions, and theorems that come before it. Grant the premises, and the result has to be true. To complete the structure he surely had to invent much of it; the result, almost without changes, remained the standard mathematical textbook for about twenty-two hundred years.

Euclidean geometry has served as the ideal model for many makers of intellectual systems since that time, though I can think of none but Spinoza who have tried to imitate it literally. Alas, the system has never worked very well. Its weakness is in the definitions. Even in geometry, one can ask whether the definitions really define anything: "a *point* is that which has no parts," "a *straight line* is one that lies evenly between its extremities," "*one*

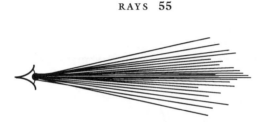

Figure 3.1. The eye sends out visual rays to sense what is in front of it.

is that according to which each existing thing is said to be one." If you don't think much of these definitions, try to do better. With great effort, mathematicians have done better; but in practice, outside mathematics, useful definitions do not have the inertia that the Euclidean scheme requires. In technical matters a definition is often a little summary of our knowledge and insights. Choose a word—"proton," "gene," or, if you will, "light"— and consider how its definition has changed as new ideas came along. Inevitably, a word means different things to different people. Experience has shown that the harder one tries to apply the Euclidean scheme in human situations, the less useful are the results. But this was the method of geometry, and there, from a practical standpoint, it worked very well. Already in the time of Pythagoras, the sixth century B.C.E., the engineer Eupalinos had saved time running an aqueduct through a mountain by surveying it carefully, then drilling from both ends and meeting perfectly in the middle. Euclid applied geometry to optics, which is the science of light and vision.

What Euclid made was only a fragment of a complete theory because he had no idea how the eye works; in particular, he did not know how an image is formed. His constructions are no more than applications of simple geometrical principles to the straight lines one can draw between a thing and an eye, but the assumptions are interesting. His theory of vision is, to begin with, Platonic: a bundle of visual rays issues from the eye and in some unexplained way senses the thing it strikes. Each ray of the bundle, considered as a line, touches only one point at the other end, but of course we see more than one point at a time. Look at one letter on this page. You see it clearly, but without shifting your eyes you can read a few words nearby. Further away, the forms are indistinct. Let us assume that the bundle consists of a finite number of rays as in Figure 3.1, dense in the middle and thinning out at the edge. This suggests that what you see would look like a cluster of points. But suppose the eye quickly vibrates the rays so that each one sweeps out a little area. Euclid assumes that is how we see; it is his principal hypothesis. Here are his first four postulates, which will be useful later (§5.4) in discussing perspective:

A. It is assumed that lines drawn directly from the eye pass through a space of great extent;

B. that the form of space included within our vision is a cone, with its

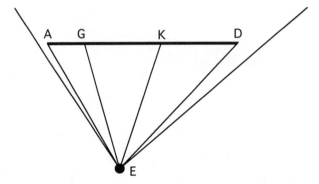

Figure 3.2. At any moment, the eye at a certain distance from the line AD sees only certain points of it.

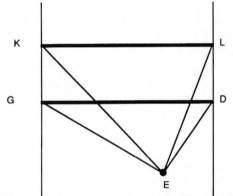

Figure 3.3. Line GD is struck by more visual rays than KL and will therefore be seen more clearly.

apex in the eye and its base at the limits of our vision;

C. that those things on which vision falls are seen, and that those things upon which vision does not fall are not seen; and

D. that those things seen within a larger angle appear larger, and those seen within a smaller angle appear smaller, and those seen within equal angles appear to be of the same size.

Starting from the postulates, Euclid then states and proves fifty-eight propositions; here are a few of them.

Proposition 1. Nothing that is seen is seen all at the same time. Figure 3.2 shows an object AD intersected by four rays coming from an eye at E. Thus only four points can be seen at any one moment, but "it seems to be seen all at once because the visual rays shift rapidly."

Figure 3.4. A distant object may fall between two visual rays and not be seen.

Proposition 2. An object located nearby is seen more clearly than an object of equal size located at a distance. Line KL, in Figure 3.3, has the same length as GD but is farther away. Every ray through KL passes through GD, but not every ray through GD passes through KL. "Therefore GD is struck by more visual rays than KL and will be seen more clearly."

Next comes *Proposition 3. Every object has a certain limit of distance beyond which it cannot be seen.* Figure 3.4 shows an object that falls between two adjacent rays and so is not seen by either of them.

I skip to *Proposition 5. Objects of equal size and unequally distant appear unequal, and the one lying nearer to the eye always appears larger.* This follows from Figure 3.3 and Postulate D. Since line KL intercepts a smaller angle at the eye E than the equal line GD, it appears shorter.

Proposition 6. Parallel lines, when seen from a distance, appear not to be equally distant from each other. We must assume that the lines extend away from the observer. The argument, using lines KG and LD in the same figure, is the same.

Euclid says nothing about representing distance in a picture, and yet the last two propositions contain the germ of a theory of perspective. For example, extend the vertical parallels of Figure 3.3 toward infinity. The angle at E gets smaller and smaller, and finally there is the vanishing point at which, in a child's drawing, railroad tracks click together. Note that even if the eye at E is off to one side and not between the parallels, the lines will still appear to meet. So much for the rails, but how to put in the cross ties? How far apart should they be? It was more than a thousand years before Filippo Brunelleschi and Leon Alberti showed how to solve that simple problem. We will get to it in §5.4, but now we must look more carefully at Euclid's theory of vision.

There are two fairly urgent questions that Euclid does not answer: what determines how many rays there are, and why do we see less clearly as night approaches? Perhaps the number of rays is determined by the level of the surrounding light—no light, no rays—but Euclid doesn't say so. One can go on asking questions. Why do some people see better than others? Why are some people nearsighted and some farsighted? How is it that we can send our visual ray to a star? On the other hand, consider how much Euclid does explain: he explains why we see clearly at the point we are looking at and only dimly at the periphery. The old theory of *eidola* cannot explain why the *eidolon* of a distant tree is smaller than that of one nearby, or why it happens to diminish in size so that just as it reaches my eye, wherever I am standing, it goes through the hole; but the ray theory assumes that the

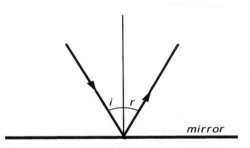

Figure 3.5. When a visual ray strikes a polished surface it is reflected with the angle of reflection, *r*, equal to the angle of incidence, *i*.

ray starts at my eye and correctly relates apparent size to distance. Looking at the diagrams just drawn, one might at first ask what difference it makes whether the visual ray goes into the eye or out of it. A line is a line. But again, when I look at a tree two bundles of rays leave my eyes and touch the tree. Reverse the process, and two bundles of rays must leave the tree and travel toward my eyes, along paths that converge toward the two pupils. But how do those rays know my eyes are going to be there at that particular moment? Are rays emitted all the time in every direction? In §6 we shall find that the Arab Alkindi says yes. But if you think not, and if you believe in the rays at all, they must start out from the eye.

Euclid's theory of vision is important for two reasons. First, like Aristotle's theory of rainbows, it is a geometrical discussion of something that happens in Nature. The older mathematicians had dealt with idealized objects which they themselves invented: straight lines, triangles, perfect circles. Surveyors and businessmen used mathematics in their work, but they rarely used those freely created abstractions. Here Euclid applies them directly. It is not just more geometry. His assumptions deal not with idealized objects but with eyes and light and the experience of seeing. He is beginning to catch Nature in the web of mathematically exact reasoning. He does not calculate any numbers in his *Optics*, but he knows how to do it if anyone wants to.

That is one reason for the historical importance of Euclid's *Optics*. The other is that the bundle of rays had an astonishingly long life. It persisted long after people had grasped that the bearer of vision goes into the eye and not out from it, long after people had learned to answer the questions I mentioned that Euclid does not answer. As late as 1800 most physicists considered a beam of light as a bundle of rays, a definite number of them, more if the light is brighter; if they followed Isaac Newton and believed that light consists of particles, they had to imagine them as moving along each ray like cars on a multilane highway. We shall see in Chapter 7 how this venerable image was ruined by the great destroyer of theories. It crumbled under the weight of facts it could not explain.

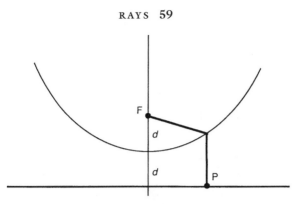

Figure 3.6. Construction of a parabola. Any point on it is equally distant from the focus F and a straight line.

Euclid's name is attached also to a work on the optics of mirrors, the *Catoptrics* (*katoptron* is Greek for a mirror). It studies the consequences of the basic law that governs the reflection of a light ray from a flat polished surface, as shown in Figure 3.5: the angles *i* (incident) and *r* (reflected) are equal. If the mirror has a curved surface, the same principle holds if one considers reflection from the exact spot the ray strikes, and the book explains, for example, what you see if you look into a concave shaving or makeup mirror from various distances and at various angles. The intellectual level of this book is below that of the *Optics*, and Euclid probably did not write it; but a century after Euclid, a remarkable work by a Greek named Diocles proved two theorems on curved mirrors that solve an important practical problem.

Suppose we need a fire but don't want to go to a neighbor's house or waste time rubbing two sticks together; we can use the sun. The idea is to focus its rays on a small spot and put something there that will burn. For this purpose people used glass lenses or concave mirrors made out of brass. Aristophanes, in *The Clouds* (c. 400 B.C.E.), says you can buy the lens at a drugstore, though because nobody yet understood the principle of lenses they were probably fairly primitive. (Strangely, not one of these household objects has survived.) But the law of mirrors was known, and mathematicians were ready to answer the question "What is the curve of the mirror that gives the highest temperature?" Diocles gives the answer, though he was probably not the one who first knew it. The original of his book *On Burning Mirrors* has vanished, but an Arabic translation turned up in the library of the shrine of Imam Reza at Meshhed in Iran, and an English version has been made from it.

Diocles's Proposition 1 says that the mirror must be made in the shape of a paraboloid. Start with a parabola, which can be defined (Figure 3.6) as a curve made up of all points P such that the distance from P to a fixed point F equals the distance of P above a horizontal line. F is called the focus

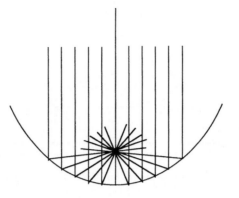

Figure 3.7. The focusing property of a parabolic reflector. All rays incident along the axis are reflected through the focus.

of the parabola, and Diocles, in a proof that takes about a page, proves that a parabola has the remarkable property that any light ray coming vertically downward onto it is reflected toward the focus, as in Figure 3.7. Now spin the parabola around the vertical axis; it becomes a bowl called a paraboloid. Make it out of brass, polish it, point it at the sun, and place a bit of tinder at the focus.

It is hard to form an accurate paraboloid; spherical bowls are easier and were generally used. How well do they focus the sun's rays? The discussion in "Euclid's" *Catoptrics* is faulty, but Diocles gets it right. His Proposition 2 proves that the circle does not focus at a point as a paraboloid does, for the sun's rays will illuminate the whole line between A and a point B, halfway to the center C, Figure 3.8. Note however that rays that strike the mirror near A are all reflected quite close to B, making a hot spot there. A spherical bowl serves fairly well. The loss of focus illustrated here is called *spherical aberration*; it happens with lenses too, and we shall see in Chapter 5 that it poses problems in the design of telescopes and microscopes.

Is light a substance or an accident? Euclid's visual rays, spurting out of someone's eye, bouncing off any reflecting surfaces on the way, and sensing distant objects, are clearly substances. Of course the sun's rays that hit Diocles's burning mirror are different physically from the visual rays, but they obey the same laws of geometry, and I think that Diocles, having burned his fingers trying to hold a bunch of dry grass near the focus of his mirror, would think the question—substance or accident?—is silly. Certainly it is a substance.

The successors of Ptolemy Soter continued to support the Museum. The first was his son Ptolemy Philadelphus; the name means that he loved his sister Arsinoe. That is a good thing, since they were also husband and wife, a custom continued by several generations of Ptolemys ending with Cle-

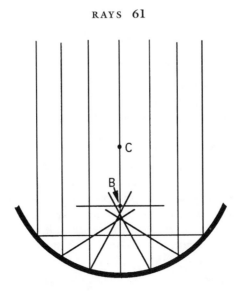

Figure 3.8. A spherical mirror does not focus as well as one that is shaped like a parabola.

opatra VI, who married two Pharaohs at once and in her extra time was the lover of Caesar and Mark Anthony. Philadelphus's successor Euergetes (he who does good works) was among the last who took an active interest in the Museum, but royal patronage, though it installed a certain number of nonfunctioning members, also assured its financial and physical safety in the troubled times that followed.

3. Hero Finds Economy in Nature

In the first century C.E. lived a very unusual character in the history of early science, Hero (or Heron) of Alexandria. We have books of his, but other than where he lived we know only one item of his résumé: that he observed an eclipse of the sun in the year 62, while Nero was emperor.

His books read like texts for students who have already sat through the elementary lectures, and they cover a variety of scientific subjects, among them mathematics, theoretical mechanics, and catoptrics. Their form suggests that he taught mathematics and physics, perhaps at the Museum. But other books he wrote go against the well-known Greek taste for intellectual structures, for example a treatise on heavy machinery and another on an instrument for surveyors. There is also a work, the *Pneumatics*, which starts by arguing from the compressibility of gases that they must consist of atoms and emptiness, and then goes on to describe a jumble of magic tricks, automata, and mechanical toys: arrangements for opening the doors of a temple or making puppets dance when a fire is lit on the altar, a mechanical doll that drinks water, a lamp that trims its own wick, and many others. If these trivia belong anywhere, it is in the other Alexandrian tradition of engineering and applied science. This was probably in its time better known and more esteemed than the kind of thinking that went on at the Museum, but the inventors left few written works, and we know only a handful of their names, so that our view of Greek science is everywhere distorted in the direction of speculative theory. That is what we have, but we should remember its incompleteness.

There is a *Catoptrics* that used to be attributed to the astronomer Ptolemy, but modern historians give it to Hero. It begins with a discussion of sight, in which the author calls attention to the speed of the visual ray. Go outside at night, close your eyes, and open them again. The ray traverses the vast distance to a star and sensation comes back again so quickly that we see it at once. "Thus it follows that the ray is emitted with infinite speed.[1] And therefore it moves without interruption, it makes no detours, but travels along the shortest way, which is a straight line."

The "therefore" isn't really a therefore, but Hero makes use of the insight as he goes on to discuss reflection from a flat surface. He asks, "What is the path of the reflected ray?" The answer has been known for centuries: the incident and reflected rays are straight lines that make equal angles with a line perpendicular to the surface (Figure 3.5). But that is not the way Hero puts it, for he seeks to relate it to the principle he has just tried to establish for a visual ray in space.

Visual rays are straight, and they reflect at equal angles. Euclid started his discussion by assuming these statements without comment. They are two of his axioms, supposed to be known facts that anyone can verify. Hero goes further and asks himself *why* these statements are true, and he finds that he can replace the two with one, that the ray follows the shortest possible path. It implies that a ray traveling between any two points follows a straight line, but it also implies by a simple geometrical argument that when a ray reflects from a polished flat surface the angle of reflection equals the angle of incidence. It does not take him long to prove the same for reflection from curved surfaces. There is, however, a catch. If the mirror is strongly curved, as shown in Figure 3.9, and the eye at A looks at the image of a dot at B, the true visual ray (if you want to talk about visual rays) is ray no. 1; but clearly ray no. 2 is much shorter. In fact, ray no. 1 is the longest, not the shortest, broken path that can be drawn from A to B. I do not know whether Hero ever noticed this, but if he had, a mathematician of his talent might have been led to some profound thoughts.

Hero's name is known to most physicists, and the reason for his immortality is that he sought the single basic principle that underlay the only two mathematical facts known in his time about how light rays behave. The principle he found was of a new kind: that Nature acts in what by some standard is the most economical way, one might say the easiest way. This

---

[1] The word "infinite" did not carry as heavy a weight then as it does now. Read: light moves with immense speed.

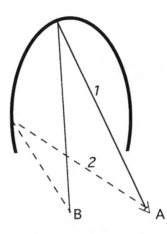

Figure 3.9. The ray reflected from A to B does not necessarily follow the shortest path. In this case it follows path 1, which is longer than any other path 2.

is not an observed fact; the principle is inferred from its consequences. It is a qualitative premise that gives rise to exact quantitative results. We shall not hear of such a principle again until the seventeenth century.

The laws of modern physics are expressed as mathematical structures that summarize what a million experiments have told us about Nature. The experiments are quantitative, the formulas are quantitative. They describe how Nature is, but they do not say why it is this way. The question "Why?" pertains to what Empedocles called the *logos* of Nature, and it seems unlikely that the answer to it would be expressed in a formula. Is it possible that if we knew enough we could answer the question in a qualitative principle? This is the program that Hero bequeaths to us, springing from an example that applies to one simple case.

Hero's short book concludes with some tricks one can play with mirrors: mount one on the ceiling so that you can sit in a room and see what is going on in the street; use a pair of large mirrors to make an image that hovers above an altar. The tricks look rather hard to arrange, flat mirrors of that size would have been very expensive, and the descriptions are probably meant only to illustrate the principles of catoptrics. After Hero there was not much more to say about mirrors; it was time to study some new optical phenomena.

4. Ptolemy Studies Optics

Though Claudius Ptolemaeus, known in English as Ptolemy, was one of the most influential thinkers who ever lived, we know nothing of his life. The name suggests a Greek-Egyptian family; Claudius shows he was a

Figure 3.10. In this illustration from Gregor Reisch's *Margarita philosophica* [1504], Ptolemy is both astronomer and king.

Roman citizen. Probably he was a contemporary of Galen, born in the reign of Hadrian and dying under Marcus Aurelius; his dates are something like 100–178. Some medieval writers thought he belonged to the dynasty of Ptolemys that ruled Egypt until Antony overthrew Cleopatra, and old woodcuts (Figure 3.10) show him wearing a crown. Today he is best known for a method of calculating planetary positions, based on an elaborate system of circular motions, that was used until the seventeenth century, but he worked in other scientific fields as well, among them observational astronomy, geography, and optics. In the middle ages he was known primarily for his astrological textbook *Tetrabiblos*. Here we look first at *Optics* and then briefly at *Tetrabiblos*.

Like Diocles's *On Burning Mirrors*, *Optics* comes to us via an Arabic version. In about 1160 a copy turned up at Palermo and was translated into Latin by the Emir (or Admiral) Eugenius, a versatile official in the Court of the Two Sicilies. Eugenius seems to have been a Greek, but his duties in that country of many cultures demanded Latin and Arabic. His copy lacked Book I (there is evidence in other manuscripts that the Arabs never had it), and it stopped in the middle of a sentence in Book V, but what remained is substantial.

Ptolemy's basic theory of vision is almost the same as Euclid's, but he carries the investigation much further. In Book I he assumes a visual cone, but it does not consist of a bundle of rays, for he reasons that if it did, in Euclid's model a very distant object like a star would tend to fall between two rays and would be invisible. (Evidently he does not believe that the rays vibrate.) His cone is a continuum, denser at the middle where vision is most acute. The visual ray strikes a surface but does not interact with it unless light from the environment is also present. The more light the better it interacts, reporting back to the eye the surface's *chroma*. This is the theoretical background, but as with Euclid, once we have drawn a line between the eye and the object it does not make much difference which way the ray is supposed to travel, and therefore nothing changes but the rhetoric if we assume that light passes from the surface to the eye.

One can tell at once that Ptolemy, unlike Euclid, is an experimenter, and this gives life to his writing, for in addition to giving propositions in mathematical form he talks about how things actually look, as seen through one eye or both. Book II contains a long discussion of optical illusions. They are called *deceptiones* in Eugenius's translation, but today many of the topics would be classed under the psychology of vision. To call them illusions shows the influence of Plato and his followers, for whom the world given to our senses is only a ragged copy of the real one. Our eyes give us contrary indications: if in the evening we walk past a house it seems to move past us in the opposite direction; but if, still walking, we look at the moon, it seems to go along with us. Try it. Shouldn't they either both move or both stand still? For another example, it is sometimes impossible to say whether a curved surface is convex toward or away from us. Ptolemy also points out that the apparent colors of things change with distance; "this is why painters choose veiled and cloudy colors for things that they wish to represent as far away."

Ptolemy's treatment of reflection from plane and curved mirrors goes well beyond Euclid's *Catoptrics*, but I skip over it because he extends optics in a new direction when in Book V he discusses how a ray is bent if it enters a transparent medium obliquely. First, he describes a little experiment mentioned in Euclid's *Catoptrics* that anyone can do and that, if you haven't, you should. You need a china bowl, a pitcher of water, and a coin.

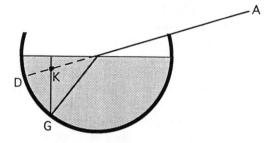

Figure 3.11. If one looks from A, a coin at G in a bowl of water appears to be at K. (This and Figure 3.12 are redrawn from those in a manuscript of Ptolemy's *Optics*.)

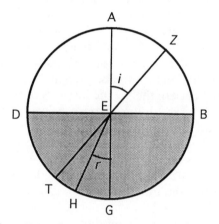

Figure 3.12. A ray from Z is refracted toward H when it crosses a surface of water or glass.

Set the bowl on the table in front of you and put the coin into it, arranging things so that from where you sit you can't see the coin without rising a little from your seat. The coin should be just barely out of sight. Now *slowly* (says Ptolemy) pour water into the bowl, and as the water rises the coin will come into sight. Figure 3.11 explains what happened. Your eye is at A, and initially the ray AD, grazing the edge of the bowl, misses the coin, G. But if there is water in the bowl, the ray bends downward where it reaches the water, and so touches the coin. Of course, when you look at things, your eye assumes that the ray is straight, and so it reports that the coin has risen to the point K. (Plate 3 shows a common effect of refraction.)

The ray bends, but how much does it bend? Figure 3.12 diagrams an apparatus Ptolemy built to find out. It is a vertical disc on which the arcs AB and DG are divided into degrees, ninety in each. Stick a marker (perhaps a bit of colored wax) at the center, E, and then mount the disc in a tub of water so that the line DB exactly coincides with the surface. Stick another marker at some point Z. "Sight along the surface of the disc until the

TABLE 3.1
Refraction of a Light Ray Entering Water

Angles measured in degrees			
Ptolemy's Data			Modern Numbers
Angle of incidence (AEH)	Angle of refraction (GEH)	Difference	Angle of refraction (GEH)
10	8		7.5
		7.5	
20	15.5		14.9
		7.0	
30	22.5		22.0
		6.5	
40	29		28.8
		6.0	
50	35		35.1
		5.5	
60	40.5		40.5
		5.0	
70	45.5		44.8
		4.5	
80	50		47.6

two markers at Z and E seem to coincide on the visual ray. Now take a small pointer and move it along the arc GD until the tip of the pointer at H is masked by the two markers. You will see that the arc GH is always smaller than the arc AZ," even though as you sight along the three points they seem to be in a straight line. Comparison of the two arcs shows how much the ray is refracted (i.e., bent) at the surface of the water.

Table 3.1 shows the values Ptolemy found for the angle GEH when he varied AEZ in steps of ten degrees, together with a little trick which shows that there is something special about his numbers. The third column shows the difference between successive values of GEH, and their regularity is obvious: the numbers increase in equal steps. This regularity, for Ptolemy, is the law of refraction, just as the rule of equal angles is the law of reflection, and he finds the same law when he does a similar but more difficult experiment on a half-cylinder of glass. Considering that they are naked-eye observations, his results are excellent; but results measured only to the nearest half-degree are not accurate enough to reveal the true relation between the values. The table's fourth column shows values he might have obtained had his measurements been more accurate—I say "might" because exact figures would depend on the temperature and purity of the water, but these are close enough. If you try the subtraction trick with them you will find it does not work; the differences do not progress in even steps. Ptolemy's numbers are the result of a measurement and a guess. He knows he cannot measure any better than about half a degree, he guesses that the numbers must have some simple pattern, he nudges his results so as to reveal one. Alas, his pattern is not really there. Does this mean that the numbers have no pattern in them? No; with better numbers he might have found the right one, but the opportunity passed. He was almost the last great scientist of the Greek and Roman culture, and it was not until about 1600 that the right one was discovered, almost simultaneously, in Holland, France, and England. It is known by the name of a Dutchman, Willebrord Snel (also Snell, but Snel seems to be more authentic). We shall come to it in Chapter 6.

Apart from its intrinsic interest, Ptolemy's theory of refraction had an important application, for when an astronomer tries to measure the position of a star or a planet, the Earth's atmosphere bends the ray to it (or from it) just as water in the bowl bent the ray going from your eye to the coin. The effect is substantial at low altitudes. If the star's apparent elevation above the horizon is 5 degrees, the ray is bent through more than 9 minutes of arc (a minute is 1/60 of a degree, and the moon's disc is about 30 minutes wide), which affects observations with the naked eye. Since Ptolemy lacked numbers he could describe the nature of the effect but could not calculate its magnitude. *Optics* discusses the problem, but *Almagest*, his treatise on astronomy, does not even mention it.

Ptolemy's *Optics* goes on to study refraction at the plane or curved surface of a transparent solid. He studies the refraction of rays passing transversely through a cylinder; next comes a spherical surface, and then the text breaks off. At this point we are halfway through a lens, but we have to get the light out through the back surface before we really have a lens. Did Ptolemy continue? Did he get it out? We shall probably never know. Perhaps his mind stopped there, in the middle of a sentence, as Bach's stopped in the middle of a fugue. If he had continued we might have had a theory of the formation of optical images then and there, fifteen hundred years before Johannes Kepler, and the histories of astronomy, medicine, warfare, and other arts would have been different. My guess is that he didn't get it out.

Ptolemy did not discover the phenomenon of refraction, but he was the first we know of who measured it. He then scanned the data to see whether it contained a general law, and found that it did. The result of a measurement is a number, but Ptolemy asked a different question: what is the *relation* between the angle of incidence and the angle of refraction when a light ray enters a transparent substance? He found approximately what it was; in about 1600 came the more exact form, and in §6.6 we shall see that a mathematician named Pierre Fermat soon showed how one could understand it by giving a twist to Hero's principle of the shortest path.

By Ptolemy's time the visual ray that Empedocles introduced to Greek philosophy had dominated the theory of vision for six hundred years, and it was still far from the end of its life. Its main competitor, since Aristotle was little known, was the atomists' *eidolon*. This interested the Epicureans, but with the exception of Lucretius most of them did not really care about scientific questions, and *eidola* were mentioned more often around the dinner table than in the lecture room. Considering the theory's conceptual difficulties, one might take this as a sign that *eidola* had no future, but we must remember that visual rays had their difficulties too, and that the people who wrote about them were mostly mathematicians, happy to explain vision by drawing straight lines on a piece of paper. For people who were not mathematicians the idea of the little image, flying through space and entering the mind of the observer, never mind how, was easy to imagine and pleasant to think about; it flew for a long time, and we find it mentioned as a matter of common knowledge among educated people in the eighteenth century.

Tetrabiblos is the abbreviated title of Ptolemy's astrological work [1940], the *Systematic² Treatise in Four Books*. It was translated from Arabic into Latin in 1138, was widely copied, and remained an authority on the subject for centuries thereafter. As late as 1553 it was issued in a new Latin translation edited by Philipp Melanchthon, who was Martin Luther's philosophical collaborator. *Tetrabiblos* and a ninth-century Arabic work called *The Greater Introduction to Astronomy* by Abu Ma'shar (known in Europe as Albumasar), which arrived at the same time, introduced systematic astrology to Europe.

Any tendency we might have to dismiss *Tetrabiblos* as unworthy of a great master should be fought down, for it and *Almagest*, which is the treatise on the calculation of planetary movements, came out of the same hemisphere of Ptolemy's brain, and they support each other. He says at the beginning that astronomy allows two kinds of predictions. We can accurately calculate the configurations of the stars and planets at any future time. That is the subject of *Almagest*, which speaks no word of what we would call astrology. *Tetrabiblos* tells about the other kind, less certain, which shows how those configurations influence earthly events. Ptolemy's list of the ways the sun and moon affect temperature, weather, seasons, and tides contains a few surprises: the lunar phase changes the flow of rivers and the growth of plants and animals. Obviously sun and moon, and presumably the other planets also, affect existent things as well as processes of germination and fruition. But how? That is a very difficult question. In principle it could be answered by comparing planetary configurations with records of actual events, but the configurations never repeat exactly, and one does not know how much variation the system should allow.

What is the use of even trying to predict these things, if what will happen will happen anyhow? Here Ptolemy's answer is unequivocal: the question's

---

² The Greek word, *mathematikés*, means something like "learned" or "systematic"; there is little mathematics in *Tetrabiblos*.

premise is false. If you let an infected foot go untreated, you may die from the consequences, but a little care makes the danger much less. If your horoscope predicts that you will catch the plague in Alexandria in your thirtieth year, you should arrange to be somewhere else. It is not that Fate seeks out wherever you are; rather it is the sky, acting in its orderly way, that determines what is going to happen. In §1.6 I mentioned that the prevailing assumption of Roman astrology was inescapable destiny. In that view the sky is only an index of decisions made at a higher level. Ptolemy thought otherwise.

Ptolemy then goes on for two hundred more pages to classify the planets and constellations according to the tendencies they show in their various aspects. Jupiter, Venus, and the moon are generally beneficial, while Mars and Saturn are harmful. He relates the various regions of the known world and their inhabitants to stellar constellations, then continues with the construction of horoscopes and the kinds of events and illnesses they can predict. Finally we study the stars' influence as life goes along, on money and career, on marriage and children, and finally, on the manner and quality of death.

Ancient Rome was full of astrologers. They were known as Chaldeans, and some probably were. Periodically they predicted catastrophe and death in the imperial family accompanied by riot and turmoil. This led to civil unrest even when nothing was wrong in the palace, and caused them again and again to be outlawed. Tacitus's *History* defines astrologers as "a class of men who betray princes and deceive the ambitious, and in our state will forever be both forbidden and retained." Revelations of how certain decisions were made in the White House under President Ronald Reagan show how slowly old habits die.

From about 1300 to 1700, astrology played a large part in almost every European activity: medicine, pharmacy, navigation, psychiatry, politics, literature, warfare, farming, administration, and many others. Churchmen from priests to popes asked the stars for advice on difficult choices, but they clashed their weapons if an astrologer claimed that stars usurp our willpower. Nobody doubted that they influence our states of mind, but the Church taught that the influence was no stronger than that of a bad cold: it may lead us to make a mistake, but we are no less responsible for it.

How does it all work, anyhow? Ptolemy has little to say. As so often in Antiquity, we wander in uncharted realms of analogy. The sun's radiation affects our environment in obvious ways, and therefore the radiations from the other planets must also have some effect. The moon pulls the tides with a tremendous force out of all proportion to its cool pale light, and therefore one would expect that the other planets act with a strength that is not measured by their brightness. A mechanism was needed to make the discussion more plausible. After a while it was provided by Arab philosophers, but first came one more Greek.

5. Philoponus, the Last Greek Philosopher in Egypt

The Museum continued for two more centuries after Ptolemy. Outside its walls the later history of Alexandria was turbulent as civil and religious disputes erupted into warfare. The library suffered damage, but scholars still edited and published the classics, and Romans still came to study them. Research continued amid riot and destruction that increased in ferocity as time went on. In 313 Constantine's Edict of Milan announced that Christianity would be tolerated in the Empire. Soon it became the state religion, but the Museum kept its pagan tradition. In 369 the Emperor Theodosius ordered the destruction of Alexandria's pagan temples, and on this hint the Museum's last buildings were wrecked by a howling mob. A cloister was erected in the ruins where illiterate monks wandered among empty shelves while the remaining scholars, taking what rolls they could carry, dispersed like leaves blown by the wind. Some went to Constantinople where they added to the flow of Greek philosophy and literature eastward to Damascus and Baghdad; from thence translated into Arabic after several centuries, the old books once more returned to universities in Cairo and Alexandria. When they were destroyed, the Museum and the library had lasted about seven hundred years.[3] For most of this time Rome was the center of military and political power, but Alexandria, the seaport of Egypt and a point through which passed much of the commerce between the Orient and the Mediterranean world, was the center of wealth, intellect, and style.

A few scholars stayed in Alexandria. One of them, a mathematician and Neoplatonist philosopher named Theon (c. 320–c. 90), put together an edition of Euclid's *Elements* that is the ancestor of nearly every manuscript of the book that has come down to us. In this he was probably helped by his daughter Hypatia, who legend says was as wise, learned, and eloquent as she was beautiful. She lectured in the schools, wrote mathematical commentaries, including one on Ptolemy, and became head of a Neoplatonic academy. Though she remained a pagan she seems to have been on good terms with the Christian intellectuals, and two of her students later became bishops. But there were also other Christians, who in 415 pulled the middle-aged woman from her carriage and tore her body to pieces in a nearby church. After this time Alexandria was a purely Christian city until 642 when the Arabs arrived, but before that happened one more Greek philosopher had something to say about light.

John Philoponus (this means John the Industrious, but the Arabs later knew him as Yahya al-Nahwi, John the Grammarian), was active c. 520–50 and one of Alexandria's last Christian professors. He was keenly interested in philosophy and viewed the Aristotelian tradition with healthy skepti-

[3] Since 1981, UNESCO and the Government of Egypt have been raising funds to establish a new library in Alexandria on a scale to match the old one.

cism. He argued that since God created Earth and stars at the same time, they are probably all made of the same substances, pointing out further that because the sun illuminates the moon just as it illuminates anything else, the sun is probably made of fire and the moon of ordinary matter. In this way and many others he confronted Aristotle, for whom the stars and the heavenly spheres that carry them are made of the fifth element, ether, and their nature is divine, and he launched a strange attack on Aristotle's theory of vision.

In Aristotle's definition, "light is the activity of what is transparent," the word "activity" (*energeia*) has a range of meanings. It can mean actuality as opposed to potentiality, and that seems to be how Aristotle uses it here, but John takes it more literally. It is impossible, he says, that light is only a state of being, because it has properties and produces effects. Light rays are straight lines, and light from the sun, if concentrated by a concave mirror, generates heat. Besides, if, as Aristotle says, all the air around us is full of the images of things, why can we see only what faces us, and why do distant things look small? John's solution is to interpret *energeia* as referring to some incorporeal substance that carries images. "The *energeiai* of the colors and light are emitted in straight lines and reflected under equal angles. For this reason images appear in mirrors, not because our visual rays are projected to the objects but because the *energeiai* of the objects are projected in our direction. And therefore we do not see things behind us, because the *energeiai* move in straight lines toward our eyes." He can explain why we see less clearly at a distance by saying that *energeiai* are weakened by distance. But these *energeiai* are not the same as the *eidola* of Democritus and Lucretius, for they are activities, not things any more than sound is a thing, and by giving them a name John spares himself the embarrassment of having to say exactly what they are.

Greek science produced three different theories of vision: rays, *eidola*, and Aristotle's images that float in the air. John Philoponus has combined three into one, but he has not brought us any closer to the central mystery: what is an image, and what actually happens inside our eyes? The next developments came from other people in other places.

6. Alkindi's Web of Radiation

The Prophet Mohammed died in 632, leaving the Arab world in a state of divine fermentation that sent forth waves of energy and change where old regimes were foundering under their own troubles. One wave traveled eastward through the Persian Empire and entered India, another swept westward across North Africa. In ten years it took possession of Egypt and the sandy remnants of Byzantine power, then swung north through Spain until exactly a century after Mohammed's death it was halted near Poitiers by a Frankish army under Charles Martel. In the Near East, Arabian tribes slowly awoke to a sense of national identity and unity, and in 749 they threw off

Persian rule and founded a state of their own, known as the Abbasid Califate. In 762 Calif al-Mansur laid the first stone of a new capital at Baghdad.
Four years later the gardens and palaces of the city were finished, and for the
first time Arabs had a court in which the arts and graces could take root.
Scholars, some of them Nestorian Christians, came from Syria and Persia
bringing Greek science and literature to enrich the culture of Islam.

In 786 the young and brilliant Haroun al-Rashid inherited the Califate.
Educated in Persia and liberal in his views, he opened a brief golden age,
celebrated in the *Arabian Nights* as a time of chivalry and music, poetry
and military conquest, but the state was administered, lacking any Arab
tradition, under the old Persian formulas of oriental despotism.

Even before Haroun, isolated scholars had been translating philosophical works from Greek into Arabic, at first a difficult process because abstract terms were scarce in the vivid and allusive Arabic language. Haroun
subsidized these efforts, and at the same time that Charlemagne and his
peers were trying to learn to write their names, he was sending agents to
find Greek and Syriac manuscripts for his translators' workshop so that he
and the scholars around him could read them and learn from them. First
priority was given to medical texts. Aristotle came early, but most of the
manuscripts were in Syriac, and by the time they had been translated
a second time for readers in Baghdad they had lost much of their Greek
specificity.

After Haroun died in 809 the usual power struggle began, and it was ten
years before his victorious son al-Ma'moun, with flags waving and trum-
pets blowing, entered the desolate ruins of Baghdad. He rebuilt them into
the most magnificent city in the world, but it rarely knew peace thereafter,
for the sheer magnitude of its wealth overwhelmed the old virtues of the
desert. Haroun had left the treasury overflowing; the *Arabian Nights*
hardly exaggerates its wild superfluity. When al-Mutadid came to power in
892 it was empty, and his court, including its retinue of poets, musicians,
and scholars, was supported hand-to-mouth by confiscatory taxes that only
increased the turmoil of the state.

It was under al-Ma'moun that Yaqub ibn Ishaq al-Kindi (c. 801–66),
the first Arab philosopher (in Arabic, *faylasuf*), perfected his theory of rays.
Al-Kindi, written Alkindi in the Latin texts, was born in Kufa, south of
Baghdad, but he spent most of his life in the capital. He learned some
Greek and was appointed editor for the Calif's board of translators. During
those troubled times he and his students kept out of sight in a world of
violence and disorder. He wrote books on philosophy, science, and music.
His book *De radiis stellarum*, or *On Radiations from the Stars* (1974),
which survived only in a twelfth-century Latin translation, shows that for
him astrology was more philosophical than practical, a study of the celestial
order rather than a way of reading the future.

> With Earth's first Clay They did the Last Man knead,
> And there of the Last Harvest sow'd the Seed;

And the first Morning of Creation wrote
What the Last Dawn of Reckoning shall read.

(Omar Khayyam, *Rubayyat*, trans. Edward Fitzgerald[4])

The Persian mathematician sings sadly of a world in which no moral effort we might make is of any use; the future is already decided. But is it really that bad? Did God give us imagination and courage and enterprise for no reason? Alkindi has thought deeply about God and destiny and speaks in favor of our ability to guide our own lives, if we know how.

To follow his reasoning, let us return to Aristotle's explanation of why anything happens at all. It is, he says, because the heavenly spheres move in response to the Prime Mover, and their motion, in turn, produces change in the world we live in. The implication is that if the spheres stopped turning the potter's hands would cease to move, and his wheel would stop also. Aristotle never says how the spheres exert their power, but Alkindi knows: it is by the rays they emit. Not just the spheres or the stars but everything that actually exists emits rays in every direction, so that the whole universe is causally bound together by a web of radiations filling all space, rays going from everything to everything else. (This means that wherever our eyes are, rays will come in.) Rays originate in substance and act on form; they are what link the two. Substance, composed of the four elements, remains fixed, but every quality except its composition changes in response to the radiations it receives.

In Alkindi's theory of vision, visual rays connect our senses to the world by changing the form of the air in front of us so that it becomes able to transmit qualities of shape and *chroma*. Otherwise his theory is essentially Ptolemy's. The rays are shaped like little cones, with apex at the eye. These cones have length, breadth, and thickness, and exist as physical things. The sun's rays allow us to see, but they do more than that. They give warmth, and if concentrated by a burning mirror they produce heat. Medicines radiate their effects throughout the body, and a magnet's radiation attracts a nail. Sound is also a radiation, but the vast majority of radiations do not reach our senses. They carry influences that we do not understand and cannot interpret; Alkindi calls them metaphysical. Every star is differently composed, and so each one acts differently on the elements in a material object. The interactions vary in strength and direction as the stars move nearer to us or farther away. A ray from the center of a star toward the center of the Earth is strongest; if it strikes obliquely its effect will be weaker, unless it

---

[4] Fitzgerald's translation is in Edwardian style. A more literal version (cf. Thompson 1907) would be:

> On the day They harnessed the horses of Heaven
> And ornamented it with Jupiter and the Pleiades
> Our lot was recorded by the Divan of Fate.
> How have we sinned, if They decreed our destiny?

But the sense is the same.

happens to be strengthened by a ray from somewhere else. We ourselves radiate; the desire and hope we experience as feelings are actual substances that act upon the world to produce changes in the forms around us.

All this was not entirely new; remember that Democritus has already said that our moods and emotions give off *eidola* that are too thin for us to detect except when we are asleep, and they enter our dreams. According to Alkindi, if we imagine a new form that we hope will appear, our desire forms an image in a vaporous substance known as spirit, which like the *pneuma* of the Stoics is stored in the brain and runs along the nerves.[5] Spirit radiates, and so if we want something, our desire acts on the whole universe to make it a little more likely that we will have it. Precisely because they are reinforced by desire and intelligence, the radiations of a voice can be powerfully effective, but just as all voices sound differently, so their radiations have different effects. When the right voice says the right words, scorpions and wolves are expelled, earth is warmed, wind is stilled, fires are extinguished, water floats a piece of iron. And one can intensify a star's helpful influence by invoking it with the right words at the right time. Prayers, magic formulas, and curses all set the world in motion, though often the effect is so small that one does not perceive the result.

The universe interacting in this way seems unimaginably complicated, and of course it is, as we know from daily experience; but for Alkindi it is not chaotic, even though it may seem so, for God has established a harmony in which every radiation and every material substance plays its part. Our free will dances to this harmony, and the result is a cosmic order so perfect that if one could only know it perfectly one could unlock both the past and the future. But this is not possible, for there must always be open water for the free soul to sail its little boat.

Alkindi's book goes on with examples of magic operations: the use of images and emblems, the effectiveness of animal sacrifices, the choice of exactly the right moment to begin an undertaking. Outwardly the world he describes is very like the one we know: unpredictable and uncontrollable; an astrologer may know where the planets are but can know nothing of the millions of other radiations that combined with planetary rays at the moment of birth to form the actual person, and so the horoscope cast at great expense often turns out to have little to do with who we are or how our life progresses. The doctor's medicine succeeds or fails in its purpose; our dearest wish comes true, or else it doesn't. Seen in this way Alkindi's theory makes perfect sense, much better sense than an astrologer who claims to know what cannot possibly be known.

If Alkindi's theory gave little useful information about the world, what was the point of it? I think it ministered to the religious sense that behind the world's chaos and suffering there is some unity, that merciful God has planned some good end to the confusion, simply because the world is One, and that One must in the end be good.

---

[5] Spirit will be further discussed in §4.8.

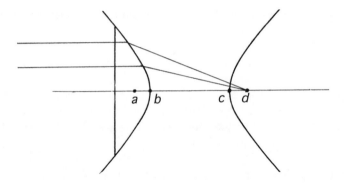

Figure 3.13. Ibn Sahl's lens is the lens-shaped figure bounded by the hyperbola on the left. If the hyperbolas are constructed so that the ratio of the distances ad/bc equals a number called the index of refraction of the glass (defined in §6.4), all rays entering the lens from the left parallel to the axis will be brought to a focus at point d, which is, appropriately, called a focus of the hyperbola.

7. Alhazen Writes a Book

Through decades of war and destruction alternating with years of gorgeous magnificence, the Abbasid Califate flourished for about a century, ruling from Egypt to the Indus River. Thereafter it slowly crumbled as local commanders began to call themselves Sultan. From 868 Egypt was ruled by a series of ambitious conquerors, and in 969 it came under the domination of the Fatimid dynasty, which like the Abbasids reigned in wealth and safety for a hundred years before it fell victim to a greedy and incompetent ruling class. William Blake says that the road of excess leads to the palace of wisdom, but in Africa, again and again, it has led only to more excess, to misery and desolation. By the end of another century of decline Moslem and Crusader armies were chasing each other around the Egyptian deserts, and in 1171 the Fatimids were swept away by the Crusaders' most admired opponent, Salah al-Din, known in the West as Saladin.

Sometime around 984 a mathematician named Ibn Sahl attached to the court at Baghdad produced a treatise on burning mirrors and lenses that has recently been reconstructed and partially translated. It discusses the focal property of a paraboloidal mirror (§3.2) and then goes from bronze to glass with the question "Suppose a burning lens presents a flat face to the sun: what must be the shape of the other face that will focus all the rays at a single point?" The answer is that the shape is a hyperboloid (Figure 3.13).[6] Ibn Sahl's proof shows he had a mastery of geometrical reasoning and also a precious piece of knowledge that had to be rediscovered long

---

[6] *Hyperboloid* is defined in the glossary of technical terms at the back of this book.

afterward: Snel's exact law that relates the angles of incidence and refraction when light passes from glass to air. He had everything that was needed to create a theory of optical instruments more than 725 years before Kepler—except, apparently, the concept of an optical instrument. He only wanted to light a fire.

Why did Ibn Sahl not use Ptolemy's form of the refraction law? The data available at the time might have shown that his new law is a little better, but they could not possibly have established it as correct. I think the reason was mathematical convenience. With the new law the geometric construction, though very difficult, finally yields a hyperbola, a beautiful shape studied since antiquity; whereas Ptolemy's law would have led into uncharted mathematical swamps. Seen in this way, the new law is only a handy approximation to the old one.

In about 965 Abu Ali al-Hasan ibn al-Haitham—in Latin, Alhazen—was born in Basra, south of Baghdad. Almost nothing is known of his life. There is a story that he disentangled himself from an irksome political appointment in Basra by feigning insanity; he moved to Cairo and established himself at Alkhazar mosque, a theological university that still exists. Here he spent the rest of a long life, supporting himself by making each year one copy of Euclid and one of Ptolemy's *Almagest*. He died about 1041 in what had become a decaying despotism, after having written something like 120 books.

One book of Alhazen's, called *Kitab al-manazir*, or *The Book of Optics*, is a classic, a masterpiece of observation and deduction. Some time around 1200 it was translated into Latin as *De aspectibus, On Vision*, but manuscript copies spread slowly in Europe, partly because of the book's bulk, 288 dense quarto pages in the printed edition of 1572. Once established, it dominated teaching and speculation on optics until after 1610, when Johannes Kepler answered some of the questions it did not. The ideas that revolutionized the theory of vision are in Books I and II.

The difference between Alhazen and his Greek predecessors is clear from the first sentence. Euclid and Ptolemy follow the familiar classical method of announcing a set of axioms: take them or leave them, they are assumed. Most of Alhazen's book draws conclusions from what he has observed. Does light (in the absence of mirrors and refracting surfaces) travel in straight lines? The fact is crucial to his theory of vision, and he verifies it in many ways, with sunlight but also with the light of moon, stars, and flames. Light is defined in terms of actual perception. What we see is things. What then is light? First, there are sources of illumination. He proposes that light is an accidental quality of a thing that is illuminated by a source. When this happens, the thing becomes luminous and is itself a source, and of course there are also some self-luminous sources. But what happens in the medium between the source and the illuminated thing? Alhazen's answer is that there are no perfectly transparent media; each point of the intervening air is illuminated and becomes a new source of illumination that spreads outward in every direction and in this way is propagated from point to point until it reaches an opaque object. We will find a version of this idea in Chapter 4, where it is termed the multiplication of species, and it appears again in §7.7 as Huygens's principle. What is the relation between light and color? Every observation shows that the quality of color always accompanies the quality of light; the two are inseparable. When we perceive the form of something, what is the nature of this form? Perhaps we see a shape approaching and think, "That is a dog." Aristotle explains that the form of the dog has entered our soul. But suppose we pick up some strange little thing from the ground and at first have no idea what it is. In that case we may think for some time before any form enters our soul. If so, what was the form doing while we were thinking, and what if we have made a mistake and have got hold of the wrong form; can Aristotle tell us where that one came from? No, says Alhazen, the incoming form is purely visual. If we have never seen a dog before we will not recognize the first one we see. Recognition is the result of memory and inference.

Early in his book, Alhazen sets out to annihilate the idea of the visual ray: "We find that anyone who looks at a very strong source of light will experience both pain and damage to his eyesight; if he wants to look at the Sun he cannot do it because of the pain." Clearly, something is coming into the eye. Alhazen tells how to produce afterimages: look at something white in the sunlight, then go into a dark place—your eyes will still register it. He notes that you can see fine details of a thing only if the light is not too strong or too weak. He has tried everything, and the weight of evidence convinces him that "there is no vision unless something comes from the visible object to the eye, whether or not anything goes out." Alhazen has physical reasons also:

> If vision occurs only through something that issues forth from the eye to the visible object, then that thing is either a body or not. If it is, then when we look at the sky and see it and contemplate the stars, there issues at that moment from our eyes a body which fills the whole space between the sky and the earth without the eye losing anything of itself. But that is quite impossible and quite absurd. Vision does not, therefore, occur by means of a body that goes out of the eye. If, on the other hand, the thing that issues forth from the eye is not a body, then it will not sense the visible object, for sensation belongs only to animate bodies. Therefore, nothing issues from the eye that senses the visible object.

The visual ray is superfluous, and he ignores it thereafter.

Alhazen is criticized for using many words when few would do, but to me he communicates the joy of someone who investigates an old subject in a new way and finds that every observation can be explained by the same simple model. He is ignoring the traditional philosophical method that bases its arguments on unquestioned principles.

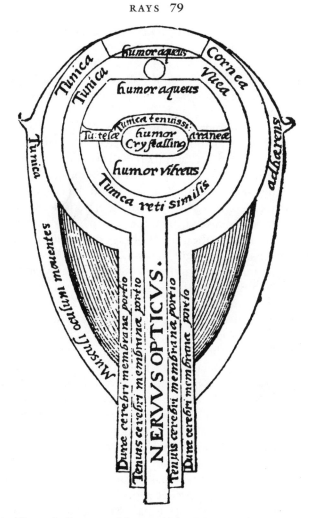

Figure 3.14. Alhazen's diagram of the eye (from Risner's edition of 1572). *Tunica adhærens* anchors the eyeball; *tunica cornea*: the tough corneal layer; *tunica uvea*: layer resembling a grape and containing the pupil (the small circle); *tunica telæ araneæ*: layer as fine as a spider's web; *tunica reti similis*: layer like a net, the retina. Aqueous, vitreous, and crystalline humors are shown, and the optic nerve, enclosed in two layers of cerebral membranes, leads to the brain.

Next in the *Book of Optics* comes the anatomy of the eye, illustrated in Figure 3.14 which is taken from the Latin version and can be compared with Plate 1. There are some changes in terminology, one of which is important. What we call the crystalline lens Alhazen calls the crystalline humor, since he has no idea of the function of a lens. And he has placed it at the center of the eyeball, whereas it should be near the front. He did not perform the dissection himself; Islamic law forbade that, but, as he says, "everything we have said about the inner layers of the eye and their composition has already been written by anatomists in books of anatomy." There is a reason for putting the crystalline humor where he did. Noses and ears can come in a variety of shapes, but in the old philosophy the eye is the noblest of the senses, and this is why all eyes are in the shape of a sphere, the shape of the cosmos; for the same reason the crystalline humor, as the most important part of the eye, must be at the center.

In Book I Alhazen summarizes his theory of vision: "If the object is opaque, then it is colored. And if it is irradiated by any light whatever, the light will be fixed in its surface, and there will arise from its color and from the irradiating light a form which extends through the air in every direction. Reaching the eye the form will affect it, and the eye will thereby sense the object." Because color is a form, nothing material leaks away from the object, and it is not diminished; one does not have to answer the embarrassing questions raised by the theory of rays or of *eidola*. With light it is different, for every source of light that we can study consumes something. When a candle has burned down it goes out; this is easily understood if we assume that the light it gives off is a substance.

Now Alhazen has to decide what part of the eye actually registers color. Following Galen, he proposes that it is the crystalline humor. Light and color enter it, but since the humor is a material object the light interacts with it, and from this, somehow, a message passes to the soul. How do we actually see the image of something? Take the simplest case, a point of light, say a star. It sends its rays in every direction, and Figure 3.15, which uses a cleaned-up version of Figure 3.14, shows the little bundle of rays that reach the pupil. If they all continue they will illuminate the whole surface of the crystalline humor, and we will see a blur, not a point of light. But only one ray of the bundle strikes the pupil perpendicular to its curved transparent surface; all the others make a small angle with the perpendicular direction. Ptolemy and his predecessors have studied refraction enough to know that if the front of the eye is part of a sphere, a ray entering along a radius passes straight through and the others are slightly refracted. Now suppose that refraction weakens those other rays so that they do not register on the crystalline humor. The single ray that is left is the one that goes exactly through the center of the eyeball and hits the surface of the crystalline humor at a single point. That, then, is perceived by the soul as a white dot, and thus we perceive the star.

So much for a point of light, but what about a tree? What happens when we see something more complicated than a star? Alhazen's answer is his most original contribution. He says simply that the problem has just been solved, for the surface of any object consists of points, and we have seen how each point produces its image on the crystalline humor.

If each point of the object corresponds to one point of the image, then the whole image will be a faithful small-scale representation of the object, and if the crystalline humor is the part of the eye that receives light and forms, the image will be clearly seen.

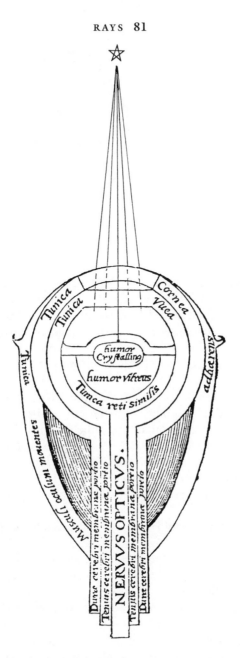

Figure 3.15. The only ray that reaches the crystalline lens from a point source is the one that arrives perpendicular to the cornea.

Alhazen's explanation raises a mathematical question that careful readers must have found very puzzling. The argument states that when I look at a mountain, every point of the scene in front of me produces an image at a corresponding point on my crystalline lens, which is perhaps a centimeter wide. The correspondence is one-to-one, meaning that to each point in the

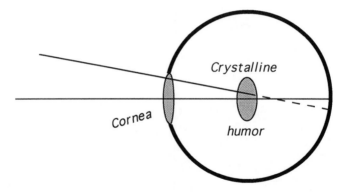

Figure 3.16. If the ray were detected at the surface of the retina, we would see everything upside down.

scene corresponds one point on the surface of the lens, and vice versa. Therefore in spite of their vastly different sizes the two areas contain the same number of points. The graphical construction seems to show that this is true, but in doing so it implies that a whole is equal to its part, which both instinct and tradition deny. The subtle point is this: the mountain's visible surface can be considered as a distribution of points. Its image in my eye consists of the same number of points but is much smaller. But what is a point? Have we any right to assume that the image of a point is another point exactly like it? If all points have the same "size," then how can the mountainside and its image contain the same number of them? If not, then we must face, and not dodge, the question of points of different sizes. Much later Galileo was intrigued by this situation and interrupted Day 1 of his *Dialogues Concerning Two New Sciences* [1638, 1914] to insert a long discussion of infinite sets of points or numbers, but it was not until the nineteenth century that Georg Cantor wrote down the definitions and proved the theorems that made infinite sets part of respectable mathematics. Mathematicians now agree with Alhazen; it is correct to say that the surface of the mountainside and its tiny image on the humor consist of the same *infinite* number of points, and it makes no sense to talk of the size of a point.

What happens when light rays arrive at the crystalline humor? Dissection, of course, shows that it is almost perfectly transparent, whence its name, and therefore it interacts very little with the rays that reach it. Why should one think it interacts at all? This is a hard question, but it has an answer. We do not see with our pupils, for the cataract operation that punctures an opaque pupil makes vision better, not worse. And if the rays of light were to pass through the crystalline humor and be detected at the back of the eye everything would look upside down (Figure 3.16). Therefore it must be the crystalline humor that detects the rays; nothing else is there. The surface behind it, which the Latin calls *tunica reti similis*, membrane like a net, and we call the retina, must just be intended to complete

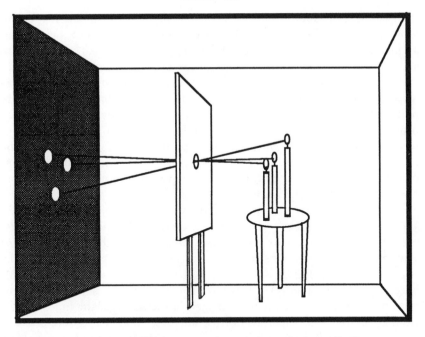

Figure 3.17. Alhazen's camera obscura. The rays from three candle flames pass through a small hole without interfering with each other.

the sphere and perform some useful service like helping to support the whole structure. (We have seen that Galen, recognizing the retina as part of the brain, came closer to the truth.)

Aristotle, in §2.5, has already said that just as a piece of wax can receive different shapes and impressions, so a sense organ takes on the property it is sensing, becoming hot, green, and so on, as the case may be. We can now see that this is what happens in Alhazen's theory: the dot of light that comes from a leaf and falls on the crystalline sphere is actually green, and the lens, being slightly opaque, is colored green at that point. How the image gets from the crystalline humor to brain and consciousness is less clear, but note that in the figure the optic nerve empties like a funnel from the center of the retina. (There is in fact a funnel, but it is small and off-center.) Evidently, as Galen had also said, the form of the image detected by the crystalline humor enters the nerve through this funnel and passes along it to the brain, where it encounters the soul.

One can dispute Alhazen's theory by objecting that the innumerable rays that crowd through the pupil on the way to the crystalline humor must come so close together that they interfere with each other. To investigate this point he invents what is called a *camera obscura*, Latin for a dark room, a device with which people amused themselves many years later. He erects a screen in a darkened room and makes a little hole in it (Figure 3.17). He then sets several candles in front of the screen, goes to the other side, and looks at the wall. There are several spots of light. The pattern they form is an inverted image of the candle flames, but this is not what interests him, and he does not mention it. The important point is that the spots are clear and distinct, so that the various rays do not bother one another as they pass through the little hole. He blows a candle out; one spot disappears. He relights the candle; it comes back. The rays behave independently of each other, and he concludes that the multitude of rays from a scene in front of us can pass undisturbed through the pupil of our eye. I suspect that the reason he does not mention the inversion of the image is that in his theory of vision the image formed on the crystalline lens is upright, and he does not want to confuse the reader.

There is no further discussion of this way of projecting images, but in another work, *The Shape of the Eclipse*, he explains a phenomenon that Aristotle or one of his students had noticed long ago, that though the spots of sunlight under a tree are usually circular, during an eclipse of the sun they become crescents (Figure 3.18). The explanation follows at once from Alhazen's principles. Figure 3.19 shows that if the hole is small, each luminous point of the crescent sun is imaged in the shape of the hole (here, a circle), but that the superposition of all these images is a reversed and somewhat blurred picture of the sun. A little thought shows that the shape of the hole is unimportant as long as it is small. Alhazen's discussion is unclear and not entirely correct, but it explains the crescents and goes on to show that if the hole is made larger the crescent images become less distinct until, if hole and image have the same size, we no longer see the crescent and the shape of the image is determined by that of the hole. This explains a mysterious remark in Aristotle that while a crescent sun is imaged as a crescent, a crescent moon is imaged as a circle. The point is that because the moon is so much fainter than the sun one has to use a larger hole in order to see any image at all. People who believed that images are carried by *eidola* had a very hard time with observations like these.

Book II of Alhazen's *Optics* analyzes the process of vision. It is basically Aristotelian, and it emphasizes that the process is not simple: "Not everything perceived by the sense of sight is perceived by pure sensation; rather, many visible properties are perceived by judgment and inference in addition to the sensing of the object's visible form." And of course, to know what an object actually is we must know something about it that enables us to recognize it. Alhazen notices that we perceive familiar things at once; there just is not time for any consecutive mental process. Therefore, he reasons, we skip judgment and inference and perceive at once by recognition. Farther on, he lists twenty properties that we perceive by sight: a few of them are distance, shape, size, number, motion, roughness, and beauty, and he describes how we infer each from the raw data. For the first time, we have a coherent picture of the entire process of vision.

Figure 3.18. Spots of light under a tree during a solar eclipse (photograph by the author).

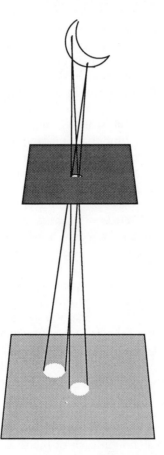

Figure 3.19. If light from a crescent moon passes through too large a hole, each point will be imaged as large circle and the shape of the crescent will not be seen.

Book III treats optical illusions, and the next three books deal with reflection from flat and curved surfaces and the distortions they can produce.
In the seventh and last book Alhazen discusses refraction from plane and
spherical surfaces. He tells how to measure refraction but shows no signs of
having done so, possibly because the experimental techniques at his disposal
were no different from those Ptolemy had used eight centuries earlier and
so there was no use repeating the work. He does not mention Ibn Sahl's law
of refraction, but even if he knew it he had no way of knowing it is correct.
He continues Ptolemy's mathematical studies of the effects of atmospheric
refraction on the measurement of stellar positions and can show, roughly,
how the bending of a star's ray varies with the star's position in the sky. But
since like Ptolemy he has no numbers, he cannot calculate the size of the
effect, and so the value of the discussion is mostly pedagogical.

Alhazen and Ptolemy were the only early students of light who thought in
ways we would now call scientific. They questioned Nature with experiments and tried to interpret its answers in ways that made sense. The inter-
pretations often involved hypotheses that could not be directly tested, but
that is true of all science—yesterday, today, and tomorrow. The other early
students of light were either mathematicians like Euclid and Hero, who
posited the existence of visual rays and then went on to make geometry, or
else philosophers who filled the gaps in their understanding with analogies
and verbal formulas. Alhazen's greatest achievement was to get the image
into the eye, impressed on the sensitive surface of the crystalline humor.
He understood for the first time that one cannot stuff the whole image of
something through the pupil; instead, we must consider the scene in front
of us as dissected into an array of points which the eye sees all at once and
assembles into an image. How do all the forms then reach the soul? Are
they transmitted point by point, or does the crystalline humor unite them
into a single form in the likeness of what we behold? The answer was beyond him, but at least it was clear why Euclid's constructions that used
straight lines for rays make sense provided one reverses the arrows so that
they enter the eye. *Eidola* are no longer necessary or even useful for understanding vision, for they raise more questions than they can answer, but it
took more than this to banish them from the minds of philosophers.

For people who believed that an optical image carries an object's form
and not just the data of its shape and color, Alhazen's analysis made no
sense. A form is an entity, and you can't break it up into little pieces. Alhazen did more than clarify the theory of vision; he demanded a new concep-
tion of what it is to see.

At the end of Chapter 2, I asked some questions about light: how does it
get from one place to another, what are its powers; I even asked if it really
exists, so as to show that though each of the three classical theories of vision explained how the image of the thing in front of us arrives at the door-

way of the soul, the role of light in the process remained obscure. Now, in the terms of Alhazen's *Optics*, we can begin to understand it. Light is a substance. It comes from the sun, or a candle, or whatever, a little of it touches an apple; then it spreads out from each point of the apple's surface accompanied by the form of the apple's *chroma* so that even if a hundred people are looking at the apple, each eye gets some. When this light and *chroma* enter an eye they create an image of the apple on the eye's crystalline lens, and from there the image passes to the soul. Alhazen's account of vision is essentially correct, but we still need to get the image onto the retina, not the lens, we need to know how it gets there, and we need to know how the color of a surface pertains to the color that leaves it. What is colored light, anyhow? We haven't finished with vision, but we have begun to make sense of it.

Alhazen's light is what we see by; Alkindi's is only one of the many radiations that bind the universe in a network of causal power. Alhazen's light descends from Euclid and Ptolemy, whereas Alkindi's descends from Plato and Plotinus. Alhazen knows the laws of optics, which explain how light carries its influences through space until it reaches the crystalline lens. These laws can be expressed in the language of geometry. If every physical influence is by way of radiation and all radiations follow the principles of optics, then these are the principles that govern the natural order of things, optics is the most fundamental of all sciences, and its principles are mathematical. Imagine a great medieval tapestry, the kind called *millefleur* that portrays ladies and huntsmen at peace in a garden or forest with every leaf and flower distinct. The background recedes into an unfathomable distance. Now turn up a corner of the tapestry and see the back of it. It is mathematics. The next chapter deals with the front of the tapestry; later we shall look at the back. But both sides are created from a single plan.

Light in the Middle Ages

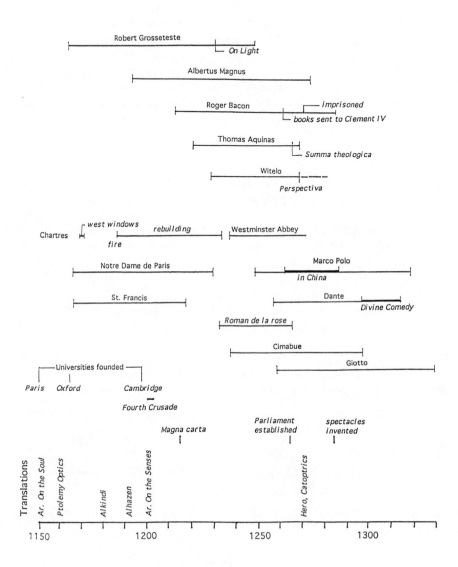

Robert Grosseteste

└─ *On Light*

Albertus Magnus

Roger Bacon ──── *imprisoned*

└─ *books sent to Clement IV*

Thomas Aquinas

└─ *Summa theologica*

Witelo

Perspectiva

Chartres ┌─ *west windows* *rebuilding* Westminster Abbey

fire

Notre Dame de Paris Marco Polo

in China

St. Francis Dante

Divine Comedy

Roman de la rose

Cimabue

Giotto

── Universities founded ──

Paris *Oxford* *Cambridge*

Fourth Crusade

Magna carta *Parliament established* *spectacles invented*

Translations

Ar. On the Soul

Ptolemy Optics

Alkindi

Alhazen

Ar. On the Senses

Hero, Catoptrics

1150 1200 1250 1300

4

THE LIGHT THAT SHINETH IN DARKNESS

[The] immortal and blessed beings who dwell in Heaven . . .
receive their happiness from the same source as we do—from
the ray of a certain intelligible Light which is the God of angels and is distinct from them, for only by this light are they
resplendent and only by participation in God are they established in perfection and beatitude.
(Saint Augustine)

1. Plato and the Meaning of Light

WE HAVE SEEN how a ray of light started, perhaps from the
mind of Empedocles in Sicily, flashed over to Alexandria where
Euclid and Ptolemy taught it mathematics, then to Baghdad
where it studied Arabic, how Alkindi gave it physical power and Alhazen
reversed its arrows. Finally, in about 1200, it came back to Europe again,
speaking Latin. But at about the same time, Europeans were seeing light
that had started from Plato and followed a different path.

Plato's *Timaeus* is a very serious book, but is hard for moderns to read
it seriously because it is written in a language we no longer use in scientific
contexts, the language of myth. It tells how the universe came to be created
by a divine craftsman from the world of Ideas. He looks out at the material
universe and finds it in a state of chaotic motion and decides to use its
material to make something that is beautiful and good. He reflects that
intelligence is beautiful and that it does not exist without soul, "for which
reason, when he was framing the universe, he put intelligence in soul, and
soul in body, that he might be creator of a work which was by nature fairest
and best. On this wise, using the language of probability, we may say that
the world came into being—a living creature truly endowed with soul and
intelligence by the providence of God." Plato had to be careful. He heard
a little scraping noise as his student Aristotle sharpened an awesome array
of critical tools with which to distinguish fact from fancy and good reasoning from bad. Plato did not want these tools used on him, and so seven or
eight times more in the *Timaeus* he cautions that the tale he tells is a probability, a likely story. But it is not a fairy tale. He is trying to tell how he
thinks the world really is, but he does it in his own way, by explaining how
the Creator formed it as the best material structure that could be modeled

on a timeless original. Because the nature of matter is to change, he created "a moving image of eternity, . . . moving according to number, while eternity itself rests in unity, and this image we call time." Plato describes the arrangement of the planets and tells how and why they move, he explains why plants and animals are as they are, he explains human physiology and how the senses work (we have already noted his theory of vision), and much more. Throughout, he tells us that the world is constructed on mathematical principles, and his blueprint is sprinkled with shapes and numbers. The element Earth is represented by a cube, the figure bounded by six squares. One can make triangles, pentagons, and so on—plane figures with any number of sides. Can you stick them together to make a solid as squares make a cube? Mathematicians of Plato's time knew there are exactly five ways it can be done: you can fit together

four triangles to make a tetrahedron,
eight triangles to make an octahedron,
twenty triangles to make an icosahedron,
six squares to make a cube, and
twelve pentagons to make a dodecahedron,

and that is all; they are known as the Platonic solids.

Plato borrowed the idea of atoms from Democritus, but he did not want his universe to be made of little pieces of unexplained matter, so he identified the first four figures mentioned here with what we would call atoms of Fire, Air, Water, and Earth respectively, while the last one gave the form of the universe as a whole. At its most fundamental level Plato's universe is made out of mathematical forms. This was a remarkable intuition. No physicist today thinks of an atom as if it were a thing in the usual sense; there are, for example, experiments in which an atom seems to be in two places at once. This is not allowed for a thing, but we understand an atom very well as a mathematical form. And the curved space-time of the universe as a whole is beyond our imagining; it is grasped only with mathematics.

When the Christians arrived on the scene, they found themselves derided as simpleminded by Greek and Roman intellectuals, and one of their responses was to formulate in philosophical language the answer to the question, "Exactly what does a Christian believe?" I have mentioned their attempts to reconstruct the Creation from the scanty suggestions of the Bible. It is hard to give any other reason why the early thinkers set themselves such a grim task; other great religions get along without so much specificity. Plato's philosophy, with its idea of a transcendent Good and a reality above the cares of this world, got woven into the new theology, but his ideas spread more widely than the books he wrote. Saint Augustine, who knew little Greek, had a Latin *Timaeus* and was impressed that he could read it as a pagan version of Saint John's mantra of Creation, "In the beginning was the Word, and the Word was with God, and the Word was

God. The same was in the beginning with God. All things were made by him; and without him was not anything made that was made. In him was life; and the life was the light of men" (John 1:1–4). All this is in *Timaeus*, he says, "not indeed in these words, but in much the same thought," and obviously the thought was divinely inspired. Augustine found the Platonic philosophers superior to all others, for "they transcended every soul and spirit subject to change in their search for God. They perceived that no determining form by which any mutable being is what it is . . . could have any existence apart from him who truly exists because his existence is immutable. From this it follows that neither the whole universe . . . nor any life . . . can have existence apart from him." This is Christian Platonism, according to which the ideas that shape the universe exist not in the "heaven above heaven" but in the mind of God. Everything we see is a changing symbol of the eternal reality; Plato, says Augustine, distinguishes "the intelligible world, where truth itself dwells," from "this sensible world, which we perceive by sight and touch." Of these two worlds, "the first is the true world, while the other is made like the true world and in its image."

In the middle ages Plato's doctrines reached Europe by several routes. Before 1150, as I have mentioned, a few scholars knew *Timaeus* through the old Latin translation, but otherwise Plato was virtually unknown. Cicero quoted him, and there were later commentators who, though they may never have seen a copy of Plato in their lives, produced their own versions of various parcels of his thought. Augustine's immense authority commanded attention to what Plato had said, but teachers found it easier to unwrap parcels than to navigate Augustine's treatises, and so it was the parcels that were read in the schools, and from the twelfth century they taught scholars to think about the world in terms of the eternal realities of number and light.

The Gospel of Saint John continues, introducing John the Baptist,

> And the Light shineth in darkness, and the darkness comprehended it not.[1]
>
> There was a man, sent from God, whose name was John. The same came for a witness, to bear witness of the Light, that all men through him might believe. He was not that Light, but was sent to bear witness of that Light. That was the true Light, which lighteth every man that cometh into the world. (John 1:5–9)

And later the same John, or possibly another, writing to Christians everywhere, says, "God is light, and in him is no darkness at all" (1 John 1.5). "True Light" is not to be taken as literary metaphor, for it denotes nothing but plain truth. John understood it this way, and so did Augustine, three centuries later, when he wrote that "one does not say that Christ is light in

---

[1] Modern translations say that the darkness never extinguished the light.

the same sense that he is called a rock, for the first statement is literal, while the second is only a figure of speech."

This is the light that shone beside the prisoners in Plato's cave and which they could not see—the light of the ideal world. It is what the eye of the soul seeks as it looks upward toward the truth. In our world truth is impossible, and what we experience as light is something different, the transitory and imperfect copy of the Light that never goes out. In the ideal world God, truth, and light are the same.

2. Three Denises and the Ghost of Plato

The civilization of Europe entered a new phase during the 1100s. As people moved into cities and commerce expanded, the number of those outside the Church who could read and write rose sharply; life ceased to be a struggle for existence, and people began to think wider thoughts. Even if *Timaeus* was out of reach, they could and did read in great numbers a simplified summary of it in Macrobius's commentary [1952] on Cicero's *Dream of Scipio*, c. 400 C.E. From this and other works some literate people began to think platonically, and when the first cathedral schools were set up the ghost of Plato hovered over them.

There were three holy men named Denis or, in Latin, Dionysius. The first was an Athenian, whom the Bible, in a single sentence, mentions as a member of the Council of the Areopagus who became a follower of Saint Paul. Eusebius, the historian of the early Church, says Dionysius became Bishop of Athens, but this is unconvincing; really we know nothing about him.

We know even less about the second Dionysius, not even his real name. He was a Christian Neoplatonist of the fifth or sixth century, probably no sort of pretender but a monk inspired by mystic visions of the Athenian. He describes being present with Saint Peter and Saint James the brother of Jesus at the death of Mary, and he writes a friendly and collegial letter to Saint John on the island of Patmos. By the eighteenth century his name had become Dionysius the Pseudo-Areopagite, now shortened to Pseudo-Dionysius and indexed under *P* as if that were the name he was given at birth. From early days well into the Renaissance he was placed just after the Evangelists as a witness of sacred history. His works were translated into Latin in the ninth century and several times thereafter, and every library held them. His tone is Neoplatonist throughout, the words recall Plotinus, and the text that helps our story along is called *The Divine Names*, an ecstatic treatise that depicts the nature of God in terms of the names by which he may be called. Chapter 4 begins,

> Let us move on now to the name "Good." . . . Think of how it is with our
> Sun. It exercises no rational process, no act of choice, and yet by the very fact
> of its existence it gives light to whatever is able to partake of its light in its own

way. So it is with the Good. Existing far above the Sun, an archetype far superior to its dull image, it sends the rays of its undivided goodness to everything
with the capacity, such as this may be, to receive it. . . .

And what of the Sun's rays? Light comes from the Good, and light is an
image of this archetypal Good. Thus the Good is also praised by the name
"Light," just as an archetype is revealed in its image.

We have already seen the power of this vision when Plotinus described
the light of goodness flowing downward from the One. Before the tempest
of Aristotle swamped every vessel of the other philosophies, the words of
the second Denis encouraged the Neoplatonic tendency of the cathedral
schools; when the storm had settled these words were taken merely as the
symbolic utterances of a mystic. Thomas Aquinas probably did not know
Plotinus, but he refers to Dionysius a thousand times, and Dante's poetic
vision of the point of light at the center of the spiritual universe may have
been inspired by *The Divine Names*.

The third Denis, according to Church history, was the saint who converted
France to Christianity, perhaps in the third century. With two companions
he journeyed as far as Paris. Here he preached and baptized and was finally
beheaded on orders of the Roman emperor, in front of the statue of Mercury on a hill outside Paris that received the name of Mons Martyrorum,
the hill of the martyrs; now it is Montmartre.[2] The story goes on to say that
after the executioner had done his job Denis picked up his head and walked
two miles to a pleasant area where he put it down and declared that this was
where his bones would lie.

Everyone in France believed that the three Denises were one and the
same person, and that the tomb covered by this abbey contained the body
of a companion of Saints Peter, John, and Paul, next in rank after the
twelve apostles themselves, one of the greatest theologians, the man who
brought Christ to France, France's patron saint. No wonder this became
the country's main center of pilgrimage and the church erected here became the richest and most splendid this side of Constantinople, and no
wonder that the shrine built to hold the saint's bones was designed as a
visible manifestation of the Areopagite's divine light.

3. Images of Heaven

In the Old Testament's apocryphal *Book of Wisdom* Solomon praises God
with the words, "Thou has ordered all things in measure, and number, and
weight," and everyone understood that there was a reason for this: it is so
that humankind can see that the world is God's handiwork. The same principle can be seen in the design of medieval cathedrals. A cathedral was more

[2] Antiquarians will tell you that long before Denis came it bore a temple to Mercury and was
called Mons Mercore.

than a place to hold services, or even a house of God, for it was imagined as a material representation of Heaven: what Heaven is in the divine order, a cathedral is in the country around it. To create this sensible analog of the world beyond the senses, art and science worked together so that the art of construction and the science of numbers combined to create an earthly index of Heaven's mathematical perfection.

The study of numerical harmonies was thought to go back to Pythagoras. Experiment, as he did, with a stringed musical instrument whose string is of length one: one foot, one meter, one cubit, it doesn't matter. There is something special about the relation of the sound of this string to that produced when a stop is introduced so that the vibrating part is only half as long: the ratio 1:2 defines an octave.[3] The ratios 2:3, 3:5, and 4:5 sound well, whereas ratios a little to one side or the other of these values are discordant. From early days the Pythagoreans and their descendants taught that these simple harmonic ratios are found not just in music but are embodied in the structure of the created world. Timaeus finds them in the shape of the universe, and Augustine writes that a work of art that ignores just proportions cannot be beautiful.

The perfection of number relates the fabric of a church to that of the world God made; the perfection of light relates it to the moral order. Psalm 104, a hymn to the world and its Creator, says that God covers himself with light as with a garment, and we know that one medieval mind imagined Heaven in the same way. The last few cantos of Dante's *Paradiso* describe Heaven in terms of the light that filled it, and his final vision of the Trinity is of three circles of three-colored light, one of which, as he looked long and carefully, seemed painted with the image of a man. When the medieval cathedral was being designed, there were prelates and builders who set out to perform the nearly impossible, to represent, in an earthly structure, that blinding effulgence.

By tradition, the Abbot of Saint Denis was chosen from the highest nobility, but in 1122, perhaps because the abbey had become a cesspool of depravity, the honor was bestowed on a commoner named Suger (1081–1151). He called himself a beggar lifted from a dunghill, but he was a man who combined piety with worldly ability and personal charm, and he put them all to work in reforming the abbey and replacing its old dark church with a structure worthy of the saint. He raised the money and superintended the design and every phase of the construction. He made a new façade, massive and beautifully proportioned, left for the moment the dark old nave, and in 1144 completed a new apse which opened the era of Gothic architecture. Construction of the Romanesque style, which established a space of prayer protected from the yelling world outside by heavy

---

[3] Aristotle gives this as an example of a formal cause: the ratio 1:2, a mathematical form, causes or is responsible for the sound of the octave (*Physics* 194b28).

walls symbolized the Church's power to defend the faith. Suger replaced this with a design that leaves as little masonry as possible. Thick walls are replaced by slender piers that support glass, and much of the stone that supports roof and towers is moved outside the building in the form of buttresses. The apse is full of painted light. Even the ancient nave became a bit brighter; as Suger wrote in one of the many exultant inscriptions with which he explains his church: "Even the old part shines, / For bright is that which brightly joins with brightness, / And bright is the noble work in its new light." The glass served several purposes. In the first place it proclaimed the dignity of light as a vehicle of God's goodness and power. Medieval images and reliquaries often have so many jewels stuck to them that their outline is obscured and they may even look a little ridiculous, but jewels gather light and were thought to preserve it; gold and silver, the noblest metals, are the brightest; a diamond in sunlight creates the colors of the rainbow. To the medieval imagination the splendor of these shining things was not in their aesthetic quality but in their association with the divine nature. Their brightness proclaimed the powers God had given them to cure illness and to sanctify the believer who came near them.

Second, the windows illustrated history. One of Suger's, for example, depicts the lineage of Christ by the convention that situates his ancestors on the branches of a tree that issues from the loins of Jesse; another shows the Israelites passing the Red Sea. And third, the windows were intended to instruct us; this was called their anagogical mission. Suger places an inscription around the Red Sea window to remind believers that though immersion in water means salvation to a Christian, it meant perdition to the soldiers of Pharaoh's army, drowned when the flood returned. In another window Pharaoh's daughter rescues Moses as he floats past in his basket, but according to Suger's inscription it also shows how the Church cherishes Christ. To the medieval mind God's plan was all-inclusive, and nothing showed this more clearly than the correspondences he implanted between the events of the Old Testament and the doctrines of Christianity.

Within the church, colored rays fell upon gold chalices and candlesticks covered with rubies, emeralds, and pearls, and surrounded by textiles worked with red, blue, gold, and silver thread. Some of the worshipers standing in the crowd in the dark nave and looking up into this blaze of glory must have remembered the lines from Revelation, "I saw the Holy City, the New Jerusalem, coming down out of Heaven from God, prepared and ready, like a bride dressed to meet her husband."

A succession of kings and queens and great prelates, including Suger, were buried in vaults below the church, but when the Revolution came their bones were shoveled into the street. Now the bones are back in the vaults, helter-skelter, and it will take all the wit of Resurrection angels to unscramble them. The fate of the church itself is nearly as sad, for within a century the nave was raised and the apse rebuilt to fit it. Suger's church is now a confused and impersonal space that functions as a museum of funerary sculpture where one pays to get in. But the new nave is very fine, and some of Suger's old glass is still in the apse.

Take the train from Paris to Chartres, walk out of the sunlight and into the cathedral that stands on the top of the hill. It is dark in there. There are low voices and the whisper of many feet on the stone paving. Your ears tell you of the height and volume around you, and as your eyes adjust to the darkness you begin to see colors. There is no daylight at all; at first you see the glow of red and bits of green and yellow and, most of all, a blue that brings a measure of pure sky into the space around you. Like the embers of a fire, the glass surfaces seem to produce their own light. There are almost two hundred of them: "windows," they are called; but windows let us see the world outside, whereas these are walls that enclose a space, walls of colored figures of men and women and animals.

We talk of beauty and make the trip from Paris to see it, but beauty was not something that people in the middle ages thought much about. The glass was praised because it communicated sacred history through the radiance of divine light. You see robed figures and men on horses and a king on his throne; nearby someone is being struck with a sword. The images are so distant that even if you know the saint and the story you can rarely make out what is happening. Are these scenes meant to be seen only by God? I do not think so, for Chartres was also a school.[4] As early as the 1000s, education had started to spread out from the monasteries. A big town's cathedral was the natural place for a school. Chartres became a local center of learning, at first in the liberal arts and then, in the first half of the next century, of an enlightened Neoplatonism. There was not much of Plato to study, but scholars assembled scraps from Latin authors like Cicero and Seneca and Macrobius; they had Calcidius's *Timaeus* and, most important, translations of Pseudo-Dionysius. These, together with Latin classics, were read in the school of Chartres for a hundred years. The aim of this education was to produce minds that were ethical, cultivated, and serene. Then came a new intellectual fashion, personified by Peter Abelard (1079–1142) and his students, who combed the forests of philosophy and theology, hunting for truth. They were skeptical of received notions and abusive of their teachers; they had no time for Platonic myths that meant different things to different people. The situation was a little as it must have been in Athens after the death of Plato more then fourteen hundred years earlier, and what happened was similar: in about 1200, translations of Aristotle started to arrive, and soon they swept the old education away.

---

[4] There is some doubt as to the size and importance of the school, but several notable scholars seem to have been associated with it.

Chartres's most famous window, whose luminous blue no one forgets, dates from about 1150[5] and shows the Tree of Jesse. Another shows the Virgin with her child on her lap, adored by angels with a dove descending on her from the heavenly city far above. She is in pale blue; behind her the background is mostly red; around her is a border of dark blue against which six angels hold candles and swing censers. Another window recounts the deeds of Charlemagne, but most of the glass shows the sufferings and triumphs of saints.

Now imagine being a child in Chartres, where a college of learned clerics buzzes around. Your eyes are young and clear, you see the images, a teacher points to the saints and tells you their stories. People come in from the countryside; you pass the knowledge to them, finally to your own children. The stories get garbled; it makes no difference, for a story that is wrong can be just as edifying as one that is right. Even for those who had no other education, illuminated glass was a school that brought the generations together in piety and learning. And perhaps some of the older clerics could still remember a voice that bade them look past the colored figures above them to see the eternal and unchanging truths they represent, existing in a Heaven whose color our eyes have never seen and which the glass only dimly suggests.

4. Robert Grosseteste, Bishop of Lincoln

In the years around 1200 the new cathedral of Chartres was going up, and a new class of administrators and businesspeople was beginning to fill the needs of a mercantile society. Their education in reading, writing, figuring, and religion left them realizing how little they knew about the world, and for those who wanted to learn, books in their own languages appeared. These were largely based on translations of Aristotle, which were beginning to become available through the slow and expensive process of being copied, letter by letter, onto vellum (one book used up the skins of a flock of sheep and could cost as much as a house). People had known, for example, nothing whatever about human generation except that it happened; now they could read about it in Aristotle or some popular version. We might say today that having done so they really knew no more about it than before, but what is important is that they were beginning to know what it feels like to be educated.

In the universities and among theologians, discussion swirled around the problem of reconciling the conflicting claims of reason and faith. Faith

---

[5] The present cathedral was started in 1194, replacing an older structure that had burned, and it was largely finished by 1220. The Jesse window, dating from about the time of Suger's death, belonged to the west wall of the old cathedral and had not been destroyed in the fire.

assumes direct access to the highest truth; reason slowly builds up its version out of Scripture and experience. Everybody understood that Adam, in that first sunny morning in the garden, had direct and certain experience of the divine, but that after he fell from grace he and his successors were left only with their senses to rely on, and these are the weakest of human powers. Humanity can only guess about the world in God's mind by looking at its material correlate. We learn qualities, not substance; appearance, not reality.

Consider that the medieval mind was occupied with three great questions—the nature of God, of angels, and of the soul. Nothing else was so important, and yet one could not directly detect any of these three or anything about them with the senses. Many educated modern people take knowledge of the material world as certain and verifiable, and tend to regard claims of spiritual knowledge with skepticism. In the middle ages it was the other way around. Certainty and proof pertained to the unseen, while the world of experience was a tarnished reflection of truth, not a source of it, and the idea that this world is a reliable source of knowledge would have seemed very strange.

When Latin translations started coming, schoolmen eagerly absorbed Aristotle's logical works, especially the *Posterior Analytics*, to help them climb out of the dark pit into which Adam had thrown them. They found they did not have to describe Nature in terms of fallible sensory data; they could describe it instead in terms of abstract categories. Here for example is a late medieval explanation of natural local motion. ("Local motion" means what we usually mean by motion, a change of place; "natural" means that the thing moves freely, nothing is pushing it. The motion of a freely falling object is an example.) "Natural local motion is motion arising from intrinsic form, as the motion of a heavy body arises from [its] gravity, which gravity indeed is the intrinsic form of the heavy body." The words of the explanation refer to concepts, not experience. It was like trying to study Nature while looking through the bottom of a bottle. The remarkable thing is that some thinkers were able to use Aristotle's formal method to develop and express new ideas. And when one of these, whom we shall now study for a while, tried to understand how the world was created, he was led to a comprehensive theory of the nature and power of light.

Robert Grosseteste (c. 1170–1253) was born into a poor family in a little town in Suffolk. Henry II, first of the Plantagenets, was king. Robert's French family name suggests that he was not born a serf, but the biographers say that when his father died his desperate mother sent him begging from door to door. Or perhaps she didn't; there is little firm information about his origins or what he was doing before he was fifty-five years old. He must have found a generous patron, for he got an education somehow and probably took an M.A. at one of the schools. Perhaps he studied in Paris, but a recent biographer has him working his way up as a diocesan administrator in England. In 1225 he was lecturing at Oxford, and in that year he entered the priesthood. From this moment pastoral duties seem to have taken up much of his time, but he continued at Oxford while tending his flock as archdeacon of Leicester.

In the next few years Robert produced a number of scientific works on the calendar, the tides, the rainbow, and light, a work called *On Lines, Angles, and Figures,* and commentaries on Aristotle's newly available *Physics* and *Posterior Analytics.* The last was widely read and was still quoted two hundred years later. Realizing that the texts on his desk had suffered through a series of translations, he studied Greek, and though he never knew it really well he produced versions, notably of Dionysius, that were the best of their time. Later, as Bishop of Lincoln, he could set up a small translation workshop and bring educated Greek monks to work in it. They produced versions of Aristotle's *Ethics* and a few other works that spread quickly over Europe. At the end of his life he was studying Hebrew so that he could at last read the Psalms as they were written.

The book called *On Light* is Robert's most interesting work for our purposes because of the intellectual problem it addresses. As we have seen, the biblical account of Creation is sketchy on details. In Robert's youth arrived the new translations that established Aristotelian metaphysics as the model of intellectual method. I think Robert set himself a project: suppose God had determined to create a world whose form and function were determined by Aristotle's principles. Moses (the supposed author of Genesis) lived before these principles were known to humankind and so was obliged to leave gaps in his account. Now, knowing the principles, Robert set out to deduce what must have happened at the Creation.

Robert has read Alkindi and continues the line of thought that binds the world in a web of radiation. What was there before God's command "Let there be light" started the process that created our world? Moses tells us that something was there, but if it was matter it lacked all the qualities of matter; this means it did not even occupy space. Before it could assume any qualities of the matter that makes up the world it had to do that. It needed to acquire what Grosseteste calls corporeity, the first bodily form. And this form, he says, is light,

> for light of its very nature diffuses itself in every direction in such a way that a point of light will produce instantaneously a sphere of light of any size whatsoever, unless some opaque object stands in the way. Now the extension of matter in three dimensions is a necessary concomitant of corporeity, and this despite the fact that both corporeity and matter are in themselves simple substances lacking all dimension. But a form that is in itself simple and without dimension could not project dimension in every direction into matter which is simple and without dimension, except by multiplying itself and diffusing itself instantaneously in every direction.

There was no way for light to spread out into three dimensions if three dimensions did not exist; therefore as light moved it carried dimensionality with it, producing not only the material of our world but also the space in which we live and move. There is an echo here of the *Hexaëmeron* of Gregory of Nyssa, in which God's Word first created the qualities that belong to matter and then, because if there are qualities there must necessarily be something for them to belong to, matter came into existence by itself.

The total quantity of the first light was finite (here the argument gets rather involved), and this limited how far it could go. When the original flash reached its limit it stopped, formed the outermost of the heavenly spheres the Bible calls the firmament, and returned inward, creating the other spheres as it did so. And finally, reaching the center, it formed the four elements that make our world.

In Robert's cosmic vision light has the causal power of Alkindi's rays and spreads goodness through the universe just as Dionysius describes, but there is also the beginning of a physical theory, signaled by the phrase "multiplying itself and diffusing itself," and if we examine the meaning of these words we can learn something about how light produces its effects.

5. Species

I have spoken as if light were a substance that moves from place to place, but for Aristotelians this is not possible, since light is a form, not a substance. Suppose I look at a candle from close by. The flame produces images of itself something like the *eidola* of Democritus and Leucippus, but because they are not things and consist only of form they need a new name, which in Latin is *species*. They do not move but spread by a process called multiplication, just as our shadow on the ground, as we walk along, is not really a moving thing but is continually being re-created in a new place. In general a species produces more species, but as they become greater in number they become weaker in effect. This is why if I stand farther from my candle it seems less bright, and so it has been with every action since the first great flash. From then until now, its species have been multiplying, with range ever wider and intensity and effectiveness ever diminishing, to produce actions and reactions in the world of matter. Light, originating at the first command and now multiplying its species downward through the heavenly spheres, is the power that moves the world.

This reconstruction raises an important difficulty. We can see the light from the stars and planets, but how can anything so faint be so efficacious? Robert's answer does not settle the question, but it shows he has thought about it. Imagine radiations beaming down from the spheres toward the Earth. As they approach the center the rays become closer together, more concentrated. If the Earth is very small compared with the spheres, the rays will be very dense near its surface and their effects much enhanced. We recognize their effects everywhere, and therefore it follows that the Earth is very small compared with the architecture of the cosmos.

One might expect that next would come a ringing defense of astrology, but though in his youth he seems to have been a believer, his mature opinion echoed the conclusion that Basil and Augustine had drawn from the very different lives led by pairs of twins. Even if the planetary influences that astrologers talk about actually exist, they would be impossible to disentangle and understand.

The word *species* was used by Augustine and other early writers as the Latin version of *eidolon,* to denote what acts on the eye to cause vision, but for Robert and his followers it was intolerable that God should have decreed more than one fundamental means of physical causation, and so species acquired a more general meaning which it kept for almost four hundred years. It was the power through which one thing acts on another, but it always had the character of light. In fact, says Robert, "The species and perfection of all bodies is light." Species are, as far as I can see, almost synonymous with the *energeiai* of John Philoponus.

As an example of how this power is exerted, consider Robert's theory of sound. Suppose I ring a bell. A bell is a three-dimensional structure, and it therefore contains some of the primeval light that made it so. As the bell vibrates the light in it vibrates also, and the species of its motion induce the vibrations in the surrounding air, which make sound.

Robert returns to the subject of the Creation in a later book to be discussed in a moment, but first I must mention a short work that tries to explain the world's mathematical ordering. *On Lines, Angles, and Figures* defends the importance of mathematics in the sciences: "All causes of natural effects must be expressed by means of lines, angles, and figures, for otherwise it is impossible to grasp their explanation."

The power of species moves along straight lines. In this way, "the action is stronger and better, as Aristotle proposes in Book V of the *Physics,* since Nature acts in the briefest possible manner, and a straight line is the shortest of all lines, as he says in the same place." The message is clear: because light, and all the other radiations that influence the physical world, act in straight lines, a scientific theory of these influences must be a geometrical theory. It was still too early to talk about laws of Nature, but one cannot understand what Nature does without the aid of geometrical reasoning. Not only has Robert Christianized Alkindi but he seeks to mathematize him also, and all his principles come from Aristotle. Sometimes, it is true, Robert refers to some known fact of experience (when he says *experimentum* he is not ordinarily talking about an experiment), but most often he is trying to establish or support a principle on which a deductive argument can be hung. The use of an experiment to disprove a theory was far in the future, and in general his best work consists of comments, sometimes brilliantly original, on works whose truth is not questioned: Scripture, Aristotle, Dionysius. "We are dwarfs perched on the shoulders of giants," wrote Bernard of Chartres early

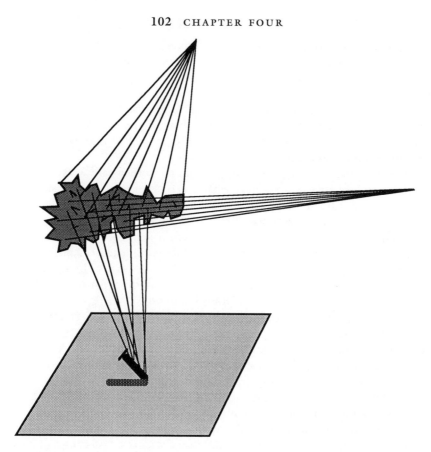

Figure 4.1. Pyramids of magnetic action surrounding a magnet. Those that reach points on a nail lift it from the table.

in the twelfth century, "and so we see more and farther than they, not because we have better vision or greater height, but because we are lifted up and borne aloft on their gigantic stature." Robert, like most brilliant men of his time, felt comfortable on his perch.

The little tract *On Lines, Angles, and Figures* concludes with some striking remarks on how species concentrate their action on small things. We have seen how they tend to expand into a sphere—a candle flame throws its light in every direction, and we have seen that this is how the heavenly spheres were formed. But there must be more to it, for consider how a magnet concentrates its action to lift a nail. To explain, Robert fills space with what he calls pyramids; these are geometrical figures generated by straight lines issuing from everywhere on the thing that emits the power and gathering at an apex which is a point of the thing that receives it (Figure 4.1). The magnet attracts a nearby nail wherever it is, and

thus infinitely many pyramids are able to issue from one surface of an agent;
as base of all these pyramids there is one thing, namely the surface of the
agent, and there are as many apexes as there are pyramids. The apexes are

located everywhere, at all the different points of the medium or recipient; and infinitely many pyramids issue in any one direction, some shorter and some longer.

Perhaps Robert imagined this magnificent hedgehog as a thing that actually exists; at any rate it reminds us that to understand the actions of nature we must learn to think geometrically. The pyramids come from Alkindi's *Optics*, where they are visual rays going outward from the eye, and in a book on the rainbow Robert defends them: "It should be understood that the visual species is a substance, shining and radiating like the Sun, the radiation of which, when coupled with the radiation from the exterior shining body, entirely completes vision."

Geometrical reasoning is of course important in modern physics, but as part of a general mathematical science of quantity as well as form. Philosophers of the middle ages and Renaissance sought to explain the world in terms of causes, but as long as they overlooked the importance of measurements and arguments focused on quantity, they made little progress toward understanding why events happen as they do.

When Robert returned to the theme of the Creation it was in his own *Hexaëmeron*. Here I will mention only the part played by light in the first scene of the drama. To follow his account one needs to understand two points that are somewhat foreign to our thinking. The first is that form has two aspects; one might say that it is both a noun and a verb. As a noun, it pertains to a substance and is defined by its qualities. As a verb it *happens*, as when it multiplies its action from one point to another. In imagining the Creation we see it both ways.

Second, we need a concept for which there is no word in English. It is intelligence pure and simple, without regard for who or what is being intelligent, or the subject to which intelligence is being applied. I shall spell it Intelligence, with the capital letter used in the same sense as in Ideas. It is a substance pertaining to the divine world, and in our material world it shows itself as light. Thus when Jesus says, "I am the light of the world" (John 8:12; Plate 2), he means that he is appearing in the material world as a manifestation of Divine Intelligence, or Holy Wisdom.

Before the Creation, says Robert, the world was empty, featureless, and dark. "Insofar as it lacked the plenitude of inclination toward action it was empty; insofar as it lacked useful action it was featureless. . . . Lacking action ordained by light, it was dark, that is, unformed and unfinished." Note that dark does not mean that you couldn't see anything. This was darkness at the root of being, and when the command was spoken and there was light, it was not the light the eye sees. That came three days later with the creation of the sun and moon.

There was always an awkward point in interpreting the Genesis story, the Bible's silence concerning two of God's greatest creations: the heavenly spheres and the angelic hosts. We have seen how Robert explained the spheres. The angels are concentrations of Intelligence, multiplied from the first flash. As Robert says: "By light we understand first, the invisible light acting everywhere for three days; second, angelic nature turned toward the contemplation of God; and third, we understand that the action of light is what compels formless matter to become formed. For every form is some kind of light, and light also makes manifest the matter it informs; as Paul says, 'Everything that is manifest is light.'" The act of creation was instantaneous, and Intelligence saw it that way, but there was a causal sequence of events during that instant, and Genesis, which is Moses's account of the Six Days, describes the sequence of events in terms that the human mind can understand.

In 1235, Robert was appointed Bishop of Lincoln, the largest diocese in England. At once he was up to his ears in conflict and administration as he struggled to preserve souls in a Church whose Roman center was daily becoming more cynical and more corrupt. In 1250, when he was about eighty, he traveled to Lyon and told Pope Innocent IV that by appointing unqualified men to pastoral positions for political and family reasons, and by preventing a bishop from correcting these and other abuses, Innocent was directly responsible for the perdition of thousands. The pope listened, and the abuses continued. Three years later a particularly offensive appointment was made, and in a letter to the pope and his court the old bishop fired his last shot: "Therefore, reverend Sirs, on account of [my] duty of obedience and fidelity I do not obey; I resist, I rebel." A few months later he died.

Robert Grosseteste was not canonized. Given the nature of his relations with the papacy this is not hard to understand, but perhaps also the necessary miracles did not happen. Still, I think the radiations from his being act very strongly on students of his life and work, so that even if he was a hard man to deal with in his time, they love him as few others are loved after seven long centuries.

Interlude on University Life

A professorial chair, in the middle ages, was more than just a chair. Often it was a black wooden tower supporting a seat into which the lecturer climbed by steps or a ladder. It was elaborately carved and backed with a high sounding board.[6] In front sat rows of students, aged thirteen and up, on backless benches, taking notes if the stream of Latin did not flow too quickly. In Paris the schools—liberal arts, theology, medicine, philosophy, civil and canon law—were scattered on the Left Bank, mostly in the space between the Pont Neuf and Mont Sainte-Geneviève. It is still called the Latin Quarter because that was the language of streets and wineshops. Not

---

[6] A beautiful example is preserved in the University of San Marcos in Lima, which was founded in 1551.

only did Latin connect the countries of Europe; there was at that time no real French or German language, and people from different parts of the same country spoke very differently.

In the earliest days of Oxford and Paris the curriculum for the B.A. covered the seven liberal arts. First in importance were the tools of language and thought: logic, grammar, and rhetoric; these were called the *trivium*. Next came the *quadrivium*, consisting of two mathematical disciplines, arithmetic and geometry, then music (often described as arithmetic in motion) and astronomy (geometry in motion). Cicero and other Romans were studied for style as well as for what they had to say about law and politics. Note the absence of subjects that one would label religious. Of course, any subject was conceived in the light of Christian principles, but theology was not a liberal art.

The method followed in these studies was what intellectual historians have called Scholastic, and it had been used in schools from the tenth century onward. First the lecturer read and explained a text, then he led the students in discussion and debate concerning the fine points of its construction, logic, meaning, and implications, studying how the argument establishes the conclusion. By its emphasis on exactness of thought and expression the method provided superb training for the civil and ecclesiastical lawyers and administrators who issued from these schools, but the texts were always old ones, the emphasis was on process rather than content, and the lecturer did not discuss the truth of what they said. In the beginning that was all right, but in the end it failed to satisfy a growing thirst for new knowledge.

In the late 1100s more texts of Aristotle began to arrive, newly translated, and by 1200 the liberal Parisian curriculum was turning into a cram course on "The Philosopher," as they called him, read, interpreted, and to some extent expurgated in the light of Christian theology. The arts student, or "artist," heard courses of lectures and attended disputations first on Aristotle's logical works, called the *Organon*, and on Priscian's Latin grammar. Then came natural philosophy—*Physics, On the Heavens,* the books about animals—and finally the *Ethics, Politics,* and *Poetics.* In addition there were special lectures on literature and philosophy outside the regular courses. A rich menu, but note that it lists no classics of Latin literature, while in the distance the mountains that were Homer, Plato, and the Greek dramatists stood wrapped in the clouds of a language that few in the West had learned to read. Aristotle's postlogical works were enormously popular with arts students because he gave them a fresh way of looking at the world around them. He started with experience available to everyone, and drew novel and not obvious conclusions. Why does a stone fall? We know that Fire rises and that the natural motion of Earth is downward. A stone contains much more Earth than Fire, therefore it falls. Perhaps this is not very informative, but it accepted the question as a reasonable one that has an answer, and it showed the student what an answer looks like. And today, if asked why the Earth produces the gravity that gives the stone its weight, one has to say one doesn't really know.

In the beginning the Church was uncomfortable with Aristotle. He taught that the world is eternal. Such a teaching could be suppressed and was, but everybody knew about it and its implications lingered as doubts: perhaps God moves but did not create, so that the world exists independently of him. Aristotle's God, the source of movement and change, is surrounded by love but seems quite indifferent to the hopes and fate of humanity. Christ's suffering is less significant in a world with no beginning and no end, in which civilization periodically rises, crashes, and slowly begins again. Theology directs the student's attention to God; Aristotle draws it away from him and toward things of this world. Church authorities banished Aristotle's *Physics* and *Metaphysics* from the University of Paris in 1210 and 1215 and again in 1228, but the three dates show what really happened: nobody paid much attention. The faculty of theology remained in the sheepfold, but the doctors and students of the arts faculty were becoming interested in exploring knowledge as well as in fitting what they knew into traditional molds. The doctors were not theologians; they had their degrees in arts, and reason pushed hard against their faith. Heresy flourished; thousands of students from all over Europe debated freely in class, in wineshops, in the street, whether Aristotle was right and Creation was a myth, whether (in Aristotle's terms) God was not the world's efficient cause but only its final cause, whether he could restore a lost virginity.[7] The rebel spirit prevailed; by 1254 all Aristotle's works were approved (with occasional corrections), and many were required.

By 1277 thought was flowing so freely that Etienne Tempier, Bishop of Paris, convened a committee of theologians which issued a list of 213 propositions that crossed the border of heresy. Students were forbidden to debate them, on either side, or even hear them debated, under threat of excommunication. Here are five examples:

> that Aristotle was right when he wrote that the world is eternal
> that if the heavenly spheres stood still fire would not burn flax, because God
>   would not exist
> that our will is subject to the celestial bodies
> that the teachings of theologians are based on fables
> that the immediate efficient cause of every form is a celestial sphere

Few people took the list seriously, and the Dominicans ignored it from the beginning. Various items were dropped from it, and time swept the whole thing quietly away. What this episode shows us is the atmosphere of intellectual freedom that existed among the arts students in medieval universi-

---

[7] This question originates in a declaration by Saint Jerome, "I will speak boldly: although God can do all things, he cannot raise up a virgin after she has fallen." In the eleventh century this was disputed by Peter Damiani, who claimed that God's power could even change the past, and the subject later became a popular one for debate. Before smiling at the question, reflect that the Church's traditional attitude toward virginity makes it both profound and difficult.

ties. As long as one did not try to teach heresy, no question was too contentious or too silly to be raised and argued in public. A contrast leaps to mind: how many questions are there that cannot be discussed openly today in the world's colleges and universities?

In the arts school the lectures were often confused and confusing. This was especially so when they expounded a text that might have been translated successively from Greek to Syriac to Arabic to Hebrew to Latin and finally, in the student's mind, to his own dialect. The translations were all as literal as possible so as not to lose a drop of precious wisdom, but how did someone render Greek subtleties of meaning in the richly allusive language of desert Arabs, or even that of the Romans, who had no tradition of such thinking? Lecturers who understood what they were talking about and explained it clearly were tremendous stars, and everyone went to hear them.

For about two centuries Aristotelian thought and language dominated most of the schools, and all went well as long as everyone thought that Aristotle's substances and forms really exist at the foundation of being. The reaction came when scholars like William of Ockham began to teach that they exist only as names of concepts, and that debates about them concern not the real world or anything that really happens but mostly terminology.

In the sixteenth century Pierre de la Ramée (1515–72) studied logic in Paris and found that the subject defined a world of its own. "Never, in all the voices of the schools where I passed so many days, so many months, so many years, did I ever hear a word, a single word, on the applications of logic." His disgust with Scholasticism entrained contempt for Aristotle, which wasn't quite fair; and at the age of twenty-one he successfully defended his thesis for the M.A., "Everything Aristotle said is false," in a day-long shouting match with the doctors of Paris. That the doctors could not strike down this extravagant claim shows how blunt their weapons were. Later Ramée published his own system of logic, which was influential in Protestant countries and built into the curricula of New England's early universities. By the end of the century the English word "trivial" had acquired its modern meaning, and after that time, though curricula changed slowly, the retarding force that Aristotle exerted in the intellectual life of Europe became much less. Remember though, the minds that pushed beyond Aristotle were minds that he had awakened. Today Scholasticism remains a small territory, mostly outside university walls, in which live those who believe that truth is already known, and that they know it.

6. Roger Bacon Rebels

Roger Bacon (c. 1219–92), an Englishman, was a rebel in a world supposed to be ruled by vows of obedience. At Oxford he took the M.A. that allowed him to teach in Paris and ride the Aristotelian wave. Most of the

Parisian lecturers were subjecting the traditional texts to logical analysis, a process whose unintended effect was to make belief in their content less absolute. Roger set out to counteract this tendency and reinforce Christian belief by careful study of all God's works, as revealed in Nature and doctrine. In an autobiographical fragment he tells how he sought the friendship of wise men and encouraged his students to study mathematics, languages, and experimental science. As for men who were not wise, he is liberal with his comments on Albert the Great (see §8), who did not know as much as he thought he did, on William of Moerbeke flooding the schools with mistranslations from the Greek, on all those who lectured from garbled texts when with a little trouble they could have learned Greek and done it right, and on those who wasted their students' time with senseless quibbles like "If a man leads a pig to market, is it the man or the rope that pulls the pig along?" He names names and considers these people dangerous to religion.

A few years later Roger returned to Oxford, where he continued his studies and made experiments on optics and vision. Some time around 1257 he joined the Franciscan order, considered to be the one in England most concerned with intellectual issues, but as it turned out this was a disastrous mistake.

Why would anyone dedicated to the purification of theology and the rightness of his Church spend time on optical studies? Considering this question helps one understand the minds of European Christians until well into the seventeenth century. First, one must realize that God has a plan for the world, in which nothing is ignored or approximated, and nothing happens by chance. "Are not two sparrows sold for a penny?" asks Jesus, "Yet without your Father's knowledge not one of them can fall to the ground. As for you, even the hairs of your head have all been counted." Not only is there a single plan but it is fundamentally simple and unified, with the sole purpose of leading humanity toward salvation. Everything we see, if we think about it, gives evidence of this plan and helps us to understand it. There is one sun, which floods us with warmth and light and supports the life of the world; if we attend to its deeper meaning, it also shows us that God is loving and merciful and that he is One. God provides that small trees do not grow in the shadow of a great one; we learn from this that a country must have a single ruler. From seeing how a mother animal cares for her young the ruler learns his duty toward his people. From studying the form of the Earth and the motions of the planets we learn that the divine plan is mathematical, including both geometry and numbers. Another clue: we know the world mainly from what we see. Gaining knowledge through light from the sun is analogous to gaining it by God's revelation, and light, as Alkindi and Robert Grosseteste have shown us, is what moves the world. Optics is therefore the fundamental science, and it is mathematical. God is telling us that if we would know him we must study the world he created.

Roger Bacon's criticisms of theology and theologians did not make his superiors happy, and he was sent back to Paris for discipline and supervision. Forbidden to write, he kept on with his experiments and, it seems, with writing as well. During this period a leading Parisian intellectual named Guy de Foulques, who had heard Roger's lectures at the university, kept in touch with him, and when Guy became Pope Clement IV he asked Roger for copies of his works, to be sent secretly in spite of orders from the Franciscan minister-general that Roger should write nothing and that no member of the order should communicate with Clement for any reason. Hastily Roger cobbled together three books known as the *Longer Work*, the *Shorter Work*, and the *Third Work*, and in 1268 he sent them to Clement at Viterbo. The papacy had recently moved to this small city north of Rome because of political turmoil and unhealthy living conditions in the capital, and for twenty years it remained the center of the Catholic world. Nothing happened. Clement died a few months later, and with him died any papal interest in what Roger had to say.

In 1271 Roger managed to publish a *Compendium of Philosophic Studies* in which again he aired his opinions of his Parisian colleagues. In 1277, the year of Bishop Tempier's broad censorship of beliefs and opinions, he was charged with "suspected novelties," tried, and sent to jail. After fourteen years the Order got a new minister-general, and Roger staggered out into the daylight, old and weary but carrying his pen. He finished part of a general treatise on theology and then, in the same year, he died. For some time the odor of heresy clung to him, but later, when his works could be freely studied and discussed, he acquired his own special name. In the middle ages Aristotle was tagged as "The Philosopher," and his most distinguished successors were each identified by an adjective. Roger was *Doctor universalis.*

After all the experiments and radical talk, Roger Bacon's work on optics is rather conventional and even, in one way, retrogressive. The main ideas come straight from Alkindi and Alhazen and Robert Grosseteste. They are clearly explained with mathematical diagrams and descriptions of experiments and remained a standard work for the next three centuries. His view of the essential role of light in all physical processes echoes Robert: "The wise and the foolish agree in this, that the agent sends forth a species into the matter of the recipient so that it can bring forth out of the potentiality of the matter the complete effect that it intends." And every physical effect has a physical cause: "Everything in Nature completes its action by its own force and species alone." Thus the stars act so as to influence our earthly surroundings, our general health, and the working of our minds. But note: they only influence; they do not control. In his criticism of the way the art of "mathematics" is practiced he writes that "in human affairs true mathematicians do not presume to certain knowledge, but they consider how the body is altered by the heavens, and when the body is changed the mind is aroused now to public actions and now to private ones, yet in all matters the freedom of the will is preserved." Thomas Aquinas, writing at about the same time, agrees. The stars, he says, influence material things directly and the intellect indirectly, for intellect is fed by the senses. The will, independent of sense, is not affected, though because most people are ruled more by passion than by will, astronomy can often predict the behavior of a group. Even on material things, however, the influence is only a tendency, not a necessity.

Roger's theory of vision is essentially Alhazen's, except that he follows Robert and retains the outgoing visual ray as well as the one that comes in. He considers the outgoing ray an obvious fact, for anything that is visible emits species all the time, and an eyeball is certainly visible. Rays are paths along which visual species multiply; but as I have said, the motion of species by multiplication is like the motion of a shadow and not like the flight of a ball. Consider for example Roger's explanation of reflection from the surface of a dense substance, solid or liquid. It is tempting to explain the action of a mirror with the analogy of a bouncing ball, but species do not move and therefore cannot bounce. Instead, "every dense body in so far as it is dense reflects species, not because violence is done to the species but because the species takes the opportunity from a dense body impeding its passage to multiply itself by another path open to it." But that is not enough to explain vision: "Vision perceives what is visible by its own force multiplied to the object. Moreover, the species of things of the world are not fitted by Nature to effect the complete act of vision because of vision's nobleness. Hence these species must be aided and excited by species from the eye." We need both rays, incoming and outgoing. Concerning light, Roger argues that the speed of light is finite, for a species considered as a causal agent must be here now and there later, otherwise it would be in both places at the same time, and cause and effect would be indistinguishable.

The air teems with species of all kinds, but the eye registers only those of light and color. As to their nature, Roger does not explain it, but he asserts that because the species that originate in a material object must convey the essence of that object, they themselves are material. The reasoning is persuasive, and until the end of the doctrine of species, though there was never agreement on how to express it, the received opinion was that light is more like a substance than a quality. But whatever the nature of species might be, Roger was careful to situate them with respect to Aristotle's four causes: the object in which a species originates is its formal cause, the medium through which it propagates is its material cause, and the process of multiplication is the efficient cause. The final one, of course, is God's plan for the salvation of humankind.

Roger devotes Book VI of the *Longer Work* to a discussion and defense of experimental science. Because science by its very definition in those days consisted in the deduction of necessary conclusions from firmly established premises, and Roger himself presents every argument in that form, one might ask what the role of experience really is. To this he answers that everyone knows the classical mode of argument is not perfect—see how often its conclusions have had to be revised—and he gives three ways in which experience comes to its aid:

It verifies the conclusions of deductive reasoning, thereby verifying the principles also.

It gives new knowledge concerning things for which science can supply no principles at all; an example is magnetism, a part of God's creation that no one claims to understand.

It leads not only to new science but also to new power, for we often invent and use new technology before we understand how it works.

There is no chance that a properly interpreted experience will conflict with that gained from revelation, because both issue from the same divine source. In fact, revelation is a necessary part of science, for experience yields only the specifics of individual cases. Science seeks general truths, and except by revelation (one form of experience), perhaps aided by experiment (another form), we can never find them.

I do not know whether Roger ever criticized Aristotle's mode of argument or the Scholastic method based on it, but with appropriate changes of language his three reasons for experimenting on Nature are good reasons today. Perhaps in restating the first one we would change its emphasis and say that an experiment can contradict the conclusions of deductive reasoning, and that if it does it shows that a principle must be false. Roger might have been willing to look at it that way, and as we move into the modern age we will see the emphasis shift in that direction.

7. How Do I See What I See?

Aristotle wrote about the physical nature of light, but he was always more interested in the process of vision, which he discussed along with the other senses at great length in *On the Soul* and other works. By the thirteenth century, stimulated by Aristotle, scholars were exploring in many directions, and much was being argued and written concerning our knowledge of the external world, what we know and how we know it. The next few pages will show how thinkers of the thirteenth century found in Aristotle and his successors the materials for a theory of perception, especially vision.

The works of Augustine (354–430) were always available and widely known. We have seen in §2.4 that he accepted the visual ray as the natural explanation of how we see. In his treatise on the Trinity he gives more detail. The act of perception involves four different species:

a species in the object itself, which produces
a species in one of the observer's senses, which produces
a species in the memory, which produces
the species in the observer's intuition.

The step involving memory is necessary. I see something approaching; memory, comparing this impression with others I remember, tells me it is a man; closer, and by the same process I see that it is Socrates. Augustine goes on to describe two other ways in which species arrive in the intuition: directly from memory, or from imagination, which can easily give me a mental picture of a bird with four feet even if I have never seen one. These species are fixed images rather than those unidentified flying objects that move along an observer's line of sight. Their nature is not further explained, but Augustine's account provides an outline of the visual process.

In about 1140, Aristotle's *On the Soul* reached Europe in a translation from the Greek. This was followed by one from Arabic; neither was very good, and copies fanned out slowly. A century later William of Moerbeke produced a better version from the Greek, and §2.5 gave a capsule version of the theory of vision it contained. At about the same time arrived translations of Galen and of Avicenna (980–1037), generally considered the greatest Arab writer on medicine. Aristotle and these two followers differed from Plato in an important respect: whereas for Plato the soul and the body moved on different planes of existence, these taught that the soul pertains to the body, that it senses and acts through physical mechanisms, and that some of it, at least, dies with the body.

According to Aristotle, "Soul is the actuality of a natural body having life potentially in it." It is the organizing, individuating, and energizing principle that distinguishes a living body from a dead one. Aristotle locates this principle in the center of the breast, in or near the heart, which is the hottest part of the body, the first organ (in the fetus) to start functioning and the last one to stop. The brain, connected to the heart by the big carotid arteries, functions only to cool the heart. Soul communicates with body by a fine, vaporous fluid called spirit. (This is Galen's *pneuma*, a word that originally meant breath.) Its presence makes semen fertile, and it was often called the first instrument of the soul. Aristotle has little to say about spirit, but it was important in the perceptual theories of the middle ages and the Renaissance.

Let us pause for a moment to consider the scientific problem that Aristotle and the two great doctors, Galen and Avicenna, were trying to solve: how does sense reach the soul? Aristotle says nothing about nerves, but Galen lists many and understands that they report sensation and control movement. Obviously they convey their messages very quickly; the question is how. Today we know the process is electrochemical, but for many practical purposes it is as if a nerve current were a fluid moving along a narrow tube. That is exactly what Galen and Avicenna imagined. Spirit is a fluid related to the ether of which stars are made, refined from blood and carried through the body by nerves and veins. It is so thin that it is almost like a vapor, so volatile that no one need be surprised if it does not show up under dissection.

Figure 4.2. The brain and its ventricles.

I mentioned in §2.5 that Aristotle gives the soul five separate powers or faculties: nutrition, appetite, sensation, movement, and thought. Of course, he knows that in man and the higher animals the soul is more complicated than this; the five are only for purposes of discussion. He locates nutrition, sensation, and movement in the heart. Galen elaborates the theory of spirit. He moves all the faculties to the brain, but its gray, amorphous tissue seems unsuitable for thinking, and so he focuses on the ventricles (Figure 4.2), which are small vessels full of fluid inside the brain that are now thought to regulate pressure. Here, he taught, spirit is refined from blood, after which it moves between ventricles by seeping through the brain tissue and communicates with muscles and sense organs through nerves leading from the two ventricles nearest the forehead. Avicenna is more specific; he invents five internal senses to match Aristotle's five faculties and distributes them among the structures of the brain. He is creating the outlines of an anatomy of thought, and since this is once more an active area of research, it may be interesting to mention some of his ideas.

Scholars of the middle ages agreed only on the general outlines of a theory to explain how the senses bring us news of the external world. The reason the texts are a chaos of principles and arguments is not hard to find: evidence was supposed to help in choosing the principles of any study, and evidence as to what actually happens in our brain when we see a rabbit was very scanty. The idea was that the species carrying a sense impression travels from the external sense organ to a corresponding internal one where it impresses itself on the spirit contained there. This spirit conveys it rapidly through the body to various organs such as the heart, where it might generate the immediate bodily movements of unreflecting passion or reaction to danger, and to the head, where common sense resides. There the process begins that impresses the species on the higher faculties, so that we become aware of what it is we are perceiving. Robert Grosseteste is careful to say that the whole process occurs by the agency of light: "Light is the substance that is most subtle and thus closest to spirit, which is incorporeal. . . . It is therefore through light that spirit acts instrumentally on all our senses."

Much later, dissection and experiment proved that thinking takes place in the tissues of the brain. But then what is the use of the ventricles? Dr. Thomas Willis, a famous student of the brain, writing in 1664, has another use for them: following Galen, he assigns them "that vile office of a Jakes or sink by which spirit is cleansed of excrementitious matter." He still has spirit, made in the brain and cerebellum and flowing through brain tissue in tiny veins. Willis does not say exactly what it does; perhaps he assumes that everyone knows, but Isaac Newton, writing in 1675, supposes that "there is such a Spirit, that is, that the Animall Spirits are neither like the liquor, vapour, or Gas of Spirit of Wine, but of an æthereall Nature, Subtile enough to pervade the Animall juices as freely as the Electric or perhaps Magnetic effluvia do glass. . . . As any of it is Spent, it is continually supplyed by new Spirit from the heart." By this time these subtle vapors were starting to evaporate from the medical literature, but it may revive your own spirits to realize that they still persist in common speech.

8. Two Doctors

The main source of medieval psychology after this point is the works of the Dominican friar Albert (c. 1200–1280), Albert the Great, as he has always been called. He was the first notable German scholar. Born in Lauingen, a town on the Danube a little below Ulm, he went early to Padua where he probably studied medicine. In his forties he was sent to Paris where he lectured to immense audiences. He was there only five years, yet today in the university quarter the Rue Maître Albert leads into a tidy little square called Place Maubert, echoing the name the students gave him. Later came another honorific: they named him *Doctor mirabilis*, which means wonderful, marvelous, admirable.

In 1248 Albert's order recalled him to teach in a new conventual school at Cologne. Here in a flood of books he tried to organize all knowledge along Aristotelian lines and write it out in Latin for the education of Europe. Among other works he produced one on gems and minerals, one called *Thirty-Six Books on Animals*, and a vast series of commentaries on (Pseudo-) Dionysius, on Euclid, and on most of Aristotle's logical and scientific works. Especially in discussing what Aristotle says about natural things, he compares it with his own observations and corrects without fanfare if he thinks Aristotle is wrong. Nobody, he says, can be right all the time. His books, because nothing comparable was available, became the rocks on which medieval education stood. Later Albert was sent off on a series of administrative jobs; he spent some time in Viterbo, then retired and wrote on, as long as he was able.

The theory of sensation that Albert transmitted to his followers is essentially that of Avicenna, with a few refinements. A sense organ becomes connected with the faculties, communication takes place, then species arrive in the brain and are processed by the five internal senses of the soul. Say that I bite into a peach. The functions of my tongue and my eye are not to report "peach" but only to register the species of the taste and of the colored shape. Common sense combines the two messages and determines that they both pertain to the same act and thing. Imagination remembers the form that these sensations comprise, and perhaps others, such as that of holding the peach in one's hand. Cogitation decides, "I have just bitten into a peach." Estimation judges, "It is not ripe; don't eat any more from this basket for a few days." Memory retains the idea as long as needed. Figure 4.3 shows an early sixteenth-century version showing six internal senses that correspond roughly with Avicenna's placements.

Albert's theory of vision follows Aristotle closely. Having refuted all theories based on *eidola* and incoming or outgoing rays, he maintains that an object changes the forms of transparent media around it, including the watery part of the eye. From the Arabs he has learned that vision occurs at the crystalline lens, whence the image is conveyed to the soul's faculties by spirit in the optic nerve.

Albert, writing shortly after Robert Grosseteste had put forth his Platonic theory of vision in *On the Rainbow*, uses Aristotle to throw down Plato; whereas Roger Bacon a few years later pays no attention to Albert and models his theory on Alhazen. Roger sat in prison, and Aristotle won, but it was a shaky victory. Clearly Roger was someone who preferred thinking for himself to scanning old texts; and if this was his habit, why should others not do likewise? In increasing numbers, they did.

Thomas Aquinas (c. 1227–74) was born into a family of South Italian nobility. At seventeen he entered the Dominican order, and as his astounding gifts became evident he was sent to Paris to study. Here he listened to Albert and followed him to Cologne, but whereas Albert's main interest was philosophy, his was theology. After a few more years he left to begin a strange, short life of travel, distraction, and unremitting toil, for just as Albert had set out to unify all natural knowledge on Aristotelian principles, Thomas planned to do the same for theology. He lectured everywhere and was the idol of students. "The goal of philosophy is not to know what men have thought," he said, "but rather to know the truth of things," and his scholarship went far beyond explaining old texts. Scripture and revelation were the only sources of undoubted truth, but even Scripture had to be intelligently interpreted before its message could be understood.

Regarding the subject of this book, the few words Thomas wrote would

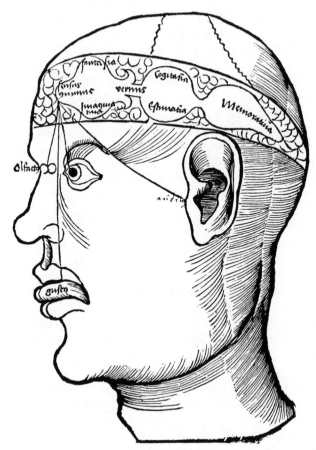

Figure 4.3. Conventional view, roughly following Albertus Magnus, of the location of the internal senses in the ventricles of the brain. The senses are *visus* (vision, not labeled), *olfactus* (smell), *gusta* (taste), *auditus* (hearing). From the left, the internal senses are *sensus communis* (common sense), *imaginativa* (identifies what is there), *fantasia* (imagines what is not there for transmission to *cogitativa* and *estimativa*), *vermis*, a passage between ventricles, *cogitativa* (thinking), *estimativa* (judgment), and *memorativa* (memory). Some authors equated *imaginativa* with *cogitativa*. From Gregor Reisch's *Margarita Philosophica* [1504].

hardly be worth reporting on except for the esteem in which he was held. He gives four reasons why light cannot be a substance:

Light and a medium such as air or water can exist in the same place.
Light seems to travel with infinite speed, and material substances cannot do that.
Light cannot be an essential property of any transparent substance, for if it were, that substance would cease to exist in the dark.

Blow out a candle. If light were a substance, the light left in the room would take some new form, but (unless darkness is also a substance) we never see this happening.

Therefore light is an accidental quality. But of what substance is light a quality? Thomas, following Aristotle (§2.5), says it is a substance called *diaphanum*, transparency, found in media like air and water, which receives the form of light from any source and conveys an object's form to the observer's soul. If light is a quality, an important conclusion follows: an accidental quality cannot do anything; therefore the beautiful light physics of Robert Grosseteste and Roger Bacon, in which light was the universal cause of change, cannot be true, and from this time that great structure begins to fall into decay.

Thomas's theory of vision starts with an Aristotelian classification. There are two kinds of changes, natural and spiritual. Natural changes are physical. If I touch something hot with my hand I perceive its heat because my hand becomes hot, but if I look at something red my eye does not become red.[8] That is spiritual perception, and thus there are two means by which a quality of the object enters the observer's soul. But what can we actually know about the material world? I have defined substance as the identity of a particular thing independent of any of the qualities by which we know it. Clearly we can talk about substance, but we cannot know it. Substances are *defined* into existence. All knowledge ultimately comes through the senses, and these tell us only certain qualities of a substance. During the middle ages it became customary to divide the qualities of a thing into two classes. The first consists of the things's *substantial form*, defined as the qualities that are absolutely essential to its being what it is; the second are its accidental qualities, those that are not fundamental to it. Thomas emphasizes that the only forms our senses respond to are the accidental ones. Aristotle taught that we know forms and not accidents, but Thomas uses "know" in a less restricted sense: we can know qualities as well. We know by our senses that a coin is round and heavy and that its color is gold, but these accidents do not tell us whether it is really made of gold. "Goldness" is what gives the coin its value, what defines it as a coin. Without it the coin is nothing: goldness is a substantial form. We may be able to find out whether the coin is genuine, but our senses alone cannot tell us. They say what a thing looks like and how it acts, but not what it is. Much later in this book we shall encounter questions like "Is an electron a wave or a particle?" Here again, to adopt medieval terminology, we are trying to identify a substantial form. Some have tried to legislate the electron into one or the other, but experiment does not tell.

The modes of thought illustrated in this chapter were developed, taught, and learned for many years. In medicine and science, experience forced cer-

---

[8] But Alhazen has correctly pointed out that it really does.

tain revisions, but in subjects like theology, the intellectual structures so beautifully adapted to giving a complete and harmonized picture of human knowledge were slow to change. As late as 1879, Pope Leo XIII directed all Catholic clergy to base their theological arguments on Thomas's Aristotelian system of thought. Things also stood thus for a long time in the universities, for a university is a good place to listen to verbal formulas, to write them down, learn them by heart, and then sleep in certainty. Aristotelian certainties continued for several centuries more but encountered increasing skepticism as time went on. In the 1340s William of Ockham pointed out that because species escape the senses we know and can know nothing about them. He then applied his famous principle of parsimony, known as Ockham's Razor (do not introduce any unnecessary complications into an argument), to the process of vision and pointed out that species are not needed; it is enough to assume that an object acts directly on the eye. We shall follow the old doctrine's further decline in the coming chapters until its inevitable end. It was a slow process, but in 1651 the end was in sight when Thomas Hobbes, at the beginning of his *Leviathan*, had this to say about species, which he defined as "a *visible shew, apparition,* or *aspect,* or a *being seen*":

> Nay for the cause of *Understanding* also, they say that the thing Understood sendeth forth *intelligible species,* that is, an *intelligible being seen*; which coming into the Understanding, makes us Understand. I say not this, as disapproving the use of Universities: but because I am to speak hereafter of their office in a Commonwealth, I must let you see on all occasions by the way, what things would be amended in them; amongst which the frequency of insignificant Speech is one.

5

THE END OF CLASSICAL OPTICS

> I think that in discussions of physical problems we ought to
> begin not from the authority of scriptural passages, but from
> sense-experiences and necessary demonstrations; for the Holy
> Bible and the phenomena of Nature proceed alike from the
> divine Word, the former as the dictate of the Holy Ghost and
> the latter as the observant executrix of God's commands.
> (Galileo Galilei[1])

1. Witelo, Also, Writes a Book

IN 1268, when Roger Bacon sent his three books to Viterbo, nothing
happened. With Pope Clement died any papal interest in what Roger
had to say, but apparently the manuscripts were available for study,
and they attracted the attention of a young Polish councillor named Witelo
(c. 1235–after 1281) who had recently arrived at the papal court. He had
taken a liberal arts degree at Paris and then gone to study canon law at
Padua, but he seems to have had time to write a treatise *On the Principal
Causes of Penitence and the Nature of Demons*, and also to study optics. He
says that seeing a rainbow in a waterfall near Viterbo inspired him to write
a book on the subject. Viterbo was a good place to work. It was quiet and
had a library, and William of Moerbeke was there. Even though Roger
judged him incompetent, he had lived in Greece and was one of the principal translators of Greek texts.[2] In Viterbo he was installed as papal confes-
sor and was willing to help Witelo with books like Euclid's *Optics* and
Hero's *Catoptrics* when he needed them.

Clearly Witelo intended *Perspectiva* to be a classic, definitive; and so it
was, for almost 350 years. It is built on a grand scale. There is a fifty-seven-
page mathematical introduction, out of Euclid and later writers, giving the
geometrical theorems he will need for discussing reflection and refraction.
Then come experiments with rays and shadows, the anatomy of the eye,
and the theory of perception, all taken from Alhazen. He explains the rule

---

[1] From Galileo's *Letter to the Grand Duchess Christina*, translated in Seeger 1966.

[2] Of course, residence in Greece was not enough to equip a scholar to read the classics, for
the language had changed in fifteen hundred years. There were Greek priests and monks in
Rome and Palermo, and entire Greek-speaking communities in South Italy, but Italian scholars who went to those speakers for help with Plato generally returned disappointed.

of reflected light—angle of reflection equals angle of incidence—by adopting Hero's principle of economy. He discusses refraction at curved surfaces: first comes a spherical crystal as a burning glass, then the rainbow, which he calls *arcus daemonis*, the demon's arc (what happened to God's covenant?), and attributes to reflection and refraction in moist air. He has repeated Ptolemy's measurements of the refraction of light passing from air to water and corrects only the first row in Table 3.1, writing 7°45' instead of 8°—a small improvement, but it spoils Ptolemy's progression of numbers and destroys the hope that they might illustrate a simple natural principle.

The book owes much to Alhazen, but when he writes what he thinks light really is and why we should study it, he reflects the views of Bacon, Grosseteste, and especially Dionysius. We find them in the Preface, in which he dedicates his book to William of Moerbeke:

> The nature of all things flows downward from the Divine nature, and all intelligibility from Divine intelligibility, and all life from Divine life. Divine light in its intelligible mode is the beginning, middle, and end of all these influences, for it determines the reason, the means, and the final manner in which things are disposed. Indeed, sensible light is the medium of the corporeal influence which wonderfully connects lower bodies (according to their different kinds and forms and potentialities) to the eternal bodies above. For light *is* the diffusion of the highest corporeal forms, according to the nature of bodily form, to the material of lower bodies.

In the sense that light is here more like a process than a thing, Witelo writes like an Aristotelian, but his book, with its assumptions about how light acts in the world, and its heavy mathematical content, is far from Aristotle. Instead, he follows Grosseteste in proposing light as the fundamental principle that underlies all the operations of Nature, and he explains why it is important to study optics:

> Just because the visual ray acts more perceptibly on our senses, we should not think that these rays do not act on natural objects that have no senses, for the presence of senses does not add to the actions of natural forms. And since every aspect of vision can be studied through either mathematical or experimental demonstrations, I will treat, as well as I can, what relates to the actions of natural forms [by studying] how they affect visible objects.

Witelo delivers his great manuscript and fades from sight. It was copied early and often. The first printed version came in 1535, and in 1572 there was a much better one, edited by Frederick Risner in Basel, which combines it with Alhazen's *Perspectiva* and lists Witelo's extensive (and unacknowledged) borrowings. Why did he borrow so much? Not for laziness or bad character; the main reason is that except for the metaphysical setting he had nothing new to say. The ancients knew the truth and wrote it down, and there was little a modern could do except arrange, clarify, and explain. The book's mathematics was all in the service of the one simple fact that when light hits a mirror, flat or curved, the angles of incidence and reflection are equal. Is there nothing more to mathematical optics? Euclid had squeezed the juice out of that orange sixteen hundred years earlier, and there was little point in kneading what was left. It was time to widen the field of inquiry, ask some new questions, and explore new phenomena so as to have something new to talk about.

2. Where Did I Put My Glasses?

Sir David Brewster, whom we shall meet again later in this book, was an expert on optics. Sometime about 1850 he was visited by Mr. Austen Henry Layard, who when not excavating the ruins of the ancient Near East served as a member of Parliament. The object Layard took from his pocket had been dug up in Nimrud, a city near the ruins of Babylon, and he dated it in the eighth century B.C.E. It was ground from rock crystal and seemed to be a lens, though a few modern skeptics see it as a crude ornament. It can be seen in the British Museum. Brewster reports,

> This lens is plano-convex, and of a slightly oval form, its length being $1^6/10$ inches and its breadth $1^4/10$. It is about $9/10$ths of an inch thick, and a little thicker one side than the other. Its plane surface is pretty even, though ill polished and scratched. Its convex surface has not been ground, or polished, on a spherical concave disc, but has been fashioned on a lapidary's wheel, or by some method equally rude. . . . It gives a tolerably distinct focus, at the distance of $4^1/2$ inches from the plane side. . . . It is obvious, from the shape and rude cutting of the lens, that it could not have been intended as an ornament; we are entitled, therefore, to consider it as intended to be used as a lens, either for magnifying or for concentrating the rays of the Sun, which it does, however, very imperfectly.

As I said earlier, Aristophanes tells us that in his time the corner drugstore in Athens carried burning lenses. Ptolemy and Alhazen and Witelo mention them as devices that use refraction, but none of these writers give them any serious attention. Euclid did not mention them; they lay in the kitchen drawer among the sieves and spoons.

In the West, Robert Grosseteste was among the first to think about lenses. Qualitatively, he understood how a flask of water brings light to a focus: a ray is refracted at each surface of the glass. To draw a proper picture he had to know how much it is refracted. He knew the law of reflection, which says that a ray makes equal angles approaching and leaving a mirror, and called it an example of Nature's tendency toward simplicity. He did not know Ptolemy's *Optics*, and so he invented his own simple law

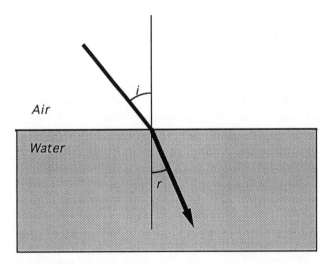

Figure 5.1. Defining the angles of incidence, i, and refraction, r, when a ray of light passes from one medium into another.

Figure 5.2. Hot spot behind a spherical flask of water in sunlight. The flask is on the left; a narrow beam of sunlight passes through it and forms a pattern on a flat screen. The fan-shaped lines arise from imperfections in the glass.

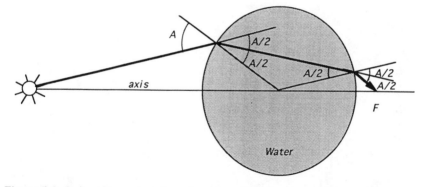

Figure 5.3. Robert Grosseteste's explanation of the hot spot of Figure 5.2. Point F is the approximate location of the spot.

of refraction, that when light enters water the angle of refraction (Figure 5.1) is half the angle of incidence.[3] Figure 5.2 shows the hot spot behind a spherical flask of water in sunlight, and Figure 5.3 shows how Robert explains it. The ray along the axis passes through point F, and so does the slightly off-axial ray I have constructed according to Robert's prescription. If all the other rays also passed through F the point would be a real focus, and it would get very hot, but a bit of trial and error with a ruler shows they do not. The light would be smeared in the vicinity of F.[4]

Robert writes about these things in a little treatise on the rainbow, together with some remarks that show him as either a prophet or someone who knows more than he says:

> This part of optics, when well understood, shows us how we can make things a very long distance off appear to be placed very close, and large near things appear very small, and how we may make small things placed at a distance appear as large as we want, so that it is possible for us to read the smallest things at an incredible distance, or to count sand, or grain, or seeds, or any sort of minute objects.

A convex lens serves as a magnifier. A concave lens is much harder to shape, and I know of none that dates from Robert's time. Such a lens does make large near things appear small, but to make distant things appear close takes a telescope, and these are made by mounting at least two lenses at some distance apart in a tube. Remember that Robert thinks we see by a ray coming out of the eye, as he continues: "The visual ray refracting through several transparent media of different natures is refracted where they meet."

[3] For small angles, three-quarters is actually a good approximation.
[4] Actually the diagram is misleading, for it puts the sun onto the piece of paper. If the distance to the sun is large, simple geometry shows that according to Robert's law of refraction the image would be in perfect focus exactly at the edge of the bottle. This doesn't happen, and so Robert's law is wrong.

Figure 5.4. An early picture of spectacles. Initial letter of the Office of the Dead, mid-fourteenth century. Bibliothèque municipale de BESANÇON, by permission.

"Several media" suggests at least two lenses, together with the air around them. If Robert possessed a concave and a convex lens, and if he tried holding the concave one near his eye and the convex one further away, he had a telescope. Any child given two convex lenses would try looking through both of them. If the lens nearer the eye is fat and the other one thin, that too makes a telescope, though the image is upside down. For whatever reason, it was four hundred years before a telescope was publicly shown and described.

Roger Bacon experimented with lenses to improve vision, but he does not seem to have thought of putting them next to the eye. That was done by some lens-maker, probably in the neighborhood of Pisa, in about 1285, while Roger was alive but in prison. Suddenly scholars and tailors and jewelers who had been forced to put down their work were able to resume it without pain or danger of blindness. Figure 5.4 shows a pair of spectacles from the mid-fourteenth century. They must have clamped onto the nose with a spring.

Sherlock Holmes pointed out that the barking dog provided a most important clue because it did not bark. Silence shouts its message, the silence of natural philosophers who for the next three hundred years paid no more scientific attention to the spectacles they wore every day than they did to the hat that protected them from the rain or to the stool they sat on. Venetian glassworkers called the pieces of glass *roidi da ogli* (little discs for the eyes); other names were used in other places, but finally a general term emerged: they were named *lente* because they were shaped like lentils. The Latin for a lentil is *lens*.

I have already suggested one reason why nobody looked for a science of lenses: refraction did not seem to be governed by any simple principle, and even if it did—even if, for example Robert's proposed law were correct— the business of philosophy was to show, from first principles, why that was so, and not to follow up its consequences. Hero's principle of economy established the theory of catoptrics, but here there was no principle. Failing that, a deductive account of refraction was impossible, and science stopped. A little further on we shall see that when the problem of principle had been solved, or rather had evaporated, the theory of lenses was worked out in a few weeks.

There was another reason for the long delay, also anchored in Aristotelian metaphysics. This states that vision is the process which brings the form of the object into the soul of the observer. I hope you will keep an ordinary magnifying glass handy to try some of the things I will be talking about, but even a cheap one will be a much better lens than any available in the thirteenth century. For one thing, the glass was somewhat cloudy and not even uniform. Clear Italian glass, *cristallo*, was not made for another two hundred years. And also, because these lenses were ground and polished on a rotating wheel, their quality was limited by the steadiness of the workman's hand. But look through your lens at a few words on this page. Unless the lens is held at the right height above the paper the image is fuzzy; unless it is held parallel to the plane of the paper the image is distorted. What is a lens? Unless properly held it is a barrier for the object's form, not a conduit. And even if it is held correctly the form is still not right, for the lens has made it too large. The form seems easier to perceive, but actually it is wrong. If you try to see something through a lens you cannot assume that the form that enters your soul is the form of the object. It was not going to be simple for Galileo to convince skeptics that the world he showed them through his telescope was the real world. But before that question arose, another question was posed and solved: how can a bit of the real world be represented on a painted surface so that it brings the same forms to the eye as the scene itself would if one were viewing it directly?

3. How to Paint Space

Words had immense weight in the middle ages. We have seen how Robert Grosseteste and Roger Bacon occasionally tried to argue from the factual thing, but for a long time after them the main currents of the liberal arts consisted of rivers, torrents, cascades of verbal formulations. A new way of saying something was a contribution to knowledge, but a new way of doing something was not. Doing was the domain of the arts. Without the "liberal," which implied that they were appropriate studies for a free man who knew Latin and not for a laborer, the arts were techniques. Building, agriculture, and hunting were respected arts; the status of painting was debated. A wall surface or a panel painted to represent a scene or point a moral was a silent substitute for words, suitable for simple minds with no education, or for the rich and lively who needed amusement.

We remember the Italian Renaissance as a time when architecture threw off Gothic tradition and looked to the classical past, when painting explored new dimensions of color and optical depth; it was also the time when, almost unnoticeably, Aristotle began to slide from his position as "The Philosopher." These changes, combining with revolutionary developments in commerce and economics and politics, produced in a few centuries a new Europe that was at many points unrecognizably different from the old one. Nothing is harder than to explain why people change their modes of thought; it is not enough to pile up examples of changes in their external circumstances. The next few pages will tell about the invention of perspective drawing and suggest how it may have played a part in dispelling the verbal smog that had so long obstructed a clear view of the world.

The First Book of Samuel, Chapter 5, tells how victorious Philistines brought the Israelites' Ark of the Covenant back to their city of Ashdod where they put it on public show, and how the Lord afflicted them with punishment that was grievous and, in the King James version, embarrassing. Plate 4 shows a fresco from the Cathedral of Anagni, a little south of Rome. It was painted about 1225, in the time of Robert Grosseteste and Roger Bacon, and represents the city and the men of Ashdod in their time of trouble; on the left side is the Ark that caused it. A hundred years ago viewers would have classed this representation as primitive. They would have pointed out, as if it needed pointing out, that the men of Ashdod are painted on a different scale than the city, and that there is no suggestion of how the city's surroundings might have looked. To show the same scene Peter Brueghel would have used a single scale, but he would have had to bring the men outside the city and into the foreground if he had wanted to show their anxious mood. The painter of Anagni keeps them inside the city, where they were and where they belong. Today, comic strips and political cartoons, as well as some gallery paintings have accustomed us once again to inconsistencies of scale, but even in these art forms there is often some effort to show where the scene takes place, to situate it in space, at least to suggest how far it is from the viewer. The next few pages will tell a little of the story of how painters in the Italian Renaissance learned to represent space, how this learning depended on optical principles, and how the new knowledge began to change our way of seeing.

The idea that is basic to drawing a scene in perspective sits on the pages of Euclid's *Optics*, where Proposition 6 states that *parallel lines, when seen from a distance, appear not to be equally distant from each other*. If you look at his proof in §3.2, you can see that he might have stated it more exactly— *parallel lines extended away from the observer appear to draw together*—and if he had been in the habit of talking about what happens at the far limit of the unbounded plane he might have added, *and appear to touch at infinity*. This is what happens to the railroad tracks that children draw at school; of course children's infinity is not mathematical infinity, but it is far enough for them and for us. The point where these lines touch is called a *vanishing point*. This is a modern term, but even the idea behind it was not clearly expressed before Leon Battista Alberti published his *De pictura* in 1435. To a mathematician of Euclid's caliber it would have been obvious at once had the problem been posed to him in the right way, but if this ever happened the news did not get back to Greece. We have a few examples of Greek perspective drawing in vase paintings of the time in which the artists use different devices to show foreshortening, but the vanishing point was unknown to them.

Marcus Vitruvius Pollio, called Vitruvius, worked in Rome in the first century B.C.E., perhaps two centuries after Euclid. His *Ten Books on Architecture* show that he is interested in history and knows some Greek, and in the introduction to Book VII he makes a few remarks on the way Greek stage designers painted scenes on strips of canvas hung from the back wall of the stage. Democritus and Anaxagoras, he says (even before Euclid) showed how to draw lines

> with due regard to the point of sight and the divergence of the visual rays, so that by this deception a faithful representation of the appearance of buildings might be given in painted scenery, and so that, though all is drawn on a vertical flat façade, some parts may seem to be withdrawing into the background and others to be standing out in front.

Here and in other remarks on the subject, Vitruvius does not show that he knows how they did it, and outdoor scenes in the frescoes of Pompeii show little improvement over the earlier Greek efforts. The back of a Roman stage was a complicated architectural construction that could be used for many purposes, and there is no evidence that Roman producers used painted scenery.

Roger Bacon hoped for a way to paint scenes as they would really look to an observer, but he had a different reason. We cannot fully receive the

Bible's spiritual message before we grasp its literal truth. We are given dimensions of Noah's ark and Solomon's temple, but what did they actually look like? "We can understand nothing fully unless its form is presented before our eyes, and therefore the whole knowledge of things defined by geometrical forms is contained in the Scripture of God, far better than mere philosophy could express it." Roger does not need to explain this remark, because everyone understood that God's creation is entirely God's, a single plan and structure, geometry is his, the physical world is his, and it is our job to understand them as minute parts of a vast intellectual scheme. Roger probably never saw Euclid's *Optics*, and if he had he might not have noticed that it would help to solve his problem; what he wanted was something more basic, that people should learn the principles of geometry so as to have an objective and scientific view of what they were doing before they started to draw. Bacon's voice was lost in the clamor of that tumultuous period of intellectual history; and anyhow, the Latin in which he wrote was not read by the painters of church frescoes or the illuminators of manuscripts. It was necessary to start almost fresh, to rediscover what was hinted in old books, and that happened two hundred years later.

In the early 1400s the city of Florence was known throughout Europe for the daring imagination of its merchant adventurers, the talent of its artists, and the splendor of its celebrations. There was, however, a mistake, a blemish, a cause of embarrassment, that was much talked about: if you entered the magnificent new cathedral and looked up you saw a great round hole that began 175 feet above the pavement and waited, as it had waited for many years, to be covered with a dome. Short of putting up an expensive forest of timber scaffolding to support the stones as the mortar dried between them, nobody knew how to proceed. The history of architecture tells how a man named Filippo Brunelleschi (1377–1446), a goldsmith who had decided to become an architect, designed the glorious cupola that stands there now and invented a scheme for hanging scaffolds on the interior of the completed portion so that masons could work in security, raising their scaffolding as they proceeded. Though born with a silver spoon and given a fine education, Brunelleschi preferred arts to sciences and making things to reading what ancients had written. In about 1413 he seems to have decided to try to make a painting that looked just like the scene it represented. He used a mirror, but nobody knows how he proceeded, and any pretended reconstruction, including this one, should be approached with suspicion.

What Brunelleschi sought was a picture that looked as real as the image of your face in a mirror, and so he started with a mirror. He set it up in the central doorway of the cathedral where it would be out of the wind and the crowds, and facing the baptistery about fifty feet away. He stood with his back to the baptistery, so that he was not distracted by the scene itself and all he could see was its image in the mirror, framed by the darkness of the

Figure 5.5. Viewing Brunelleschi's painting of the Florentine baptistery.

cathedral interior. Perhaps he put up a stick to help locate his eye at a point in front of the mirror so that the scene would always be the same. He then set up an easel with a blank wooden panel about a foot square next to the mirror and started to reproduce what he saw. There was nothing new or unfamiliar about the process; it was just like copying a painting, which all students of the art were taught to do.

Finished, Brunelleschi turns around and compares his little painting with the scene in front of him. It is perfect except that it shows a mirror image. Its perfection can only be perceived by someone who sees the painting itself in a mirror, and Brunelleschi thinks of a dramatic way to demonstrate what he has done. First he covers the picture's sky area with a thin sheet of silver and polishes it till it reflects beautifully. Then he bores a little hole, "the size of a lentil," in the lower part of the panel. He installs his mirror on the easel turned around the other way, as in Figure 5.5, and invites his friends to come to the cathedral at the proper time of day, stand in the doorway holding the panel, and look through the little hole at its image in the mirror. They see the painted baptistery and the reflected sky with clouds and birds moving across it. Then, while they look through the hole, he takes the mirror away. What they see is unchanged. He has reproduced the appearance of the scene before them, a thing not done since ancient days, if then.

And why the silver plate? Why did Brunelleschi not paint the sky blue with white clouds, or however it looked at the moment he painted it? The sky changes from moment to moment, and it would have spoiled the illusion if the panel had shown the sky as it was on the earlier day when he made the painting.

4. How to Draw Distance

Filippo Brunelleschi did not write much, and contemporary accounts are not very clear, but it seems that though he knew some geometry and thought about what he was doing, he did not try to understand it in an exact way. That was done by another Florentine, Leon Battista Alberti (1402–72), who became a leading architect, though his university education had trained him as a lawyer. His little book *De pictura*, *On Painting*, is the first rationalized explanation of what Brunelleschi had done to fulfill Roger Bacon's demand for a representation that tells the literal truth and nothing more. Truth about what? Brunelleschi reproduced a scene in front of him, but that technique would not help anyone make a picture of Noah's ark. How does one make a truthful representation of an imagined scene? The Latin text of *De pictura* appeared in 1435, and an Italian version, intended for painters, came out in the following year. Alberti's account is reasonably clear, and here is a brief summary of its simplest construction.

Alberti begins, "I inscribe a rectangle, as large as I wish, which to me is an open window through which the subject [*istoria*] is seen. Then I determine as it pleases me the size of the men in my picture." The rectangle, of course, is the border of the picture, but he thinks of it as a window somewhere in front of him—perhaps at arm's length, perhaps further away. This imaginary window defines what is called the picture plane, and though it does not appear in the picture it is essential for the construction to follow. The picture plane is Brunelleschi's discovery; it was his mirror, even if perhaps he didn't realize why he needed it. Determining "the size of the men," of course, amounts to choosing a scale for the painting.

Artists of the Renaissance were fond of situating their "men" on a tessellated marble floor. As an exercise I will put them there. My imaginary window is made of glass. I stand behind it with my feet on the ground and look straight out, so that the visual rays going out from my eye (or coming into it, Alberti doesn't know and doesn't care) extend parallel to the surface of the Earth. With a pencil I draw a line representing the horizon and make a dot on the window where the ray from my eye parallel to the tessellations passes through it. This is the vanishing point. It is the end-on view of a line at the level of my eye, and its distance above the ground represents the height of an average man. Now I start drawing the pavement. The lines extending away from me are parallel to the visual ray, and therefore, according to Euclid as emended above, they will all appear to converge at the

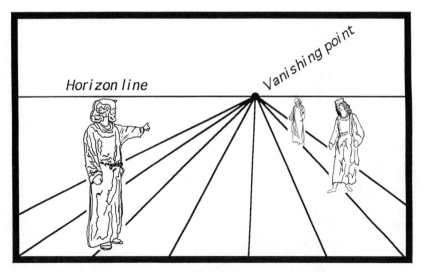

Figure 5.6a. Starting a perspective construction.

Figure 5.6b. Horizon line erased.

vanishing point. The horizon line through the dot defines the height of an average man anywhere in the picture, and now I can draw some average men (Figure 5.6a) at various distances from my window. The horizon line has done its job, and I erase it (b).

To finish the picture I must draw the cross-lines that make squares on the pavement. How far apart should they be? Alberti says to take another piece of paper and make a sketch of me and the picture plane and the floor,

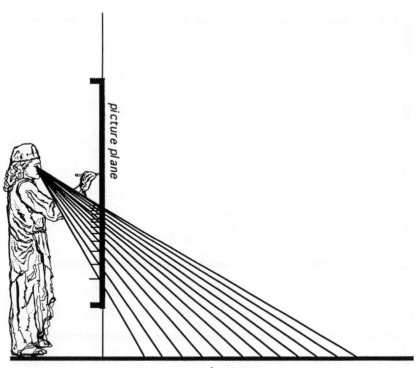

equal spaces

Figure 5.6c. Construction showing how the horizontals are drawn.

Figure 5.6d. The paving is done.

Figure 5.6e. A few extra touches.

viewed from the side, and now comes a question that did not arise earlier: how far am I standing from the picture plane? The natural distance is determined by the length of my arm—I have drawn it that way, and also shown where the successive rows of squares on the pavement are. Finally, I draw the visual rays from my eye to the squares and note where these lines cut the picture plane (c). This is where they should be drawn (d), and when the construction lines have been erased and a few details added (e) the picture is finished. To make the diagrams clearer I have pretended that the picture is actually being drawn on the picture plane, but of course one copies the image in the picture plane onto a piece of paper on an easel at one side.

You might be interested in working out the changes in this simple procedure if the painter is standing twenty feet behind the window, if the window is on the second floor, or if the vanishing point which I have located close to the center of the picture is over to one side.

Brunelleschi showed us how to reproduce a scene in front of us (actually, in back of us); the little scene just drawn came from the imagination. It needed no model, for it was based on principles that apply to vision in general. Alberti's construction explains itself with visual rays, but what of the other classical theories of vision? Think of Brunelleschi's picture as seen in the mirror, and try to explain it by *eidola*. Difficult, since those of the painting are flat and yet they look the same to the soul. The same difficulty arises if we explain vision in Aristotle's way: here again, two very different forms enter the soul, and the soul can hardly distinguish them. Far from weakening support for Aristotle, this and other puzzles forced his defenders to build their walls higher and set up heavier cannon. The painted panel, the magnifying lens, and soon the telescope—all of them are liars: they show us

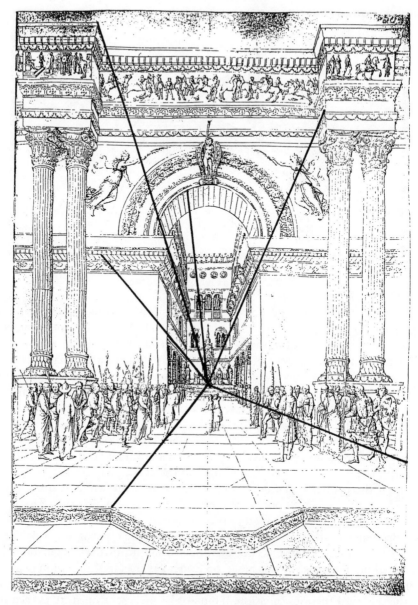

Figure 5.7. Drawing in one-point perspective by Jacopo Bellini (1424–70) with a few construction lines added.

a fictitious world. This is especially true of the panel, on which someone can paint a four-footed bird or an event that never happened, forms that are falsehoods. As long as that is understood, these things have their uses, but in no case have they any authority against philosophic truth. We shall see what happened when Galileo, almost alone, stormed that high fortress.

In the years following the publication of *On Painting*, Italian artists played with perspective like children with a new toy. Figure 5.7, in which only a few of the perspective lines have been drawn in, shows a single-minded concentration on the kind of one-point perspective just explained, but it depicts the open space of the courtyard in a way that was inconceivable to the painter of the Anagni fresco, two hundred years earlier. Evidently two things have happened: artists have developed a new technique of representation, and they are also looking at space and volume in a new way.

There was more of common sense than of Euclid in Alberti's diagrams. Euclid's only contribution was to point out what is obvious to anyone who looks at a straight road over level country: parallel lines seem to converge to a definite point. It can't have taken Leon Alberti long to work it out, though one has to think a little harder when the layout is less simple. Why didn't anyone do it earlier? There is of course no one-sentence answer. As I indicated earlier, the Renaissance was a time when perspectives in every direction of life and thought were changing rapidly, and no one would expect that painting would stay the same.

Let us start with a sentence in the beginning of *On Painting*. "The painter," Alberti writes, "has nothing to do with things that are not visible. The painter is concerned solely with what can be seen." Alberti has just made a radical statement, for it contradicts the Scholastic belief that visible form is only one index, often faulty, of the true nature and substance of whatever one wishes to portray. On the other hand, if one seeks only literal truth, Alberti's statement clarifies an artist's task. In depicting a face or a body the painter must still, as always, portray the truth that lies below the surface, but the setting, even though it too may be entirely imaginary, can be represented according to the formal laws which Alberti then proceeds to explain. Because pictures made according to these formal laws, when one got used to looking at them, looked like the world they represented, it followed that there is a harmony between the principles on which God's world is constructed and those that mathematicians and painters created by their work. Humanity has moved one step closer to an understanding of God's principles and is thus made nobler. For those educated in the old Scholastic tradition, this was new knowledge.

At this time of rapidly changing world outlook, the old question "What is reality?" reasserted itself with urgent force. A new way of seeing requires a new metaphysics. Aristotelian theory, considered as a means of answering the question, had retreated to monasteries and conservative universities. There was no other language of discourse to use in this quest; even the basic words used today in our own efforts were lacking. Only art's wordless language and its laconic commands, "Stop talking, look here, see this, do that," were able to guide people's thinking into new directions.

What, then, is vision? Alkindi, Alhazen, Witelo have prepared the answer for us: vision is the process in which the eye and brain receive an optical image carried by rays of light and translate it into a mental image. The first part, at least, is a purely physical process, as Democritus had claimed long ago. Eye and brain receive images, not forms. The idea of the senses as receivers and decoders of information transmitted by physical means was a new one, and we shall soon see how it upset old and deeply engraved patterns of thought.

5. Porta Asks a Question

A magician is someone who makes things happen by means that science cannot explain. Science, in the sixteenth century, was still treated as a collection of facts that had to be true, that could not be otherwise, since they follow necessarily from fundamental principles. But look around you. How many of the things and processes you see can be brought back to some principle that is obviously true? Everything else in those days—effects of cough medicines, spot removers, beauty creams, poisons and love potions, the chemical procedures that made paints and dyes, unexplained phenomena like magnetism—were examples of what was sometimes called natural magic. This in turn lived in the vast domain of magic generally. Not only did this domain contain things and procedures; it also was inhabited by armies of spirits busy on the orders of their two commanders, God and the Devil. Everyone had bits of white magic that inclined good spirits in their direction; everyone knew the danger that a charm misspoken or written down wrong could become a weapon in the hands of a demon. Natural magic relied not on the work of unknown spirits but on the orderly machinery that God had set in motion on the first day, that one could rely on even if one did not understand it.

The book *Magia naturalis* of Giovanni Battista della Porta (1535–1616) gives a recipe to use against a wound caused by a weapon, which I shall quote in the English translation of 1658. It starts with the weapon:

> Take of the moss growing on a dead man his skull, which hath lain unburied, two ounces, as much of the fat of a man, half an ounce of Mummy[5] and man his blood: of linseed oil, turpentine, and bole-armenick, an ounce; bray them all together in a mortar, and keep them in a long streight [narrow] glass. Dip the weapon into the oyntment and so leave it: Let the Patient in the morning, wash the wound with his own water [urine]; and without adding anything else, tye it up close, and he shall be cured without any pain.

Why does this recipe work? It is an example of magic, and in his introductory chapters Porta talks about what the word means. For some, it is simply "the practical part of natural philosophy, which produceth her effects by the mutual and fit application of one thing unto another. . . . But I think that magick is nothing else but the survey of the whole course of Nature."

---

[5] The dust of an Egyptian or Syrian mummy. This important item was becoming almost unobtainable in Porta's time, but substitutes were available. Human fat could be found at the nearest gibbet.

Study Nature, not old texts. Experiment, don't use fancy language to cover your lack of understanding.

Porta was born into a wealthy Neapolitan family that welcomed philosophic conversation, but from the beginning he seems to have disliked it and shown a passion for strange facts and new ways of doing things. He dabbled in medicine and astrology and founded a little society called the Academy of Secrets, which accepted only those who had persuaded Nature to reveal one of hers. It met in his house, and much of what was published in the first edition of *Natural Magic* [1558] had been discovered by its members. To the public these secrets smelled of illicit dealings with dark powers, but when he was hauled before the Inquisition at Naples, afraid for his life, Porta explained that they were learned by experiment, not necromancy. The authorities closed his academy and let him off. My reason for quoting the remedy against wounds is to allow you to judge whether all those discoveries had really withstood experimental trials.

In 1589 appeared the second, much larger edition of *Natural Magic*. It is a collection of recipes for generating new kinds of plants and animals, for cookery, preserving fruits, counterfeiting gems, tempering steel, beautifying women, and other such arts, together with a few chapters on natural wonders such as magnets and lenses. You will recall that scholars had been writing about scientific matters for three hundred years without ever mentioning the lenses that rode on their noses; here Porta breaks the taboo, for in Book XVII, *Of Strange Glasses*, he writes, "Many are the operations of a Lenticular Crystal, and I think not fit to pass over them in silence. For they are Concave and Convexes. The same effects are in spectacles, which are most necessary for the use of man's life; whereof no man yet hath assign'd the effects, nor yet the reasons of them." Here, in a book read by thousands, was a challenge. As we shall see, Porta tried hard, but it was Johannes Kepler, a few years later, who solved the puzzle.

Porta goes on to discuss the magic of lenses and curved mirrors. He knows that a convex lens serves as a burning glass and shows how the process can be turned around to form a straight beam of light if a candle is placed near the "point of burning." Then comes a description of something that he cannot possibly have tried, "For setting your eye in the Centre of it behind the Lenticular, you are to look upon a thing afar off, and it will shew so neer, that you should think to touch it with your hand." For this and a few other remarks Porta has been hailed as the inventor of the telescope, but after due allowance for his tendency to exaggerate, it does not seem that this lover of the marvelous had seen or made one. Conceivably he was recalling what Robert Grosseteste had said. He did, however, put a lens into the opening of a camera obscura to produce bright images (upside down, of course) on a screen at the back wall of the room.

The principle of the camera obscura is then used to teach an important lesson: "From this it is clear to philosophers and students of optics where vision takes place, and the old and much-vexed question of incoming rays is also answered, nor is there any better way to settle both these questions.

The *simulacrum* enters through the pupil in the same way as it enters through the hole in the camera obscura, and the crystalline sphere in the middle of the eye acts as the screen; I know this will please ingenious people." Obviously Porta has studied Witelo's big book without believing all of it, for like many others he was not ready to contemplate breaking up the optical image of an object into infinitesimal bits that would pass through the pupil and then reassemble themselves into a much smaller image on the surface of the crystalline lens. He is still trying to squeeze the whole *simulacrum* through the pupil.

Having issued his challenge, Porta himself took it up in a book called *De refractione, On Refraction*, in 1593. In this curious volume (of which only two copies are listed in the United States) Porta shows that he has been experimenting and thinking since he put out the new edition of *Natural Magic*. He has not forgotten *simulacra*, for he defines a ray[6] as the line "along which the light of a source arrives, or a *simulacrum* moves through a transparent substance," but I think the words are to be read in the same sense that we draw a line on a map to tell where a car goes. He has absorbed a little more of Alhazen and Witelo but is writing for people who have not, for whom the *simulacrum* is real.

In Book II Porta discusses refraction by a curved surface as Alhazen and Witelo had done, and shows (Proposition 7) how to locate the image that is seen when one looks through a glass sphere at a small object. His optical diagrams make some sense, but he cannot interpret them because he does not understand that when we use a magnifying glass the lens is forming an image for us to look at, just as we look at a thing. He thinks in terms of *simulacra*, which are different. In that theory we do not look at a *simulacrum* but perceive it only when it actually enters our eye. Book IV further discusses *simulacra* but adds nothing new. It is pure Democritus: they are surface layers of the thing seen that peel off under the action of light and get smaller as they approach our eyes. He has learned nothing about vision from Alhazen and Witelo.

Book VIII, on *specilli*, or lenses, enters new territory. "Heretofore we have studied refraction but have not attempted a theory of lenses, a task that is difficult, admirable, useful, delightful, and not attempted by anyone before." In fact Porta has no idea how to find the point along a ray at which an image is formed. His conclusions tend to be facts that one can establish in five minutes of playing with a lens, supported by faulty mathematical reasoning, but it is not all wrong. Proposition 15, for example, reads, "Through a concave lens an object will always seem smaller," and it is clear from the proof that Porta has learned a little of the lesson of Alkindi. What comes into his lens is not a *simulacrum* but the rays from two points on the object's surface, and he shows correctly that they look closer together when seen through the lens.

---

[6] Porta says "line," not "ray"; I have changed the word for consistency with the vocabulary of this book.

Porta lived for many more years, well known as a dramatist and magi-cian, scattering his energy over a flock of books on cryptography, distilla-tion, agriculture, physiognomy, and other arts, and visited by people who hoped that magic would solve their personal problems. By 1615 telescopes had been invented. Many people owned them, and we shall see in the next section how he proposed to make them more powerful. Porta's optics don't amount to much, but *Natural Magick* must have persuaded many people that there was more to Nature than could possibly be approached by deductive reasoning and that questions are especially interesting when one has no idea what the answer is.

6. Galileo the Investigator

It is hard to write anything about Galileo Galilei[7] (1564–1632) without producing another book about him. The broad range of his work, the con-troversies that interrupted it, his suffering for disobeying the Church's command not to defend the Copernican theory—all this is important for intellectual history, some of it is very dramatic, and most of it is beside the point of our discussion. I will therefore deal with Galileo's biography in the sketchiest fashion and concentrate on one strand of the story: how astron-omy was transformed by the telescopes he built and the observations he made with them.

Galileo was born in Pisa in 1564, the year Michelangelo died and Shake-speare was born, to a family of minor nobility. His early education was in Pisa. He never took a degree at the university but at twenty-five was ap-pointed Professor of Mathematics without it. Two years later he moved to the larger university in Padua, again as Professor of Mathematics, where he stayed for eighteen years and did his best scientific work. It was during this time, in 1609, that he heard about a spectacle-maker in Holland who had put two lenses together to make an instrument that made distant objects seem closer. If you try this with a couple of magnifying glasses you will be disappointed. The image is only a little larger and it is upside down. After thinking about the question for an evening (so he says) Galileo mounted two lenses in a tube, a convex one in front and a concave one next to the eye, and found that this worked better: the image was larger and not in-verted. It was, however, fuzzy and distorted, but Galileo guessed that lenses good enough for eyeglasses were not good enough for a telescope (the word seems to have been invented by Porta's patron Prince Cesi) and set about grinding some for himself out of the excellent glass, clear and uniform, that was (and still is) produced by makers of fancy glassware on the Venetian island of Murano.

In a few days Galileo had made a telescope that magnified about eight

---

[7] In Tuscany at this time it was customary to baptize a family's oldest son with the family name; thus for example Michelangelo's elder brother, Buonarroto Buonarroti.

times. If you have a pair of "opera glasses" they work on the principle of a Galilean telescope, though for convenience the lenses are now put somewhat closer together, and they have about the same magnifying power. You can easily measure the power by looking at a brick wall while using the glasses at one eye and not the other. The two images will be superposed in your vision, and you can count how many courses of bricks seen with the unaided eye correspond with five, say, seen through the lens.

Two of Galileo's telescopes and the front ("objective") lens of a third have survived and were recently studied in Italy. The three objectives are almost optically perfect, but they suffer from two faults inherent in their nature which are called spherical and chromatic aberration. I have already mentioned spherical aberration in describing Diocles's concave mirror in §3.2, and the same thing happens in a lens. A glance at Figure 3.8 reveals that although rays arriving near the center of the mirror are quite accurately focused at B, those that arrive near the edge are not. It is the same with a lens ground with spherical surfaces, and in order to get a clear image Galileo was obliged to screen off the edges of his lenses. The largest and best of the objectives, belonging to a telescope that according to Galileo magnified thirty times, was stopped down to an aperture of only 38 millimeters, or 1½ inches, and so it did not let in much light, and its images cannot have been very vivid.

The other aberration, called chromatic, occurs because a lens bends light in the manner of a prism, especially near its edge, and thus produces rings of color that confuse the image. Stopping the light from the edge of the lens reduces this confusion, but the larger the lens the worse it is.

Galileo's instrument produced a sensation in the maritime capital of Venice. He showed it, he says, to the entire senate "to the infinite amazement of all; and many gentlemen and senators, even though they are old, have several times climbed the stairs of the highest campaniles to observe sails and vessels so far out to sea that if they had headed for port under full sail, it would have been at least two hours before anyone could see them without my telescope."

Galileo seems to have been more fascinated by his new toy than curious as to how it worked. If he had been curious, the big volume containing Alhazen and Witelo was available and would have started him thinking about lenses, but it seems he had such distrust of his predecessors that he never even looked at the book, for in the nineteen volumes of his collected works he nowhere mentions Witelo. His telescope formed an enlarged image, and that was enough. He made better designs and better lenses, and his instrument that magnified thirty times is nearly the best that can be done with that simple combination of lenses.

At about the same time Galileo seems to have asked himself, "If my lenses make ships seem larger, what can they do with an insect?" and a bit of trial and error produced the ancestor of today's compound microscope. In one of his letters he refers to something one can see "using a telescope

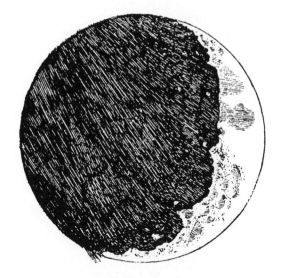

Figure 5.8. Galileo's drawing of the moon. From *Sidereus nuncius* [1610].

adjusted for looking at very nearby objects," but no general interest was ignited. Evidence from telescopes was answering hot questions in astronomy and philosophy, but nobody thought of applying microscopes in medicine. Some drawings of bees were published, but nobody was wondering about the nature of animal tissue or the difference between the material of plants and animals; nobody thought there might be a whole world of living creatures too small to see. I mentioned in §1.3 that Aristotelian philosophy always dealt with the qualities of a thing as a whole and did not care to have them explained in terms of atoms, which have their own qualities. I think that mental habits inherited from Scholasticism made it hard for people even to wonder about the finer structure of things around them. Microscopes amused a few people with enlarged views of familiar objects, while telescopes showed things that were absolutely new and played a major part in the long Copernican drama.

From the beginning, Galileo had turned his telescope toward the moon and the stars, and in 1610 he published a little book called *Sidereus nuncius*[8] telling what he had seen. He describes the moon's surface and writes "I have been led to the opinion that the surface of the moon is not polished, even, and exactly spherical as a great number of philosophers believe it and the other celestial bodies to be, but is on the contrary uneven, rough, and full of cavities and prominences, not very different from the Earth's own appearance, which is marked by mountain chains and deep valleys." Since the moon rotates only once a month its dawn comes

---

[8] I use the Latin title because it has two translations, *The Sidereal Messenger* and *The Sidereal Message*. Because it would have been presumptuous for Galileo to announce himself as a messenger from Heaven, it seems likely that he meant message, but Johannes Kepler, who was fond of verbal play, read messenger, and that is the translation that has generally been adopted.

slowly, and Galileo describes how the morning light first touches its highest peaks and then spreads into the valleys. Figure 5.8 reproduces one of his drawings.

The sidereal message was not welcome to the "great number of philosophers," because however they may have differed on the exact workings of the astronomical universe, they were agreed on what it was made of: changeless, unblemished ether. The world of fire and water and dirt, of generation and corruption, lies below the sphere of the moon; above is the gigantic mechanism of the cosmos, installed by God for his own inscrutable purposes. It is true that even to the naked eye the moon does not seem specially ethereal. There were some who imagined it as a flat disc mirroring the Earth and fancied that they could see a reflection of the Italian boot. Others suggested that the moon's apparent roughness lay beneath a smooth surface of transparent crystal. Yet others spoke of a perfect crystalline sphere in which the density varied from place to place so as to give the appearance of mountains. Three hundred years earlier, in Canto 2 of the *Paradiso*, Dante had described the sensation of actually being inside the moon, made of ether and not rock. Its material is luminous, solid, dense, and clean; what seem to be variations in density are really variations in luminosity. There are different kinds of divine power pouring down through the heavenly spheres, and the moon's influences on tides, weather, and so on originate in different places inside it which, along with their influences, radiate light of different colors and intensities. In Galileo's time most people were not sure what they saw and tried to keep an open mind. The trouble was, Galileo urged reasons why they must now close their minds, and of course most of them did, in favor of tradition.

At the end of his book Galileo describes his most remarkable discovery: that Jupiter has four moons and that, far from taking a month to circle the planet the way our moon does, they have periods measured in days. This was a second shock for the traditionalists: they were expected to look through a tube and confess that not every celestial body circles the Earth. (Of course, the new satellites circle Jupiter, and in the Ptolemaic system Jupiter circles the Earth, but these were unique in that their obvious motion was around Jupiter.) For years, Galileo had taught Ptolemy to his students in Padua while quietly nourishing the conviction that Copernicus was right. In a letter to Kepler in 1597 he admits to being "intimidated by the fate of our teacher Copernicus, whose fame will be immortal for some, while an infinite number (for such is the multitude of fools) laugh at him and reject him." Galileo's original reason for being a Copernican may simply have been scientific intuition, but later he had convinced himself (by reasoning that turns out to be fallacious) that the tides are an effect like the sloshing of coffee in a cup carried from the kitchen; they could not occur in a stationary cup. And here was a demonstration model of small celestial bodies circling a large one. In the dedicatory letter that introduces *Sidereus nuncius* he announces the new satellites and finally takes his stand: "Moving around Jupiter, noblest star of all, as if they were his own children, they circle in their orbits with marvelous speed, while at the same time with one accord they execute their huge revolution every twelve years around the center of the world, which is undoubtedly the Sun."

Galileo's little book sold out at once and became the talk of all intellectual circles. His salary at Padua was doubled, but he wrote letters and paid visits and managed to get himself recalled to his native city as Chief Philosopher and Mathematician to the Grand Duke of Tuscany. The name "Philosopher" denotes a big increase in rank. A mathematician casts horoscopes and calculates how things move; a philosopher teaches what things are and why they move. Galileo's position and achievements gained him the right to be called Your Lordship, access to the most influential figures of the court, and a very large salary. He could also expect that his noble patrons would try to protect him from any trouble that might be created by the Church.

Galileo's Copernican conviction was further strengthened by his observations of Venus. As expected, it turned out to have phases like the moon, but what was more striking, its diameter appeared to be always about six times greater when it was in its crescent phase than when it was full. Every observation but this one could be explained with a little ingenuity by Ptolemy's supporters. If Ptolemy was to survive, it was Galileo's telescope, not just his reasoning, that must be broken.

There were several reasons based on common understanding why one should not believe Galileo. First, and most generally persuasive, was that seven has to be the number of the planets (Sun, Moon, Mercury, Venus, Mars, Jupiter, and Saturn), because, as we have seen and everyone knew, the number seven is the sign that a list has been completed. As Cicero had written long ago, the number seven might almost be called the key to the universe, and it was impossible to imagine why God would have made planets in any other number. Another reason was that astrology, which was regarded as a perfected science, had no room for four new planets, even small ones. One astronomer wrote to Galileo that he would not believe in their existence until he saw definite evidence of their effects on Earth. Galileo's reply was to ask him to examine whatever predictions he had based on the motions of Jupiter and see if they had all been correct, and if not, to consider that he might have left something out. Actually, as Galileo clearly understood, the astronomer's complaint opened a can of astrological worms, for it encouraged a suspicion that any failure of prediction might be due to other stars that had not been discovered and might never be, planets too small to see but too important to ignore. Obviously, the sun and moon affect what happens on Earth, and Galileo uses the familiar word "species" when he discusses their influence, but he does not, as astrologers do, assume that species from other planets also affect us. Skeptic or agnostic? Skeptic probably, but it is hard to tell, for he rarely mentions astrology.

Figure 5.9. Title page of Hevelius's *Selenographia* [1647], showing the moon as it appears and the sun as it undoubtedly is, if we could just see it properly.

While public consciousness was waking up to the implications of his exciting new discoveries and the opposition was preparing its responses, Galileo made a journey to Rome, where he demonstrated his telescope to learned academies and explained what he had found to Pope Paul V and leading figures of the papal court. Especially since Roman astronomers had seen the new stars in a telescope they themselves had made, the stars' existence was taken as a fact. The visit was a great success, and when dissenters had their say there is no sign that they changed many minds.

Galileo always tried to keep a safe distance between himself and theological dogma, but it was not possible to insulate himself completely. When he said that there are more than seven planets he did not directly contradict anything in the Bible, but when he said the sun does not move, he questioned the story in which Joshua commanded, "Sun, stand thou still upon Gibeon," so that he could continue to destroy his enemies, and the sun obeyed, that was a serious matter. The remarkable thing is that in 1610 the papal authorities had not begun to sense the complications that would result from Galileo's vivid and enthusiastic reports of what he had seen.

The following year Galileo and others discovered sunspots. These, like the new satellites, were hard to reconcile with Aristotle, but astronomers tried. The Jesuit Christoph Scheiner proposed that what seemed to be spots on the sun were really black clouds floating above the surface, but the fact that the sun seemed to rotate as a whole, carrying the spots with it, made this suggestion hard to believe; Scheiner later decided that the sun could be allowed some spots to ornament its otherwise monotonous surface but that of course they were neatly and tastefully arranged, and it was only Galileo's lenses that made them seem changeable and irregularly spaced. Figure 5.9, from Hevelius's famous book on the moon published many years later, shows a compromise with which people could be comfortable: the moon is a miniature Earth, nothing could be done about that, but sunspots, portrayed truly, are cosmetic.

It is time to say good-bye to Giovanni Battista della Porta. In 1614, almost eighty years old and recently ill, he writes Galileo a cheerful letter:

> As my health returns I turn again to my old fancies. We are building . . . a new kind of telescope which will be a hundred times more powerful than the present ones. With those one can see to the eighth sphere;[9] ours will reveal the Empyrean. If it pleases God we shall see what is going on up there and will issue a *Message from the Empyrean.*

Porta was working on a manuscript *De telescopio* [1962] that was never finished. It shows that in twenty years since writing *De refractione* he has learned nothing; his idea seems to be that if two lenses in a tube make a

---

[9] Above the spheres of the seven planets the eighth sphere, in the Ptolemaic system, was the one that carries the fixed stars. Above that, in the middle ages, was the Empyrean sphere, the abode of God.

good telescope, five lenses will make a better one. Four months after the letter to Galileo he made his own journey to the Empyrean. I hope he is happy there.

Galileo's advocacy of the Copernican system, and especially his habit of insulting the intelligence of his opponents, brought him into conflict with the Inquisition. Finally, in 1633, there was a trial "on vehement suspicion of heresy." He was shown the instruments of torture, and in the end he recanted his claims and was returned to Florence to spend the rest of his life under house arrest. The story of his trial and defeat and the eventual triumph of his ideas has been told often and well enough so that it needs no reprise here, and I can jump to the end of it, which occurred in Rome on 31 October 1992 when Pope John Paul II told the Pontifical Academy of Sciences that both sides had made mistakes, and the Church had been wrong to condemn Galileo for holding that the Earth is not at the center of the world. Theologians, he said, should not think that the literal sense of Scripture explains the physical world, and thus the Galileo affair was closed forever.

Galileo had indeed made a mistake. He had continued to explain Copernican astronomy after the Church commanded him to keep quiet about it. But consider what he was trying to do. As a good Catholic he believed that there is only one truth and that God sometimes shows it to us more clearly in Nature than in Scripture. He knew, better than anyone, that the evidence supporting his views was overwhelming and that there was an overwhelming probability that if they were rejected and condemned, some pope, sometime, would have to stand up in public and eat his biretta. And that is exactly what happened.

7. "Light Is Created"

The dispute over planetary theory raised issues deeper than the question whether the Earth moves or stands still, for it caused Catholic doctrine, that great creature that slept uneasily in the cellars of the Vatican, to groan and stir and open its eyes. What disturbed it was the suggestion that the principles behind a scientific explanation of the solar system, and natural phenomena generally, were mathematical in nature. This is the dream that began with Pythagoras, and we glimpsed it in Grosseteste's little treatise *On Lines, Angles, and Figures*, though Robert had no evidence for such a claim beyond the elementary principles of optics. In 1609, as we shall see, Johannes Kepler gave a wonderfully simple and precise description of planetary orbits in mathematical terms. Galileo stayed with the older and more primitive scheme that Copernicus had used, but this theory too was mathematical, and it enabled astronomers to calculate planetary positions with reasonable accuracy. A character in one of his dialogues says, "I believe that [mathematical] understanding equals the divine understanding in objective certainty, for it understands the necessity of things, and nothing can be more certain than that." This opinion must have crystallized from a hundred calculations, simple ones on a scrap of paper that describe what actually happens, and others not so simple, such as the ones that tell how a beam bends when it supports a load. But there was also a big reason for believing in the mathematical necessity of things, for Galileo had discovered the law according to which a heavy body increases its speed as it falls. An important point is involved here. The reflection of light from a mirror as well as the bending of beams are static phenomena, in which nothing happens in time. A beam of light makes a spot on the wall, a piece of wood sags under a load. But a mathematical theory of planetary motion, even a false one, tells where a planet will be at a given time, and a theory of falling bodies makes the same prediction for a stone that is dropped. Certain natural processes happen according to mathematical laws that make their outcome predictable. Do they all?

But now, without intending to, we have landed on theological territory. Pope Paul died and was replaced by Urban VIII Barberini, an old friend of Galileo, who saw the point at once. When Francesco Niccolini, the Tuscan ambassador to Rome, tried to help Galileo's affairs by remarking that God at least *might* have built the universe on the Copernican plan, Urban would not listen. "Red in the face, he replied to me that the blessed God is not to be necessitated, and I, seeing him so enraged, did not wish to dispute him on matters of which I knew little." While Galileo stood up for freedom to discuss scientific issues, the Pope was battling the heretical notion that God had created a system which now runs independently of his will and pleasure. That possibility suggests questions, each more dangerous than the last: are human beings, as radical astrologers were suggesting, so much a part of the cosmic machine that every choice they make is programmed in advance? And if the machine runs without God's aid, why could it not also exist without him? That Galileo had continued to teach that the planets circle the sun was not the only reason that Urban wanted him to stop talking.

In 1623 Galileo produced a polemical work called *The Assayer*. It arose from a controversy over the nature of comets (in which, by the way, his opponent was more nearly correct than he was), and in the course of the argument he wrote down his mature thoughts on the role of mathematics in science and also on light. His words about mathematics are often quoted, but once more will not hurt:

> Philosophy is written in that great book, the universe, that forever stands open before our eyes, but you cannot read it until you have first learned to understand the language and recognize the symbols in which it is written. It is written in the language of mathematics, and its symbols are triangles, circles, and other geometrical figures without which one cannot understand a word, without which one wanders through a dark labyrinth in vain.

But these eloquent words pose a challenge to the scientist as well as the theologian. Look out the window. The sun is setting, its light turns red, wind sounds in nearby trees, and the air is suddenly cooler. This is how the world appears to us, but how can these sense impressions be explained in the language of "triangles, circles, and other geometrical figures," or even of numbers, which Galileo does not mention? The challenge was not new, for the Greek atomists had faced a similar question. If atoms have only simple properties like position, motion, size, shape, and weight, how can they ever account for all the different qualities we sense in the world? Democritus has already given us his answer: "Sweet exists by convention, bitter by convention, color by convention; in reality there are only atoms and emptiness." Impressions of color, sound, and temperature originate in us, created by our own senses in response to simple atomic qualities and changes in our environment. Galileo agrees and mentions tickling as an example. There is nothing like a quality of "tickle" that resides in your finger and is transferred to the sole of someone's foot; all you do is touch it. "I think that if ears, tongues, and noses were taken away, then shapes, numbers, and motions would remain but not odors, tastes, or sounds, which, except as applied to living creatures, I take to be nothing but words."

Galileo considers the sensation of heat, which he imagines produced by the arrival of very small particles that penetrate the skin. Particles of fire are of this kind, and we can make them ourselves if we rub something hard enough, for the particles of matter are not perfectly hard. First we rub off little chips, which we detect as heat, and finally, if the process keeps on, "it liberates atoms, the smallest particles of all; then light is created and expands through space by virtue of its—I do not know whether to say subtlety, rarity, immateriality, or some other property different from these that has no name." The tentative words express Galileo's perplexity. His scientific method has always been to contemplate the experimental situation and then think his way through it. He knows something about what light does, has even performed a little experiment that showed that it moves very fast, but in a letter written at the end of his life he admits he is still in the dark;[10] he does not know what it is. He does not say like Democritus that light consists of atoms. The qualities he suggests for it are not the "shapes, numbers, and motions," that pertain to atoms; whatever they are, they pertain only to light. An immaterial substance of some kind—as we shall see, Johannes Kepler has already thought along this line, and Isaac Newton soon will, but Galileo adds words we should remember, that light pertains to the world of atoms. It will be three centuries before that egg hatches.

Though Galileo looked through his telescope and saw what no one had ever seen before, he was not an astronomer. Astronomers measure and calculate, and the subject's appalling difficulties become clear only when

[10] A sad little joke, because by this time he was blind.

someone tries to do that. The atmosphere bends a light ray so that where the stars seem to be is not where they are. Looking at a point of light in the sky gives no idea of its distance. And no matter how carefully you observe, what you see does not show why anything happens. Perhaps the reason Galileo did not publicly endorse Kepler's theory of planetary motions was that he had never tried, and failed, to make theory agree with observation; perhaps the reasons were more personal and complex, but however it may have been with him, the impulses that led to a mathematical physics and produced the first new ideas in optics since Alhazen came from those intractable problems of astronomy. The problems were technical rather than conceptual and were best approached by following Alhazen's path rather than Alkindi's, by considering light as a part of the external world, by separating it from questions of vision and then by studying vision, too, as objectively as possible. A new concept of science was gaining ground, which demanded a mind ready to examine a question without seeking to relate it to human beliefs and perceptions. The science they were seeking must be objective; it must be abstract.

It Must Be Abstract
Wallace Stevens
From *Notes toward a Supreme Fiction*

Begin, ephebe, by perceiving the idea
Of this invention, this invented world,
The inconceivable idea of the sun.

You must become an ignorant man again
And see the sun again with an ignorant eye
And see it clearly in the idea of it.

Never suppose an inventing mind as source
Of this idea nor for that mind compose
A voluminous master folded in his fire.

How clean the sun when seen in its idea,
Washed in the remotest cleanliness of a heaven
That has expelled us and our images.

The death of one god is the death of all.
Let purple Phœbus lie in umber harvest,
Let Phœbus slumber and die in autumn umber,

Phœbus is dead, ephebe. But Phœbus was
A name for something that never could be named.
There was a project for the sun and is.

There is a project for the sun. The sun
Must bear no name, gold flourisher, but be
In the difficulty of what it is to be.

6

A NEW AGE BEGINS

> I rather wish I had a telescope at hand, with which I might
> anticipate you in discovering two satellites of Mars (as the re-
> lationship seems to me to require) and six or eight of Saturn,
> with one each perhaps for Venus and Mercury.
> (Johannes Kepler[1])

1. Johannes Kepler, the Emperor's Mathematician

I T MIGHT BE plausible to say that the most creative minds are those
least bound to the past and therefore freest to explore new thoughts
and move ahead. Then one could use Galileo as an example. His studies in Pisa immersed him in traditional natural philosophy, and his early
notes on mechanics and cosmology are cast in the mold of questions and
answers on ancient texts, with no mention of anything that actually happens. They show how thoroughly he understood and absorbed the Aristotelian method and therefore how much he had to throw overboard as he
matured. He learned largely through his eyes and his hands, and his later
writings (except where he refutes someone's arguments) are remarkably
free of references to the philosophic tradition. He never mentions that excellent optician, Witelo; he ignores astrology, an omission that must have
seemed strange in a culture saturated with it. And though he was a devout
Catholic his religious beliefs seem to have little connection with his work
except that he was concerned to bring the Church's thinking up to date.
He knew very well that giants loomed behind him, but rather than sit on
their shoulders he preferred to feel the solid ground moving smoothly beneath his feet.

Johannes Kepler (1571–1630) had a very different mind, and we shall
not understand his work unless we realize the depth of his religious beliefs
and the strength of his attachment to a philosophy that descended from
antiquity. Among these was a conviction derived from Pythagorean teachings that God designed the world according to principles of mathematical
beauty, and that the more deeply one looks at natural forms and arrangements the more mathematical relations one will discover. Kepler found
mathematical relations that express profound truths concerning the structure and movement of the solar system, plus others that we now under-

[1] From Kepler's letter to Galileo (April 1610), published as *Conversation with the Sidereal
Messenger*. Mars did, in fact, turn out to have two satellites. Translated in Rosen 1965.

stand are only coincidental and have no special meaning. He tried to explain the motions of the planets with a mathematical theory but failed because, as we shall see, he based it on a proposition of Aristotelian dynamics that was wrong. Plato's explanation of the ideal world as a single organism designed according to mathematical principles was at the core of Kepler's imagination; he was intoxicated with numbers, but he knew very well that to understand any of the great design he must learn from the forms that Nature presents to our sight.

Kepler is best known as the founder of modern astronomy, for he created a mathematical version of the Copernican model and discovered the three fundamental laws that describe how planets move. The pages to follow will mention only as much of that story as is necessary to understand Kepler's thoughts about light, for he also founded modern optics.

Johannes Kepler was born to Lutheran parents in a Swabian town now known as Weil-der-Stadt, a largely Catholic community with the status of a free city of the Holy Roman Empire. His father was a mercenary soldier, one of those who joined whatever army would pay him to fight. His mother, the daughter of an innkeeper, had six children of whom three grew up, about the usual proportion. Little Johannes almost didn't make it; when he was seven smallpox left him with weakened eyes and hands too clumsy for fine work. Supported by people who recognized his talents, he was sent to schools where he learned Latin and Greek and read the classics. This opened the Western world's literature to him and allowed him to think thoughts that could not have been expressed in the rather primitive German of the time,[2] which was only beginning to be enriched with foreign words.

The University of Tübingen, where Kepler arrived in 1589, was a center of Protestant theology. Here he followed the arts course for two years and heard the lectures of Michael Mästlin, a competent mathematician and astronomer who remained a friend and adviser throughout his life. Copernican astronomy was beginning to penetrate the European mind, and feelings pro and con were mounting. Mästlin was pro. He did not feel that his position on a theological faculty allowed him to say so openly but he mentioned Copernicus often in his lectures. To Kepler it seemed obvious that the sun, and not the Earth, belonged at the center of the world, and he began to read about the new astronomy.

When Kepler was sent to Tübingen it was assumed that he would become a Lutheran minister, but his developing interest in astronomy and his tendency to think every thought in his own way produced an impression that he was not the man to be turned loose on a congregation of simple burgers. Instead, he was offered a job teaching mathematics at a seminary

---

[2] As a result his German writings are so thickly larded with words and phrases in Latin that the effect becomes comical. The following is his German summary of a section of his *Tertium interveniens*, a work that condemns astrologers but warns that the baby should not go out with the bathwater, for a correct astrology is possible: "*Causae remotes et generales* geben in *Medicina et Astrologia* auch ihre *demonstrationes*." (From Kepler 1937, vol. 2, p. 152.)

in Graz, where he arrived in 1594. His job was not all teaching, for he was expected to be the local surveyor and settler of boundary disputes and, in addition, to put out an annual almanac. Such almanacs listed phases of the moon, eclipses, planetary conjunctions, and other useful data for the coming year, and then went on to predict events of weather and politics that would follow from these data according to the laws of astrology. For the next thirty years he helped support his family by issuing an annual almanac.

What kind of man was this anyhow, and why did a young mathematician trained in theology turn toward the stars? In a letter written in 1599 he describes himself:

> With me Saturn and the Sun operate together. . . . Therefore my body is dry and knotty, not tall. The soul is faint-hearted, it hides itself in literary nooks; it is distrustful, frightened, seeks its way through tough brambles and is entangled in them. Its moral habits are analogous. To gnaw bones, eat dry bread, taste bitter and spiced things is a delight to me; to walk over rugged paths uphill, through thickets, is a feast and a pleasure to me.

His whole life was an uphill walk.

As to astrology: I have said it is hard to understand how anyone in those days with an a priori belief in the unity of Nature under a divine plan could fail to believe in some form of it. If they had no practical use, what were the moon and the planets doing up there? The sun keeps us alive with warmth and light. The moon is connected with moisture, which is just as necessary. It governs the tides, but also, as Kepler wrote in 1601, "It has been proved by experience that all things uniformly moist swell with the waxing moon and subside with the waning moon. This one fact is the reason for most of the choices and predictions in domestic economy, agriculture, medicine, and shipping." It is just as Basil of Caesarea had said in the fourth century (§3.1). The moon does control the height of tides. I do not know whether anyone notices the other inflations today, but they seemed real enough in the sixteenth century. Figure 6.1, a woodcut made by Albrecht Dürer early in the century, shows (perhaps) Urania, the Muse of Astronomy, and zodiacal influences streaming inward toward the Earth. But how do these influences actually work? Kepler does not know, but he wants to find out. What special purpose do the planets serve? The Bible says they are there "for signs and for seasons"; but specifically, what do they do? They help us plan for the future, say Kepler and most people of his time. Having little idea of quality control, they pointed to the astrologers' successes and explained the failures; they believed that the successes raised astrology out of the quagmire of dubious science and established it as a branch of mathematics. The main problem was to get the calculations right, and this is why the *Mathematicus* of the Archduchy of Styria was responsible for the annual almanac.

Galileo was a *mathematicus* also, and about this time the dowager Grand Duchess Christina of Tuscany asked him to cast a horoscope for the Duke,

Figure 6.1. Woodcut by Albrecht Dürer (1471–1528) showing planetary influences from the houses of the zodiac.

then approaching sixty. He obliged on an optimistic note, forecasting many more active years, but the Duke followed other stars and died shortly afterward. The heavens were kinder to Kepler. His first almanac foretold cold weather and an invasion by the Turks; both happened. Much later, in 1618, he predicted that the month of May would bring great upheavals. On the 23d, delegates from a seething Protestant assembly in Prague marched up to the Hradcany Castle and threw two Catholic regents and their secretary out of a window, igniting a war that for thirty years devastated much of central Europe, and among the thousands of lives it crippled or destroyed were those of Kepler and his family. Astrology was not his main interest, but it brought him income and reputation, and his successes tightened the belief he started with, that all Nature works together, and mathematical principles govern it. There is no reason to think that Galileo ever seriously entertained such a belief, but the fiasco of the Duke's horoscope would not have encouraged him if he had.

Kepler aimed at something new, a fusion of mathematics and physics and astronomy into one great body of knowledge. He wrote all the time, twenty-four tall volumes in the Collected Works. A work from 1601, called *On the More Certain Foundations of Astrology*, explains the science's principles in fifty-one propositions, and then lists twenty-four more that predict the weather for the next year and a few of the consequences that would follow from it. Kepler has no use for zodiacal astrology. He ridicules the notion that a planet's influence changes abruptly at the instant it crosses the boundary between two arbitrarily defined areas of the sky. His astrology is lunar and planetary, but it is in some sense a physical theory. He imagines several ways in which planets influence our fate. I will discuss one that is physical and one that is mathematical, or in Aristotle's terms, one efficient and one formal cause.

The sun exerts its physical causes by the species of its heat and light. These species emanate from the sun but do not carry anything away with them; Kepler illustrates this by the Christian analogy of the Son who emanates from the Father. Species act directly, as everyone can see, but also via anything that shines by the sun's light. The reflected rays derive their natures partly from the sun and partly from any planet that reflects them, and this leads Kepler to a fanciful discussion of planetary surface textures. Saturn, he decides, is white and dry because the water in it is frozen, "but when it works to make summers rainy and winters snowy it deserves to be called humid." Jupiter is like a ruby, Mars is like a glowing coal and very dry, Venus is amber and moist, and Mercury is like a sapphire. The rays of reflected sunlight pick up these properties and, depending on the direction from which they reach us, produce significant effects on the weather.

The mathematical or formal cause depends not on any physical action but on geometry. How can geometry cause anything? It can't, but if you believe as Kepler did that everything that exists in the world is formed according to God's mathematical principles, it follows that a new soul also is formed that way, and forever afterward it retains the form. Choose a pair of planets and imagine lines drawn from them to the Earth at the moment a child is born. Now study the angle between the two lines. If it is one of the angles that flow from simple and ancient geometry, then the planetary influence is strong and predictions can be made. Figure 6.2, from his later book *The Harmonics of the World*, shows how geometrical figures fixed to the Earth pick out two stars separated by a particular angle. The first three figures produce strong influences on the weather as well as on the nascent soul. The last one, which looks equally efficacious from a geometrical point of view, has failed over a period of twenty years to show any perceptible effect on anything and therefore must be judged *imbecillus*: weak, ineffective, though it is not clear why. Trees, Kepler says, do not reason but put forth their twigs according to geometrical schemes. Five-sidedness occurs again and again in fruits and flowers. Nature is full of symmetries that we can see and wonder at, but whose reason we can only surmise:

> We believe that there is in the Earth a vegetative animate power, and in the animate power a certain sense of geometry that is formal and self-sufficient. . . . In the same way that the ear is stimulated by harmony to listen carefully and thus to hear much more (seeking pleasure, which is the goal of feel-

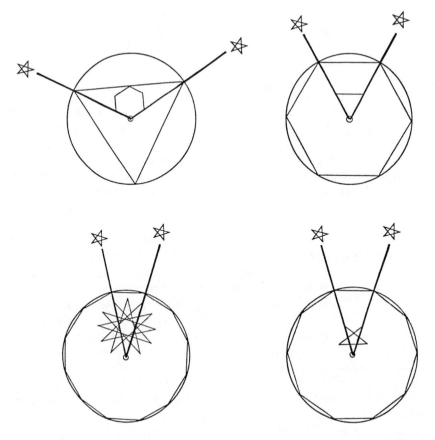

Figure 6.2. Efficacious planetary configurations (the Earth is at the center). From Kepler, *Harmonices mundi* [1619].

ing), so the Earth is stimulated through the geometric concourse of vegetating rays, . . . so that it applies itself more diligently to its function of vegetation and exudes a copious supply of vapors.

The Earth's soul responds to the harmonies of angles, but what about the human soul? In the letter of 1599 that I have already quoted, Kepler writes,

How does the face of the sky affect a man at the moment of his birth? It affects the human being as long as he lives in no other way than the knots which the peasant haphazardly puts around a pumpkin. They do not make the pumpkin grow but decide its shape. So does the sky; it does not give the human being morals, happiness, children, fortune, and wife, but it shapes everything in which he is engaged.

Rays shape a life but do not determine outcomes, for choices are ours to make, and how we make them determines our fate on the Last Day. However kings and prelates and judges may have used astrology in conducting their lives and affairs, they all understood this, and only unseasoned radicals ever adopted the extreme position that the stars, not we ourselves, are responsible for what we do.

Kepler's position as a member of a Lutheran minority in a Catholic state was always precarious. In the next few years he found himself forbidden to worship in Graz and finally, at one day's notice, even to teach there. In 1600 he found a job as assistant to the great Danish astronomer Tycho Brahe, then living in exile in Prague under the protection of Emperor Rudolph II. Tycho died a year later and Kepler was appointed *Mathematicus Cæsareus*, Imperial Mathematician, in his place. In the next few years he worked mostly on the project for which he is the best remembered, establishing the orbit of the planet Mars. There is no occasion to discuss this work here, but I will state the problem and describe the solution that Kepler obtained in 1605 and published in his *Astronomia nova* four years later.

> *The problem:* We stand on the Earth, circling the sun in an orbit whose shape and speed are only roughly known, and measure changes in the apparent position of a reddish dot in the sky that we call Mars. The shape of Mars's orbit and the speed at which the planet moves in this orbit are also roughly known; it is required to find them exactly.
>
> *The solution:* Mars's orbit is a perfect ellipse having the sun at one focus; the planet moves in empty space under the action of some unknown force such that the closer it is to the sun, the faster it travels, and a mathematical formula governing that relation is given.

The concept was absolutely new. All previous planetary systems had filled the sky with things, solid but invisible spheres or rings on which the planets were mounted, each planet turning with the cosmic machine like a spot of paint on a wagon wheel. Kepler emptied the sky of all but the planets and then put in ellipses, mathematical entities that exist only in imagination, and found that the planets move along them from night to night.

Creating this picture out of Brahe's many years of observations was made difficult by the fact that as we have already seen, the measured angular position of a star[3] does not tell what its direction actually is, for the light ray from it to the observer is bent when it enters the Earth's atmosphere. Tycho's observations were accurate to about 1 minute of arc (the full moon is 30 arc minutes across). One minute is about how much the ray is bent if the planet is 45 degrees above the horizon; the bending is greater if it is lower. If we don't know how to make the correction we don't even know where the planet is. The problem of correcting the observations is hard and becomes much harder if one has no clear theory of what is happening. Tycho for example, the best observer who ever lived, thought that the rays from sun, moon, and planets were all bent by different amounts, which is not absurd if you consider how much they differ in brightness. Kepler stopped his work on Mars to clear up these questions, he says, but

---

[3] Recall that "star" also includes sun, moon, and planets.

in dedicating his work to the emperor he mentions another reason: "I understood that certain optical theorems which seem quite unimportant contain the seeds of great discoveries, and that I must study them more deeply and get to the heart of them." Three centuries after Witelo, it was time for someone to go back to work on optics.

2. Paralipomena

Kepler's first work on optics (1604) has a baroque title that starts *Ad Vitellionem paralipomena*, which can be translated as *Supplements to Witelo*;[4] it goes on to announce the book's purpose, which is to treat the optical part of astronomy. Astronomers were bothered by several optical questions. There was, as already mentioned, the bending of light as it passes into the atmosphere. Were starlight, sunlight, and moonlight all bent by the same amount? Tycho Brahe thought not. More puzzling: one can measure the apparent diameter of the full moon as it sails in the sky or as it passes across the sun's disc in a total eclipse; the results turned out different. Kepler was not thinking about telescopes, for it was just as the book appeared that people began to hear of the gadget from Holland. At that moment, the optical part of astronomy concerned what can be seen with the naked eye.

Kepler announces that he is going to present his material in a series of propositions. They will not follow in the manner of logical proofs but will be illuminated by experience and imagination. Next, he launches into a praise of light in terms that will recall Robert Grosseteste and do not sound in the least like Galileo. It is, he says, the most excellent part of the corporeal world, the origin of the animal faculties, the link between the worlds of matter and spirit. Light is the faculty by which the sun communicates its power to every corporeal thing; thus the sun sits at the center where it belongs, radiating equally and perpetually into the whole universe. Everything that receives that light imitates the sun, and our reasonings start from this fact.

Here are the first few propositions of Book I.

1. The flow or projection of light carries it from its source toward a distant place.
2. Light from any point flows along an infinite number of straight lines.
3. Light, by its own nature, can flow infinitely far.
4. The lines along which light flows are straight and are called rays.
5. Light does not move in time but in an instant.

That is, light moves with infinite speed. The argument establishing this proposition is based on Aristotle's principle (§1.5) that everything that moves is moved by something. No force, no motion. Aristotle suggests also

---

[4] The Greek word means "things left out." It was familiar to Kepler and his readers because in the Latin and Greek Bibles *Paraleipomenon* is the title of the book of Chronicles, so named because it collects history omitted from the books of Kings and Samuel.

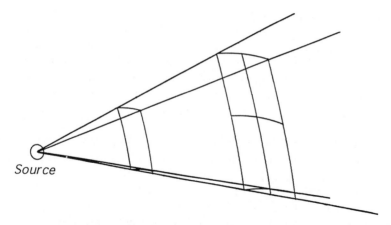

Figure 6.3. The inverse-square law. The illumination at the second surface is one-fourth as bright as at the first one.

that there is a quantitative side to this: more force makes something move faster; resistance limits its speed. The heavier something is the more resistance it offers, and the more slowly it moves; thus it is harder to throw a heavy stone than a light one. Kepler argues that because light has no mass there is no resistance to its motion and therefore no limit to its speed.

Aristotle's proposition describes experience so clearly that people tended to overlook some situations where it works less well. If I throw a ball, for example, what keeps it going after my hand no longer pushes it? Some of the ancients suggested that the air moves behind it to push it along, but as Galileo pointed out, this would imply that you can throw a wadded-up piece of paper farther than you can throw a ball. Some medieval philosophers, following John Philoponus, suggested that throwing the ball gives it a kind of force called *impetus* that resides in it and keeps it moving, but in Kepler's time people were becoming suspicious of using verbal formulas to get around conceptual difficulties. Still, everyone knows light moves very fast, and Galileo's experiment, which Kepler may have known about, had verified it, and so Kepler was not encouraged to question the Aristotelian principle that ordinary objects will not keep going without a continual push. We shall see later how this principle led him to a serious error.

Paralipomena goes on to describe the motion of light. It moves along its rays (Proposition 4), but the rays do not move and are not light. Rather, as Porta had defined them seven years earlier in *On Refraction*, they are just lines in space. What moves is a surface, perpendicular to a ray (Figure 6.3). Think of something that creates a sudden flash in the darkness. The light expands with infinite speed on the surface of a sphere centered at the place of the flash, moving outward along radii of the sphere. Suppose the light has traveled one mile and is spread out on the surface of a sphere with that radius. One mile further on, the radius is twice as large and the surface is four times as large, but the amount of light is the same, so that on any area of this larger surface it is only one-fourth as bright. This is the first statement of what is now called the inverse-square law: the intensity of light varies inversely with the square of its distance from its source.

Kepler's theory of color originates in Ptolemy but is explained in a new way. Color and light are of the same nature. They are not the same, but color is potentially light and resides inside a transparent object as well as on its surface until it is liberated by light from the sun (Propositions 11 and 15ff.). Every material is transparent, even the material of a brick; the reason we cannot see through a brick is that its broken-up texture diverts the rays in different directions so that they get lost inside it (Proposition 17).

Finally, in Proposition 32, Kepler confronts the nature of light. It is a form of heat, he says. A heated solid gives off light, but this light brings heat with it just as it also brings color. Animal life depends on the heat of the sun's light, and every living creature contains some light, as you can see in the shine of glow-worms and, occasionally, of bits of rotting wood. The heart itself has been compared to a kind of fire, and we might some day be able to detect the little flashes[5] that come from it as it beats. Even the distant planets send us heat along with their light, too weak for us to detect, but this heat enters the Earth and causes it to emit the vapors through which planets affect our climate.

Take away the fanciful conjectures and there is nothing very striking so far, but now Kepler takes up vision. Why is it, he asks, that when we look at something in a mirror it seems to be located just as far behind the mirror as it actually is in front of it? To answer he develops a theory of visual perception that Galileo may have known about later, when he fought his own battles in this arena. The image that forms in our mind when we look at something is a subjective (Kepler says "intentional") entity that the mind creates in response to the visual rays that enter our eyes. It is not something that travels along the rays; there are no *eidola*. It is the mind that creates the image and locates it at a point determined by the convergence of rays from the two eyes. If the rays are reflected (Figure 6.4), the angles presented to the eyes are unchanged, and the mind locates the image behind the mirror where it would have located it if the rays had been straight. Even if we use only one eye we can estimate distance approximately, since the pupil has a certain width and receives a sheaf of rays that are not quite parallel.

In Book V Kepler returns to the question of vision. He has obtained an eyeball and scraped away its back wall except for a thin layer, then held it up and seen that an inverted image of the scene in front of him was projected onto the retinal surface. He realizes the relevance of Porta's compar-

---

[5] The word is *lucula* and refers to the smallest quantity of light, some sort of immaterial flash that travels outward along a ray, though Kepler did not believe in atoms. It occurs in the book's optical part 1.15, and in 1.32. In total darkness things can still give off *luculae*.

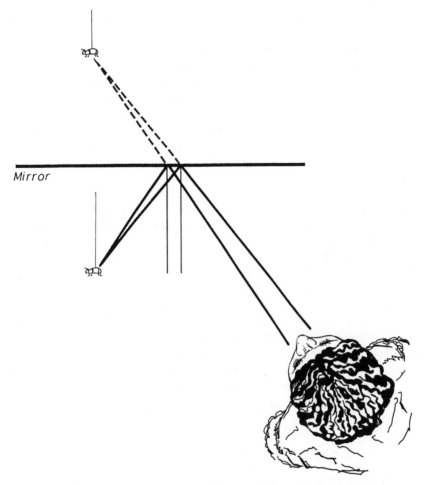

Figure 6.4. Why we perceive the image as located behind the mirror.

ison of the eye with a camera obscura, and proposes that the crystalline lens is responsible not for receiving the image but only for focusing it onto the retina. This disposes of Alhazen's mysterious refractive process that was supposed to select the single ray that arrives perpendicularly at the surface of the pupil and damp out the rest, for now every ray that starts from a given point and enters the eye is focused onto the same point of the retina. The whole image is formed on the inner surface of the retina, Kepler says, and anyone who chooses to perform the experiment can see it there. Finally, and obviously, the same experiment shows that visual rays are as superfluous as *eidola*.

Kepler understood that what forced Alhazen and Witelo to insist that the image is formed on the crystalline lens and not on the retina is that the image on the retina would be upside down; but with his new insight into the image as a mental construct, he sees that it is not a problem, or at least not his problem. "How this image connects with the visual spirit dwelling in the retina and the optic nerve, . . . I leave for physicians to decide."

Now Kepler approaches the question of refraction, and analyzes the focusing of light by a transparent sphere. It is the same question that Robert Grosseteste had treated (Figure 5.3), but whereas Robert had assumed a simple though incorrect law of refraction, Kepler had no law, only Witelo's table of empirical data. The treatment is very thorough, and at the end of it he proves what is illustrated in Figure 5.2—that a narrow beam passing through the center of the sphere is brought to a good focus. Next comes a tentative discussion of the use of lenses to correct vision, but he is not yet ready to discuss lenses in general.

The second half of *Paralipomena* is essentially a handbook covering about all that can be said, if one has no telescope, about angles, sizes, distances, and atmospheric refraction, about the intensity of light from different sources, about the texture of the moon's surface, about the information obtainable from observing eclipses, about why the eclipsed moon is reddish, and much more.

But all that was prelude, for just as the book came out the first telescopes were being made. They used lenses, those little pieces of glass that people stuck next their noses to help them to see better but that no scientific principles explained. For three centuries they had fallen through philosophy's net, and now they had arrived in an instrument that transformed astronomy. They could no longer be ignored, and someone had to bring them into the realm of knowledge.

3. Lenses and Species

Now, gentle reader, I ask you to put down this book and spend a few minutes playing with a magnifying glass. Try it on this page; the letters will look about twice as large and, when you hold the glass at a height that makes your eye comfortable, a little farther away. Next, darken the room as much as convenient, leaving an opening so you can see outside. If you hold the lens with one hand and a piece of paper with the other you will find that you can focus an inverted image of the scene outside on the paper (or, if it is dark outside, try your reading light). Kepler called this kind of image a picture; it is now called a real image. Think what is happening as Alhazen taught us to think. From the point at the top of a tree outside the window a little cone of rays reaches the lens (Figure 6.5). The lens bends each ray so that they all come together at roughly the same point behind it. If you put the paper there you will get a dot of light from the point. But each other point of the tree is sending its rays too. The rays do not get tangled up; each is imaged separately, and the result is an inverted image of the tree. Figure 6.6 shows how a lens of this kind (called a convex lens) actually works.

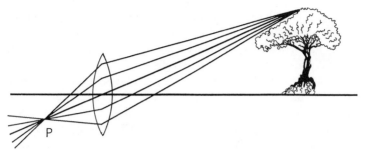

Figure 6.5. Formation of a real image, point by point (only one point is shown).

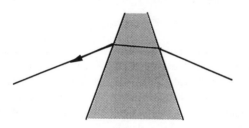

Figure 6.6. A ray is bent as it passes through a prism or part of a lens.

When the light ray goes from air into glass it is bent toward the perpendicular to the surface, and when it leaves the glass it is bent away from it. These two deflections bend the ray back toward the line called the axis that passes through the middle of the lens. Figure 6.6 shows that the further the ray is from the axis the more it is bent, and this enables the bundle of rays to converge at a point.

In Figure 6.5 I have extended the rays a little beyond the point of convergence (it is called the focal point), and you can see that they diverge from that point exactly like the rays that diverged from the treetop on the left. If you could see those diverging rays when you looked at the tree, you ought to be able to see them as they diverge from the focal point. Put down the paper and hold the lens so that the focal point is a comfortable distance in front of your eye, and you will see an inverted image of the tree, or whatever your object is, hanging in space between you and the lens. Kepler called this a pendular (hanging) image; it is now called a virtual image. A real image can be caught on a screen, but if you put your eye where the screen is you will not see any image. A virtual image is one that you can look at but not catch on a screen. Real images are formed by converging rays, virtual ones by parallel or diverging rays.

I have shown you how to produce a virtual image of something outside your room. Suppose you want to see it better. Does it not occur to you to go find another magnifier, and use it to examine the real image behind the first lens? If you can hold everything steady—or better, if you mount the lenses in a tube—you have just made a telescope. But perhaps you find that your telescope is making things look smaller instead of larger. Remember, that is what happened when you were a child and tried looking through your parents' binoculars from the wrong end. Just turn your telescope around and look through it the other way. The stronger lens should be next to the eye, and the more the two lenses differ in strength (essentially, in fatness) the more your telescope will magnify. It would not be much good for watching a horse race, because everything is upside down, but it is fine for looking at stars. It is not like those made by the lens-grinders of Middleburg and demonstrated to the senators by Galileo, for no one would pay money to see everything upside down, but potentially, if the first lens is very thin, the tube very long, and the second lens very fat, it is a much more powerful instrument.

In 1610 when Galileo introduced the moons of Jupiter to the world he ran into heavy opposition. Because he could not hope to make much headway against the Church or the universities he turned for support to the rich and powerful. This man who for years had been teaching Ptolemaic astronomy to his students even though he didn't believe much of it was not looking for trouble, and he knew that if any arose he would need support anywhere he could find it. Therefore he built several telescopes of good quality and distributed them to potential allies, and one who happened to be in Prague lent his to Kepler for ten days. Kepler had been skeptical of the reports from Florence. He had tried to make a telescope for himself but had given up, defeated by uncooperative lens-grinders, his own clumsiness, and the poor quality of the available glass. Now he was overwhelmed by what was revealed to him. His myopic eyesight could be corrected by refocusing the instrument, and for the first time in his life he really saw the stars. There were the four moons of Jupiter, which to him and Galileo were a visible model of planets revolving about the sun. There was the Milky Way, partly revealed as a cloud of faint stars, there was the surface of the moon, on which he saw mountain ranges and even, in his wondering imagination, huge structures that the moon's inhabitants might have built. It did not look like a sphere of heavenly ether.

Kepler understood that Galileo needed support. To be Chief Philosopher and Mathematician to the Grand Duke of Tuscany was a distinguished honor, but when trouble was in sight any help from the Emperor's Mathematician and friend would be welcome. Therefore Kepler quickly responded to Galileo's *Sidereal Messenger* with a little book called *Conversation with the Sidereal Messenger*, written in a tone that talks down to the ducal employee but enthusiastically endorsing his findings. And Kepler explicitly compares the new satellites (his word) with planets in the Copernican system.

Kepler's mind at this time was full of astronomical speculations but also tormented by the knowledge that nobody seemed to know how a telescope actually works. Out of the discomfort, within a few weeks, came a book

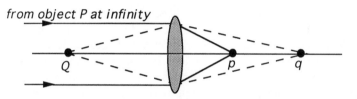

from object P at infinity

Figure 6.7. If the object is at infinity its image is focused at p. If it is at Q, the focal point moves out to q.

Dioptrice,[6] which explains how lenses form images, how the eye works, and how its vision is corrected by spectacles. It lays the foundation of the optics of lenses, now called geometrical optics, and though methods of calculation have been refined since then, the essential ideas that students learn today are still as Kepler set them down.

Dioptrics is severe in its format, consisting of 141 numbered definitions, axioms, propositions, and problems. It starts with material familiar to readers of optical works from Ptolemy on, except that it emphasizes the quantitative measure of refraction. To be sure, he does not know the exact relation between the angle of incidence and the angle of refraction (Figure 5.1), but in a lens system these angles are small, and he bases his work on an approximation that is good enough for most purposes: when light goes from air to glass, the angle of refraction is about two-thirds the angle of incidence.[7] With computers we can now do it exactly, but nothing changes much. Soon (no. 39) he has calculated that if the two spherical surfaces of a convex lens belong to spheres of radius R, and one holds the lens up as in Figure 6.5 to form a real image of a distant object on a piece of paper at P, the distance from the lens to the paper will be about equal to the radius R. Further, he knows that if one looks at an object less far away the image will move out a little, as in Figure 6.7.

No. 57 begins a section on vision in which Kepler repeats the ideas put forth in *Paralipomena*, but this time presents them more clearly because he is in control of a theory of lenses and focusing. He suggests that a normal eye adjusts its focus to near and distant objects by changing the shape of the eyeball or perhaps even changing the density of the fluid inside it, but he does not guess the principal mechanism: it does it by muscles that change the shape of the crystalline lens. This does not prevent him from giving a correct theory of eyeglasses to correct near- and farsightedness, and before rushing on he has time (no. 64) for profiles of people who will need them: a life spent sitting indoors, bent over a book or a fine manual task, leads to nearsightedness; whereas the person who drinks and sleeps too much, who is given to idleness and daydreams, who ignores what lies

---

[6] Many words Kepler invented are still used: *focus, nova, satellite, conic section. Dioptrics,* intended to accompany *catoptrics,* got as far as the Oxford English Dictionary but seems to have died there. Even so, I will call his book *Dioptrics.*

[7] This is good enough for his purposes, but the number varies from one kind of glass to another. Robert Grosseteste set it equal to $1/2$.

before his feet or under his hand, whose gaze is usually directed into the distance, will find that he can no longer see clearly what is close to his face.

Finally Kepler reaches his goal, the telescope. He explains (no. 86) the kind made by putting one convex lens behind another, as we did a while ago, and shows how its magnifying power can be calculated and controlled; then (no. 89) he shows how the same two lenses can be readjusted so as to project an inverted telescopic image onto a screen. In no. 107 he explains Galileo's telescope, which has a convex lens in front and a concave lens next to the eye, showing that this combination produces an upright image. Now that he understood how to calculate what will happen when two lenses are put together he could have gone on to analyze Galileo's compound microscope and add to his glory, but he probably never heard of it, and it was only a century later that people stopped using it to look at wasps and bees and began to ask some of the important questions that a microscope answers.

As Kepler charged into the future with his vision of a mathematical physics and his new ideas on astronomy and optics, it is easy to imagine him as someone breaking every link with tradition. But astrology was dear to him, even if he regarded most of its practitioners as fakes and crackpots, and his astrological writings speak of species, like postal couriers, carrying their messages through interplanetary space. Now we shall see how species got woven into his thoughts about planetary motion.

I have mentioned the *Astronomia nova*, in which he represents the solar system as consisting of planets (together with their satellites) orbiting the sun in elliptical paths. Something very precise is happening here, for they repeat the same orbits again and again without the slightest variation.[8] Ptolemy's great imaginary machine that mounted planets on revolving spheres explained this fact. It represented planetary motion in a way that disagrees only slightly with observation, but there was no way to correct it; Kepler's ellipses fitted the facts, but ellipses are only mathematical figures. What makes a planet move accurately along such a curve? Remember that according to Aristotle's mechanical theory, everything that is in motion must be moved by something. Something is pushing the planets all the time, a force that exercises such perfect control that at the end of the planet's year it returns exactly to where it was. What is this force? How is it exerted?

Clearly force comes from the sun, which is at the hub of every planet's orbit, and in III.33 of *Astronomia nova* Kepler compares the visible species of the sun's fire to the species of its physical power.

> Although the light of the Sun cannot be the moving power itself, I leave it to others to see whether light may perhaps be a kind of instrument or vehicle, of which the moving power makes use. Just as light, which lights the whole Earth, is an immaterial species of that fire which is in the body of the Sun, so

[8] To forestall objections, I should mention that precise measurements and calculations reveal a small drift in these orbits, too small to be detected in Tycho's observations or Kepler's interpretation of them.

this power, which enfolds and bears the bodies of the planets, is an immaterial species residing in the Sun itself, which is of inestimable strength, seeing that it is the primary agent of every action in the universe.

If he really means "every action," he is telling us that Aristotle's Prime Mover lives in the sun. In the next chapter he continues,

> For it may appear that there lies hidden in the body of the Sun a sort of divinity, which may be compared to our soul, from which flows that species driving the planets around, just as from the soul of someone throwing pebbles a species of motion comes to inhere in the pebbles thrown by him, even when he who threw them removes his hand from them.

Note the "species of motion." Robert Grosseteste has already mentioned this species in §4.5, and it seems to be the same as the *impetus* mentioned earlier. Kepler's analogy compares the motion of a planet with the flight of a pebble. But there is a cloud in this bright picture. If species radiate outward from the sun, how do they compel the planets to circle around it? Kepler answers that there is a way this could happen: if the sun rotates, the rays bearing its species will revolve like the rays of a lighthouse and sweep the planets along. One year later, in 1610, Galileo and Scheiner noticed sunspots and started to make maps of them; at once they found that the spots drift uniformly across the sun's surface and deduced that the sun does indeed rotate, with a period of about twenty-four days. Kepler rejoiced. And what keeps the sun spinning? He has provided for that: it has its own soul and internal dynamism. After thoughts of such inconceivable magnitude, going so far beyond the available evidence, he leaves the matter there. What more could he say?

At about this time Galileo was busy in his laboratory studying the motion of little balls, and coming to the conclusion, suggested but not required by his experiments, that if a perfectly smooth and level track could be provided, a ball would roll around and around the Earth without ever stopping. From this he deduced a general principle that motion in a circle can keep on indefinitely if nothing resists it, and this perhaps explains why, as I mentioned in §5.7, he stayed with the primitive Copernican model of the solar system rather than adopting Kepler's theory. The Copernican model explains planetary motion in terms of uniform circular motions, whereas the motions in Kepler's theory were neither uniform nor circular. If a planet slows down for a while it must speed up again, and this is something a ball rolling on a level surface would not do. If Kepler had lived to read Galileo's *Two New Sciences* he might have seized on the truth that things tend to keep going unless something stops them, but the book appeared only in 1638, too late.

History, which crowned the work, slowly destroyed the man. Threat of religious persecution forced him to leave Prague for the provincial town of Linz, where he found employment as a minor functionary but continued to enjoy the imperial title. His wife died, and the emperor himself came to his second wedding but, caught up in political turmoil, omitted to pay the salary that went with the title. Kepler lost children, and in the middle of it all he had to travel to Prague to defend his mother against a charge of witchcraft. The Thirty Years' War broke out, and ignorant armies swept back and forth across the Empire. After he had lived fourteen years in Linz a new government expelled the Protestants. He moved to Ulm and finally on to a town in the north called Sagan. Now it is in western Poland and called Zagan; it was then still in the Empire and therefore theoretically Catholic, but for the moment he was safe. All the time he was working, working—book after book—and slowly completing an immense project, a compilation of astronomical data to be called the *Rudolphine Tables*, based on Tycho's observations. The publication took all his money, nothing came from the emperor, and poor, ill, and without hope he set out on an old horse to ride through November weather across Germany to Regensburg, where the emperor was meeting with a general council of nobility, to beg at least a part of the salary that had been promised him. There, in a small house on a side street, fever took him and he came to the end of his journey. He was buried in the Protestant cemetery outside the city, but a few years later an army destroyed the place and denied him even that dignified repose. Nobody knows where he lies, but we have the Latin couplet that he composed for his epitaph:

> Once I measured heaven, now the darkness of earth.
> The mind belonged to heaven, the body's shadow lies here.

How can one summarize a life such as Kepler's? How many lives have shown such energy, such imagination, such sense of purpose, such ability to see facts in an entirely new way, such willingness to change one's mind? In the beginning Kepler thought that the planets were moved by angelic Intelligences, but as he discovered the laws of planetary motion he realized that with all the planets moving in empty space, angels would have no reference points by which to orient themselves with sufficient accuracy. Thus he came to the idea that the sun exerts the force which keeps the planets in their orbits and that the form of these orbits is determined by the laws of mathematics rather than angelic guidance. Loyal to Aristotle's physics, he got his planetary theory wrong, and it was not until eighty years later that Newton corrected the false assumption and set it right. But Kepler bequeathed the idea Newton started with, that the world's great motions are governed by mathematical laws. Until then, those who saw a mathematical order in the world expressed it in Galileo's geometric language of triangles and circles, which does not deal with time. The idea of a science of dynamics was his invention.

He could have prolonged his life by changing his religion as the Catholic emperor asked him to; he might have found other work so as not to die like

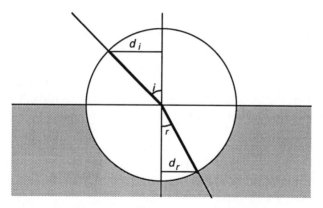

Figure 6.8. Illustrating Snel's law: a ray passes from air into a denser medium. For any angle of incidence, i, the angle r is such that the ratio d_i/d_r, known as the index of refraction, is always the same.

an old dog, but he followed his uphill path until the end. Was he content with the choice? Two years before his death he wrote a letter to his prospective son-in-law that was later printed in a small edition. In it he says, "When storms rage and we fear the shipwreck of the state, there is nothing nobler for us to do than let down the anchor of our studies into the peaceful ground of eternity."

4. Refraction: Willibrord Snel Improves on Ptolemy

The law of reflection was given by Euclid, and most probably it was known long before: the angle of reflection exactly equals the angle of incidence (Figure 3.5). Ptolemy also measured the refraction of a beam going from air to water and from air to glass. He too found a law, illustrated by the regularity of the third column of Table 3.1, but the numbers still have something arbitrary about them (Why does the second column begin with 8?), and the numbers are different for glass. Also, Ptolemy's law is not exact. Robert Grosseteste made a guess that was good enough for his purposes: the angle of refraction is half the angle of incidence. Kepler gave a formula that is both complicated and inexact, and his reasoning is unconvincing. We need a statement that tells exactly what happens when light is refracted, and then an argument showing that it must happen in that way and no other.

The correct law was announced in about 1621 by Willibrord Snel (or Snell, 1580–1626), a Dutch astronomer, physicist, mathematician, and surveyor. He did not publish it and his explanation has disappeared, but it was soon widely known through the correspondence of interested people. It is illustrated in Figure 6.8 and states that if one draws the incident and refracted rays and a circle (of any size) as shown, the distances d_i and d_r always have the same ratio. Their ratio (if the light ray enters from air) is known as the index of refraction. For air to water it is about $^4/_3$, and for air to glass it is about $^3/_2$, though it depends on the kind of glass.

The ray in the figure can go either way. Suppose it is traveling from the lower medium to the upper one. (I will not bother to interchange the labels i and r.) Make angle r larger; then i will also increase, but it cannot increase to more than 90°. That is where refraction ends. The largest possible r is about 49° for water, 42° for glass. Then what happens if you tilt the beam still further—say, to 50°? The answer is not obvious; you have to try it. When you do, you find that the surface acts like a perfect mirror; the beam is reflected down again. This fact has interesting applications.

If you shine light into the end of a glass rod it will go down the rod reflecting from side, even around curves, without leaking out (Plate 5). And the rod can be made as thin as a hair. In the 1980s and 1990s the United States was being covered with a network of fiber-optic lines replacing copper wire to carry its traffic of communications. An optical fiber is a hair of very clear glass, typically $^1/_{200}$ inch thick and many miles long, down which travels a light signal that bounces from side to side of the fiber, always at an angle that forces it to be reflected. The Information Superhighway is an almost invisible thread, and what travels along it is light.

So much for Snel, but I think he was not the first to know how light is refracted. Thomas Harriot (or Hariot, c. 1560–1621) was hired in his youth by Sir Walter Ralegh to teach him mathematics and be his assistant in various enterprises. He spent a year in the new colony of Virginia, of which Ralegh was a sponsor, to report for the London investors on its organization and commercial possibilities. Later Ralegh introduced him to the Earl of Northumberland, a potentate who enjoyed learned conversation and gave lifetime pensions to Harriot and a few others so that they could live near his mansion. From about 1590 they experimented and theorized, but they published nothing. When he died, Harriot left some ten thousand pages of notes, which show that he had been one of the best mathematicians in the world and that before 1601 his optical measurements had led him to the law later discovered by Snel. He shared the results of his measurements with Kepler but left him to find the law. Kepler never found it.

Harriot investigated refraction by different liquids and glasses and measured their indices of refraction; this was a fairly obvious thing to do, but he also asked a question whose answer contained the seeds of a great discovery: does a substance's index of refraction depend on the color of the light? He found that it does. Isaac Newton, some sixty years later, found the same thing and drew momentous conclusions concerning light and color, but those ideas slipped through Harriot's fingers. Harriot's health was always delicate, and he had many other interests.

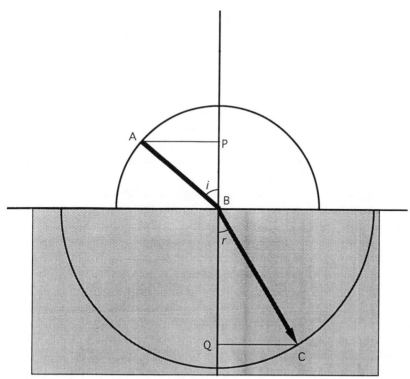

Figure 6.9. Descartes's version of Figure 6.8. Light traverses the two heavy lines in equal times. The lower one is longer because light moves faster in the medium, but AP and QC are equal. This is another way to state Snel's law.

The first to actually publish the law of refraction was René Descartes (1596–1650). We will meet him in the next section; here I mention only the argument by which in his *Dioptrics* (1637) he showed that the law must be true. He pretends that light moves like a ball; in fact the figures in his book show a little man with a tennis racket. Figure 6.9, a simpler version without the little man, represents Figure 6.8 in a new way, and shows how Descartes imagines a flash of light to move along a ray in a very short moment of time. Divide the moment into two equal parts. In the first half the light moves from A to B; in the second, from B to C. It moves further in the second half because, he claims, light travels faster in a dense medium like glass or water than it does in air. But at the same time, its speed measured in the direction parallel to the surface has not changed: the distances AP and QC are equal. This is Snel's law in a slightly disguised form (just redraw the lower half of Figure 6.8 on a larger scale), but now the index of refraction has a physical meaning: it tells how much faster light travels in the dense medium than it does in air.

What is right about this argument is the conclusion, Snel's law; what is wrong is the meaning of the index of refraction. Actually it tells how much *slower* light is in the dense medium than in air. In 6 we shall meet a mathematician who got it right, but of course the only way to prove it was to measure the speed of light in water and glass, and that was not done for another two centuries.

Only the most dedicated Cartesians were happy with the idea that light travels faster in a medium that one would expect to impede its motion. I suspect that Descartes invented implausible hypotheses to explain a result that was already well known. He himself had made careful measurements, and the opinion at the time was that he had already heard about Snel's law, though as was his custom he claimed originality. But his work in optics was only a small part of a much larger project to reform all of human thought, and I must devote a few pages to it.

5. René Descartes Invents a World

Galileo Galilei, like Johannes Kepler, was brought up in an intellectual tradition that had changed little since the middle ages. René Descartes, a generation younger than Galileo, was educated from the age of eight in an excellent Jesuit school, but he breathed a different air. The new science was showing, beyond any doubt, that even though the schools went on in the old pattern, there was more to knowledge and education than the study of old texts. As he studied what was set before him, Descartes began to see that it was unproductive of new ideas, and his whole life afterward can be seen as an effort to do something very difficult, to find a new way of thinking.

He left school to study law at Poitiers, where he received the degree in 1616. He lived for a while in Paris and then, desiring to see something of Europe, wandered off and spent a while as a noncombatant volunteer with the Prince of Nassau's army in a war against Spain. Wars were slower then. Armies paraded at sixty paces to the minute, and combat slackened when the weather turned cold because officers preferred not to share the hardships of their troops. The winter of 1619 found him in a town on the Danube near Ulm, where he was able to find a small room with a stove in it so that he could be both comfortable and alone, an uncommon arrangement in those days, and here he arranged his thoughts about human knowledge. He had been studying mathematics in his plentiful spare time, and on 10 November 1619, a day of intense concentration, he began to imagine an intellectual strategy resembling mathematics which would replace the one in which he had been brought up. He noted that mathematics (his model was Euclid) was a system of propositions which could be arranged in an order such that the proof of each one depended only on those already proved. No theorem stands alone; any one may require the use of several others to prove it. On this day it came to him that all of human knowledge might be brought into the same kind of order, atomized into a sequence of propositions, each one simple enough so that the proofs could be made free from error provided the proper starting point was found. But what an immense task; for as in geometry, but on a huge scale, each proposition might and probably would depend on a number of others, from many fields of knowledge. As he wrote many years later in his *Discourse on Method* (1637),

> These long chains of reasoning, each of them simple, that geometers commonly use to attain their most difficult demonstrations, have given me an occasion for imagining that all the things that can fall within human knowledge follow one another in the same way and that, providing only that one abstain from accepting anything as true that is not true, and that one always maintain the order to be followed in deducing one from another, there is nothing so far distant that one cannot finally reach it or so hidden that one cannot discover it.

The lazy and casual dilettante, having imagined this gigantic project, went to bed, and that night he had three dreams. The first represented him in a state of fear and perplexity and unable to move freely; the second included a great clap of thunder and a rain of sparks, and the third, quiet and peaceful in tone, concerned books and mentioned a line from the Latin version of an ancient Greek poem, "What is the path my life should follow?" Interpreting the thunder as a sign of divine inspiration and the line of verse as an invitation, he decided to build a new philosophy. Or rather, since the intellectual structure he imagined could not be built by one person, to create a new method.

He decided, and set out again on his travels. "And in all the following nine years I did nothing but wander here and there about the world, trying to be more a spectator than an actor in the plays that were being performed out there; and reflecting particularly in each matter on what might render it suspect and give us occasion for error." He spent a while in France and then in 1628 moved to Holland, where he lived for most of the next twenty years, moving from place to place as restlessness and the desire for privacy overcame him. Here he started on his great project and by 1633 had completed a book with the modest title of *The World*. He was about to publish it when he learned that old Galileo was being hauled before the Inquisition and punished for teaching that the Earth goes around the sun. *The World* made the same claim and so, having resolved long before to avoid conflicts with authority, Descartes abandoned it. Later he revised it, and it finally appeared fourteen years after his death. Its subtitle, *A Treatise on Light*, shows that an analysis of light was intended to be its centerpiece, but in the final version it talks of matter and sensation in general and explains the planetary system but says little about light. This was the first flower of his method, and one opens it eager to see how the great project unfolds.

Descartes begins with a warning: the old idea that we sense something when its form enters our soul is no explanation at all. He gives examples. The sound of words that bring us the idea of a thing does not in the least resemble the form of the thing itself, and he borrows Galileo's example of tickling, in which the sensation has nothing to do with the form of a feather. He needs this introduction because the real world is not at all as we sense it to be. It is made of three elements, Earth, Fire, and Air; these consist of particles, those of Air being much smaller than those of Earth, and Fire much smaller still. The world has no empty space at all, every gap has something in it, but we perceive only structures made of Earth; thus a vessel full of gold or lead actually contains no more matter than the same vessel when we would say it is empty. Why is the world he describes so different from the one we know? In a later book, the *Principles of Philosophy*, which appeared in 1644, he explains why there can be no vacuum:

> From the sole fact that a body is extended in length, breadth, and depth, we rightly conclude that it is a substance, because it is entirely contradictory for that which is nothing to possess extension. And the same must be concluded about space which is said to be empty: that since it certainly has extension, there must necessarily also be substance in it.

And there is another reason for filling space with matter: when I see a distant mountain, some influence comes from it to my eye, but, Descartes says in Chapter 2 of *The World*, "it does not seem possible to imagine that one thing can move another if it itself is not moving." He claims that all interactions are by contact, one bit of matter pushing another, and so there must be matter everywhere along the line between me and the mountain. But of course push is not the same as motion, and so the conclusion does not follow from the premise. We will find this confusion permeating his discussion of light.

The same hypothesis accounts for the motion of the solar system, but now Descartes is in a quandary because the model he wants to explain is Copernican. Therefore he announces he will describe not our world but another. He asks the reader, "Allow your thought to leave this world for a little while to see a new one which I shall cause to be born in an imaginary space while you watch." It is differently constructed, but our senses would perceive it as exactly like ours. He then goes on for many chapters to create stars surrounded by whirlpools of invisible matter that could drag planets around them, if any exist, and to derive, more from philosophy than from observations, the principles of motion and change. In Chapter 13 he arrives at light, but because he discusses it again in his *Dioptrics*, I will try in a moment to combine the two accounts.

In 1637 appeared anonymously a book containing four long essays. The first, which introduces the rest, staggers under the title *Discourse on the Method for Rightly Conducting One's Reason and for Seeking Truth in the Sciences*. The others concern dioptrics, meteorology, and geometry. The last one, introducing a form of what is now called analytical geometry, was a first-class contribution to mathematics. Though it is not relevant to optics, I digress for a moment to say why.

The part of classical mathematics called plane geometry dealt with a limited repertory of figures that can be drawn with simple instruments on a piece of paper: straight lines, polygons, circles, spirals, and a few more curves like the conic sections (ellipses, parabolas, hyperbolas). These figures had names, and geometers studied their properties. Descartes showed that figures are also defined by algebraic formulas in two variables. For example, if you plot $x^2 + y^2 = 4$ on a piece of graph paper, y against x, you create a circle of radius 2. How many algebraic formulas in x and y are there? An infinite number, as many as your pen can write down, and every one defines a shape of some kind. Their properties can be studied by the manipulation of algebraic symbols, which is usually easier than inventing a geometric construction. Except for what was known about a few figures defined in the old way, geometry begins with Descartes.

The *Discourse on Method* is a classic of philosophy. The *Meteorology* gives a crucial and far from obvious argument that is necessary to explain rainbows (see §7.8); the rest of it, together with the *Dioptrics*, can be dealt with quickly.

In the *Discourse* Descartes announces that he does not propose to tell anybody else how to think but only to tell how he thinks, why he came to think this way, and what kind of results he obtained. He praises his early education for its languages, its poetry, and its mathematics, but not for its philosophy. Even after hundreds of years, he says, there is nothing in philosophy that is not disputed. He saw philosophy as a body of opinion with no certainty and no plan, with nobody responsible for it, that has grown incrementally through the centuries, and he decided to study how its shaky doctrines can be replaced with certainties. The first step was to get out of his mind all the beliefs that he had accepted from birth. Next he adopted four rules:

> The first was never to accept anything as true that I did not clearly know to be true.
> The second was to divide every difficult question into parts which could be dealt with separately.
> The third was to think in an orderly way, starting with what is simplest and easiest to know.
> And last, to write down what I learned so that nothing would be omitted.

The bedrock of his knowledge, the unshakable foundation on which all his deductions would be based, is the axiom "I exist." He knows this is true because he is thinking: *Cogito, ergo sum.* I think, therefore I am. He could be mistaken about his name or what clothes he is wearing or where he is, he does not know anything with absolute certainty about his body or his house, but that he is conscious proves that he exists. His second axiom is a general rule that is less often quoted but much more conspicuous in what follows: "The things that we conceive very clearly and very distinctly are all true, but ... there only remains some difficulty in properly discerning which are the ones that we distinctly conceive." The second half of this rule ranks as one of the greatest understatements of all time, but he needs it for the first step of his deductions, which is the existence and nature of God. The *Discourse* continues, but let us turn to the *Dioptrics* to see how the method works.

We are in the year 1637, and we start with the nature of light. Descartes does not know any more than anyone else about what light actually is, and so he proposes to discuss the question in terms of three analogies (he calls them comparisons) that will help the imagination to deal with it.

The first fact to be explained is light's great, perhaps infinite speed, which he demonstrates with an argument that had escaped Galileo. Consider an eclipse of the moon, imagining it as the ancients would have, with the Earth not moving. If light moves with infinite speed, totality will occur at the moment when sun, Earth, and moon are in a straight line. If not, think what would happen. The sun's light arrives at the Earth and passes beyond it to the moon. We observe the progress of the eclipse by the light which the moon reflects back to Earth. Suppose it takes the sun's light half an hour to pass between Earth and moon. We would then see the eclipsed moon an hour after the light we are seeing left the vicinity of the Earth. But where is the sun at this point? It has been moving for an hour since that moment, and the sun, Earth, and moon are no longer in a straight line. But they are always (when one takes account of atmospheric refraction) observed to be in a straight line; therefore the time for light to traverse the 240,000 miles to the moon and back is very short, and Descartes takes it as essentially zero.[9] He never says the speed is infinite but claims that light reaches the Earth from the sun "in an instant."

The first of the three comparisons, inspired by this fact, is a blind man's cane. He touches something with it and *immediately* feels the contact in his hand. The cane scarcely moves as it delivers its message. No forms or qualities pass along the stick, and nothing passes directly to the soul; there is only the sensation of touch. Light, says Descartes, is a pressure, not a motion, in the densely packed medium of apparently empty space, transmitted instantaneously from particle to particle and finally to the eye. "And by this means your mind will be delivered from those little images fluttering through the air, the *species intentionales*[10] that burden the imaginations of philosophers."

How does the pressure originate? Let us go to the second comparison. Imagine a tub full of grapes partly trodden out, with the grapes and the

---

[9] Actually it is about 2½ seconds.

[10] These are "species that convey meaning," and the remark shows that the optical discoveries of the past thirty years had produced no ripple in the received doctrines of philosophy.

juice together. Open a small hole in the bottom. Juice (small particles) will flow out, while the grapes (large particles) remain behind. But suppose there is no hole. The juice still has a *tendency* to flow out (he refers to this ill-defined concept also as *action* and *inclination* and finally settles on *intention*); this tendency exerts the pressure in the air which constitutes light. But what produces the tendency? Descartes tells us in *The World*: the disposition of small particles to form whirlpools around a star does not stop at the star's surface. Inside it they whirl faster than ever, and the pressure we perceive as the sun's brightness is the outward push of their centrifugal force.

I have already mentioned the third comparison in discussing refraction: light travels like a tennis ball, faster in glass or water than in air. The three comparisons seem not only different but irreconcilable. Pressure is an accident; canes and tennis balls are substances. Further, a tennis ball does not travel from the sun to the Earth "in an instant." We can see, though, why he needs three analogies (or something like them) if he is to explain the two very different ways in which we experience light. The first is as illumination, proceeding from some source, that fills a room with light. It is possible to imagine this as some sort of pressure or tendency to move. The second is as a ray through a hole in the shutter, or else the kind of ray that writers since Alkindi had used to explain vision. It is hard to think of such a directed thing as a pressure; much easier if it is a stream of tennis balls. Descartes insists in his correspondence that the models are not irreconcilable, and he invents a physics of tendencies which assumes that tendencies (whatever exactly they may be) extend themselves through space along the same paths that tennis balls would follow. The argument is Aristotelian: the tendency to move is potentiality, and motion itself is actuality, but actuality is contained within potentiality, and there is no difference in their laws. And the notion of light as a tendency propagating through space without any actual motion recalls Roger Bacon's multiplication of species. They move, if you remember, as your shadow moves beside you as you walk. Armed with such reasoning, there was little that Descartes could not explain. In fact, the analogy of the tennis ball pays a dividend right away. Hit the ball so as to give it a spin. In light, he claims, the combination of forward motion and spin determine the color—a notion supported by no evidence whatever.

In explaining light by three comparisons that have nothing to do with one another and only leave a careful reader in the dark, Descartes shows his early training. I have mentioned in §4.6 how the medieval mind saw the whole of creation as a system of analogies intended to teach humankind how to live and how to know God. An argument based on an analogy was considered as more than a way of expressing oneself vividly because it related, tacitly or explicitly, to a cosmos that was founded on analogy. I am not claiming that Descartes would have justified his discussion of light by arguing in this way, but analogy is woven into his thought. Still, he could

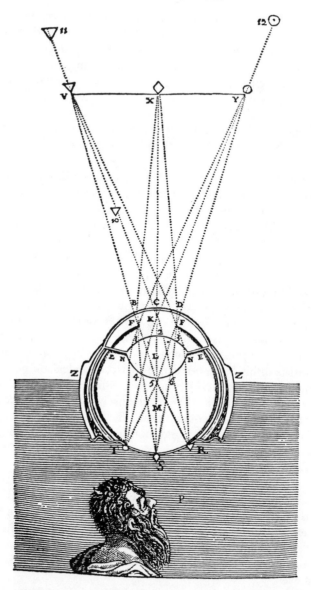

Figure 6.10. Rays of light from V, X, and Y pass through the pupil and lens and are imaged on the retina. From Descartes's *Dioptrique*.

have profited from a wise and simple remark of Aristotle's, "In inventing a model we may assume what we wish, but should avoid impossibilities."

Descartes's theory of vision rests on solider ground. His own experiments, on top of those of Scheiner and Kepler, told him the functions of the pupil, the lens, and the retina; and Figure 6.10, showing an unidentified savant studying the image formed on a retina freed from its surrounding tissue, needs no explanation.

Dioptrics continues with a discussion of lenses to help vision and improve the clarity of telescopes. After Kepler's work (which he does not mention), the first problem to be solved was spherical aberration, and with a display of virtuoso geometry Descartes designs some lenses. One of them, with a flat front surface and a back surface that is part of a properly shaped hyperboloid, focuses light from a distant source on the axis of the hyperboloid onto a single point. He has in fact rediscovered the lens that the Arab Ibn Sahl (§3.7) invented more than 650 years earlier, and except for minor details his mathematical proof is identical to Ibn Sahl's. But knowing this does not advance us very far. Nothing is said about the focus of light from directions off the axis, and though Descartes writes at length about machines for grinding lenses of nonspherical profile, he was not able to design one that worked, and it remains a difficult job.

The optics in the *Dioptrics* is not especially original, and its practical side is not very practical. Experts are not very enthusiastic when they discuss Descartes's other essays in philosophy. Why, then, has he such a high reputation in the history of European thought? For one thing, he wrote simply and clearly; he used no jargon. More than that, the *Discourse* is a charming if disorganized mixture of personal reminiscence and philosophical argument that has been enjoyed by thousands of readers. But the main point is that he had a clear idea of what needed to be done. Arguments based on authority had run their course. The debates that enlivened scholarship in the middle ages were becoming recognized as pertaining largely to language—that is, to the sounds of words and not anything real—but in universities and ecclesiastical circles the old pattern continued because no one knew how to reason in any other way. It was Descartes's merit that he realized that our idea of the world is a mental construction based on sense perceptions of sight, sound, touch, taste, and smell rather than on participation in the world's forms. Our mind, the part of us that reasons, looks out on the rest: on our body, on our society, on the physical world. Truth does not enter it from outside; rather, it constructs truth from what it observes and deduces. Little of what Descartes said about the world from this new point of view was really new, but he said it simply and in universal terms, and he supported his generalities with specific explanations of how things actually work. One wonders, seeing the logicality of his program, that he got so much of it wrong, so very wrong that French investigators remained at a standstill for many years after his death while they slowly disentangled truth from error. Perhaps the weak point was in the first of his rules of thought, "never to accept anything as true that I did not clearly know to be true." How was he supposed to know it? He spent months on anatomical dissections and in experiments on optics and mechanics. It was when he went beyond the conclusions he could draw from his observations, when he had to explain in a fundamental way how something worked, that he tended to fall back on reasoning by analogy, and an analogy, as anyone will tell you, cannot be proved true or false by experiment.

He announced that thought must be reformed and convinced people that reform was possible, even if he himself succeeded only partially. He was the first since Aristotle to try to organize knowledge of the world into a general scheme and show how it fits together. Modern philosophy, its defects and its merits, start here.

6. Fermat: A Mathematician Looks at Optics

After what has been said about Descartes's geometry, it may cause surprise to be told that Pierre Fermat (1601–65) has an even higher reputation among mathematicians. His famous Last Theorem, which he wrote in the margin of a book with a note saying that he had proved it, has inspired mathematicians for three and a half centuries to try to find his proof or any proof,[11] and his simple but subtle methods and theorems are in the toolbox of every mathematician interested in the areas in which he worked. The two giants of mathematics confronted each other over the heads of a hundred mediocrities and exchanged opinions on scientific matters. Fermat saw himself in the position of a defendant who says yes against a witness who insists on saying no; doubtless Descartes saw the situation in some other way.

When Descartes's *Dioptrics* appeared in 1637 it carried an invitation for readers to correspond with him on points that seemed obscure. Among those who replied were Thomas Hobbes and Fermat, but Descartes was more inclined to repulse their suggestions than enter into serious discussions. In private letters to his friend Marin Mersenne he described Hobbes as extremely contemptible and referred to Fermat's criticism on a mathematical point as *stercus*, which I leave in decent Latin obscurity.

Fermat, in his letters to Descartes, objects first of all to the physics that treats *intentions*, which I have called tendencies, as if they were real forces or motions. Because there was no way of observing or measuring such things, or even demonstrating their existence, there was not much that Descartes could say. The correspondence became repetitive, and Fermat broke it off. After Descartes died, discussion resumed with those who followed after him, and in about 1664 Fermat wrote a letter to a nameless gentleman, usually known as M. de ***, in which he focused on two other difficulties and said he had a theory that overcame them. First, he argued, it shocks our common sense to suppose that light travels more quickly in a resisting medium. But suppose it does: then why would the velocity parallel to the surface of the resisting medium, given by the line QC in Figure 6.9, not decrease also? Descartes had tried to explain this by a tennis-ball argument: imagine that as the ball representing light enters the water or glass a racket taps it straight downward. This, he says, will increase the overall speed by a required amount without changing the speed in the horizontal direction. But as Fermat pointed out, this can be done only with a carefully calculated tap that

---

[11] I understand that a proof announced with loud fanfare in 1993 has not entirely satisfied some discriminating minds.

will be different for every ray, for its force must depend on the ray's angle of incidence. Where there is no tennis racket, Nature must have installed some mechanism that has the same effect, and it is all very odd. And what have tennis rackets to do with a mathematical theory? Fermat complains to M. de *** that Descartes "bases [his argument] on a comparison, but you know that mathematics has little time for such tropes, since comparisons are even more odious in mathematics than in the ordinary dealings of the world." As Fermat saw the problem, it was this. The law of refraction being known, is there a way to derive this law mathematically, from clear and definite principles, without using tennis rackets or the rhetoric of analogies?

Fermat seems to have thought about this question for some years without finding the right way to approach it. After Descartes's death he continued to correspond with the Cartesians, and in about 1662 he found not only the right principles but also the mathematical method needed for his proof. As we would explain the calculation today, there are three principles:

Light, in air or in a uniform transparent medium, travels in a straight line at a definite speed.

Light travels faster in an almost weightless medium such as air than in a heavier medium such as glass.

A light ray traveling between two points follows the path that gets it from one to the other in the shortest time. (This is the same principle that Hero of Alexandria used long ago to explain reflection.)

These assumptions, plus some clever mathematics, lead to Snel's law. The index of refraction is still the ratio of the two speeds but now the ratio is turned upside down; it is the speed in air divided by the speed in the medium. The energetic reader may want to try to finish the discussion by deriving Snel's law from these principles—but no fair using calculus, because Isaac Newton had not yet invented it.[12] The method Fermat used was its immediate precursor.

I explained the problem as we would explain it today, but Fermat was a very clear thinker and would not have accepted such reasoning. His argument is geometrical—he draws lines and compares their lengths—and geometry has nothing to do with time. There had to be some way of posing the same problem in geometrical terms, and he did it by defining a "resistance" to motion in a medium in terms of the length of a line and then seeking the path of least resistance between two given points. Only later, in a private letter, did he make the obvious connection: the greater the resistance the slower the motion.[13]

[12] It was in these same years that Isaac Newton, in his early twenties, was inventing differential calculus, but as he wrote later, "The beginnings of this method were thought of by Fermat." (*Correspondence*, vol. 2, p. 182.).

[13] In case this cautious approach seems fussy, it should be noted that neither Newton nor Leibniz nor their brilliant contemporaries ever introduced time into their mathematical analyses of motion. Leonhard Euler, in 1750, was the first to take that step, which we expect schoolchildren to take in their lessons in elementary physics (See Park 1988, chap. 13 and n. F).

Note that Hero's principle is a qualitative one: Nature tends to act economically—in this case it is an economy of time. Note also that the principle does not require us to say we know what light is. Substance or accident—it does not matter, as long as something moves.

The idea that one can identify some general characteristic of Nature that can be represented by words or a simple formula and use it to calculate what actually happens is a very attractive one for theoreticians. Snel's law by itself is an accurate description of something that happens, but it is nothing more; Fermat's principle suggests the reason why Nature does it that way. It was natural to ask whether the same principle explains any other phenomena such as the flight of a ball. No; I can make the ball arrive more quickly by throwing it harder, whereas the speed of light is not under my control. But the inventors of the science of mechanics sought and found principles of economy that govern the ball and other motions; they are generally called principles of least action (still suggesting some kind of economy), and all the great dynamical principles of modern physics are expressed in this way.

Fermat's argument was not popular among the Cartesians. One named Claude Clerselier admitted that Fermat's proof was a proof, but complained that his assumption was "only a moral principle, not physical, which is not and cannot be the cause of any natural effect. . . . It cannot be because we would have to assume knowledge in Nature; and here, by 'Nature' we understand only the order of things, and that law established in the world which acts without foreknowledge, without choice, and by a necessary determinism." There is, he says, no use in ascribing knowledge and choice to Nature, for we truly understand a motion only when we know the forces that produce it. Fermat has to assume that the light starts at A and arrives at B; that is, it has to know where it is going to end up before it starts. Descartes specifies the force that determines where the light will go, just as the racket drives the ball to a certain place. Descartes deals with physics; Fermat gives us metaphysics instead.

Let me try to answer Clerselier's objection. A principle like Fermat's can be used to calculate what happens in some particular case, but it does not really explain why it happens. Instead, it identifies a general characteristic of Nature that covers a very wide variety of situations and can be stated in simple language. Does this mean that the fundamental principles of physics, when and if they are finally understood, will have this character? Who can say? Laws of economy may lie on the outskirts of a profounder truth, but one hopes that that truth, also, will be in the form of a simple statement that anyone can understand.

The Last Four Centuries

THE RISE OF OPTICAL EXPERIMENT

I saw Eternity the other night
Like a great *Ring* of pure and endless light
All calm, as it was bright,
And round beneath it, Time in hours, days, years
Driv'n by the spheres
Like a vast shadow mov'd, In which the world
And all her train were hurl'd.
(Henry Vaughn, 1650)

1. Ole Römer and the Speed of Light

AFTER THE TIME of Descartes and Fermat there was a great rise in scientific activity in Europe. Universities had not yet become centers of research, but little societies like Porta's Academy of Secrets proliferated, and some larger ones were supported by kings and noblemen. Private citizens with time and money to spare set up laboratories, and by the end of the century people began to realize that it was not enough to observe and report a new phenomenon; they must measure it, for when several explanations were offered, measurement could often allow a choice between them. The next few sections will address the work of several investigators who were active at the same time. I will introduce them in order not of their birth but (roughly) of their main contributions. The first, presented a little out of this sequence, produced a number.

Ole Römer (1644–1710) was born to humble parents in the old Danish town of Aarhus. By 1671 his talents had been noticed, and he was invited to join the Royal Observatory that Louis XIV had recently established in Paris. There he learned of a puzzle relating to the moons of Jupiter: they do not seem to circle the planet at a constant rate. With a rhythm of about four hundred days they seem to speed up and slow down, all at the same time, though their average stays constant. The number four hundred is a clue. As the Earth circles the sun there is a moment when Earth is closest to Jupiter. The next time this happens is about four hundred days later. Therefore it seemed possible that the Earth's motion around the sun has something to do with the misbehavior of the little moons. The observatory's director, Jean-Dominique Cassini, toyed with the idea that light's finite speed might be the culprit, but influenced by Cartesian philosophy and the uncertainty of the observations, he dropped the idea and insisted that there must be some other explanation. To Römer, younger and fresh from Denmark, the thing seemed obvious.

Assume for a moment that light takes a while to cross a planetary orbit. If I look at Jupiter and its moons I do not see them as they are at this moment, for the light from Jupiter takes some time to reach me. As the Earth goes around in its orbit, much more quickly than Jupiter, the distance to Jupiter varies, and so does the time delay. Therefore if something is happening perfectly regularly in the region of Jupiter, something like the regular revolution of a satellite, it will not look regular to me. That is really all. Römer was able to deduce from the accumulated readings that light takes 11 minutes to pass from the Earth to the sun. Jupiter's innermost satellite, Io, takes only 42½ hours to circle the planet, and so an advance or retard of 11 minutes is appreciable.

The argument did not play well in Paris, for Descartes had spoken: light travels from the sun to the Earth "in an instant." Even when Römer used his new insight to correctly predict the exact moment at which Io would be seen to pass behind Jupiter, the unconvinced remained unconvinced, and we shall see that it was a while before their minds changed.

But if you believe Römer, what is the speed of light? As soon as somebody tells us the size of the Earth's orbit we will know.

While these discussions were going on in Paris, surveyors were measuring the planet Earth. Snel had tried it as early as 1617, and in 1669 Jean Picard, of the Paris Observatory, made a very careful survey and gave the value 7,920 miles, about 6 miles smaller than the present value of the diameter as measured through the equator. The next problem was to establish the scale of the solar system so that the distance to the sun could be known. This is a very difficult question. The best figure available at the time came from simultaneous measurements of the apparent position of Mars made by Cassini in Paris and Jean Richer during an expedition to Cayenne in Guyana: the distance to the sun came out to be about 11,000 Earth diameters, or 90 million miles. If light goes so far in 11 minutes, it is traveling at something like 130,000 miles per second. Everyone must have worked this out on a scrap of paper when the numbers became available, but the first person to publish a figure for the speed of light was Christiaan Huygens, a Dutch physicist and astronomer, who in his *Treatise on Light* [1690] did the calculation assuming that the distance to the sun is 12,000 Earth diameters. He seems to have deduced this figure from a reconsideration of Cassini's data, and it is better than Cassini's. The resulting value for the speed is 144,000 miles per second, to compare with the modern number, about 186,000. Not much was made of the result. Partly this was because nobody could imagine any other kind of observation that might confirm or refute it, and in fact it was almost sixty years before that happened. Mostly I think it was because the figure was so ridiculously, impossibly big that no one could take it seriously. There is something comical about huge numbers—when Marco Polo returned from China with tales of vast cities and gigantic wealth the Venetians laughed and named him *Il Milione*. Robert Hooke, whom we shall meet later, remarks that according to Römer "the Motion of a Cannon Bullet is as much slower than this of Light, as the Motion of a Snail is than that of a Cannon Bullet." If it moves that fast, he says, it might as well be instantaneous, which is actually easier for the mind to grasp. But if Römer was right, Kepler and Descartes were wrong; the number is finite, and it is measurable.

The position of a Protestant in France was not entirely comfortable, and Römer's good judgment took him back to Denmark in 1681, four years before the massacres that followed the revocation of the Edict of Nantes. There he entered public service as Astronomer Royal and was successively Master of the Mint, Mayor of Copenhagen, Prefect of Police, Senator, and head of the State Council. He died in Copenhagen in 1710, more honored for serving his country than for his number.

2. Father Grimaldi Avoids a Quarrel

In 1665, the year Fermat died, two books appeared that in different ways affected the future of optics. The first was the work of Father Francesco Maria Grimaldi (1618–63), a Jesuit professor of mathematics in Bologna; the second will be discussed in §3. Grimaldi's passion was experiment, refined and very exact, and he saw things never seen before. The last decade of his short life was taken up with these experiments and the preparation of a long book with a title almost as long, which begins *A Physical and Mathematical Thesis on Light, Colors, the Rainbow, and Other Related Topics* [1665]. He knew that ecclesiastics and professors who saw his work would skip through it to see whether he agreed with received ideas. We have seen in §2.5 that Aristotle held that light is a quality of a transparent medium and in §4.8 that Saint Thomas had agreed, but he also realized that this hypothesis does not explain even so simple a phenomenon as reflection by a mirror, and so there is more to be said. Grimaldi used the question "Substance or accident?" to organize his work. It consists of two books "of which," he writes in the introduction, "the first introduces new experiments and reasoning that support the substantiality of light. The second refutes the arguments of the first and teaches that it is probably possible to sustain the Peripatetic thesis of the accidentality of light."

He pleads for a little space unobstructed by dogmatism in which to conduct his discussion. "Even though only the blind do not know light, our discourse concerning it still leaves the subject in darkness. To explain its nature is a most difficult task." The more difficult, he says, because when discussions of light tend to combine and confuse its physical and spiritual natures, nothing gets settled. In this situation, "I do not think it need be

Experimentum Primum

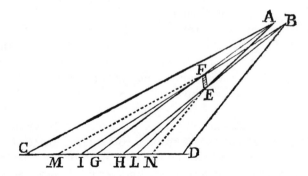

Figure 7.1. Grimaldi's demonstration of diffraction, discussed in the text. From Grimaldi 1665.

called daring if someone tries to advance these studies." He knows that experiment is challenging many teachings, and pleads gently that repetition of old formulas is not the way to deal with new facts. "Henceforth we do not intend to argue the Aristotelian opinion concerning the nature of light directly from first principles, but only to clear it of objections which several new experiments seem at first to raise against it. Whether our work contributes anything useful will be judged by the prudent and benevolent reader." Then in sixty propositions, many of them based on his own observations, he argues the case for substantiality.

Then comes Book II, "in which is established whether and for what reason one can uphold the Perapatetic opinion CONCERNING THE AC-CIDENTALITY OF LIGHT, though it does not follow from this that permanent colors are in any way different from light or that they reside in bodies when light is not there, as is commonly thought." This is an important point: color is not a quality of an object that gets brought along with light but a quality of the light itself. Book II contains only six propositions, aimed to refute the conclusions drawn in the first sixty. Father Grimaldi casts himself as a referee rather than a contestant on one side or the other. At the end he is saying, diplomatically, that though the Aristotelians are in one sense right, it is for the wrong reasons; but when he has finished, the old question still remains, wrapped in cloudy terminology, in the middle of a book of new observations and conclusions—a question that cannot be answered if it is not clearly posed: what kind of substance, what kind of accident?

Now let us look at the experiments. Grimaldi opens with his most important contribution: *Proposition 1. Light is propagated or diffused not only directly and by refraction and by reflection, but also in a fourth way, by diffraction.* Then he describes an experiment to show what he means by this new word, "diffraction."

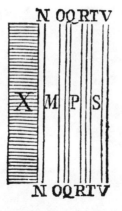

Figure 7.2. Grimaldi's drawing to show bands of colored light on each side of the shadow area (compare Plate 6).

Figure 7.1, copied from Grimaldi's book, needs a little study before it communicates anything. A and B are the top and bottom edges of a pinhole in a shutter high above the floor of an otherwise dark room; FE is a small opaque object placed in the narrow beam of sunlight that comes through; CD is a sheet of white paper on the floor.[1] Without the object FE, light from A spreads out in the cone ACL, and light from B spreads into BID. With the object in place, the doctrine of geometrically straight rays asserts that it will cast a shadow GL in the beam from A and a shadow IH in the beam from B. Thus the region IL should be shadowed, with the region GH in the middle of it perfectly dark. That is not what is observed. "The interval IL is significantly larger than it should be if all the light moved in straight lines inside a cone, interrupted by the opaque object EF." In fact, the shadowed region extends from M to N.

> Also, in the brightly illuminated regions CM and ND there are bands of colored light such that the center of each is pure white, whereas at the edges there is color, always blue on the edge nearer the shadow MN and red on the farther side. These bands depend on the size of the pinhole AB, and they do not appear if it is made too large.

The outer bands are narrower and fainter, and in some experiments there was also light barely visible within the shadowed region. Figure 7.2 is his drawing of what he saw.

Grimaldi's experiments showed another curious feature. Look at Figure 7.1. If one doubles the distance from the opaque object to the floor, one would expect that the distances MI, IG, and so forth would double also, the same diagram on a different scale, but in fact MI and LN do not double; thus the boundary ray that defines the edge of the shadow (if one may call it a ray) is not a straight line.

---

[1] Anyone proposing to repeat this experiment should have good facilities and plenty of time and patience. It is very difficult.

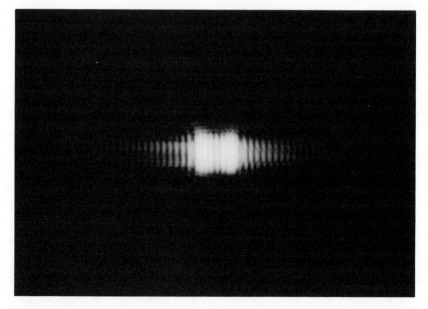

Figure 7.3. Grimaldi's experiment performed with monochromatic light and re-corded photographically. This is the shadow of a piece of wire in laser light.

That is what the eye sees. The camera (Plate 6) sees more, for a long exposure detects light that is very faint, and Figure 7.3 shows what happens when the sunlight Grimaldi used is replaced by a laser that gives a narrow beam of red light. One can see a line of light going down the center, but one can no longer even guess where the geometric shadow ought to be.

What is happening in these experiments? Grimaldi uses the familiar language of analogies to explain what he has seen. There is, of course, no vacuum; Aristotle has seen to that. Space is full, and "it seems that light is a kind of fluid [ether?] that moves very fast and sometimes passes through a transparent body in the form of a wave." We can think of sunlight as something that leaves the pinhole like a jet from a hose, but we do not see it until something interrupts its motion and causes ripples: "And just as a circular ripple is nothing but water quickly gathered together, heaped up, and followed by a double furrow, the aforesaid luminous bands consist of light unevenly distributed by the violence of the diffusion and separated by darker intervals." Visible light is here explained as the ripples in the jet, not the jet itself, and of course the ripples are accidental qualities of it. The impact of the jet itself explains the heat we feel in the sun's light. In the same vein Grimaldi explains how light goes through a piece of glass. Glass has fine holes in it through which the jet moves like a stream of water going through a sieve. The stream reforms itself on the other side, but the sieve causes a few ripples, and this is why even though most of the light goes through, one sees a little reflection from the polished surface. Note that we are not yet talking of waves that form a repeating series. Periodicity comes later.

Grimaldi also experimented with color and found that if he scratched the surface of a metal plate, the reflected light was colored; and in his book he wisely compares this effect with the iridescence of pigeon feathers, which produce color even though the fine parallel hairs of the feathers have no color themselves. Thus light can be colored in three ways: by diffraction, by reflection, and (as in a prism or a jewel) by refraction. Grimaldi argues that color is not a quality of the surface reaching us through a transparent medium as an *eidolon* or *species intentionalis* but rather a wavelike motion of the light itself which, in one way or another, it acquires at the surface. This was of course unacceptable to traditionalists who believed that light is itself a quality, for it was against the rules to talk about a quality of a quality. To the Aristotelians who surrounded him, color remains when light is taken away; Grimaldi here proposes that each color is a special motion of the ethereal stream reaching the eye. It is a question of definition. Most of us would naively say that a rose petal looks red because it *is* red. Grimaldi says no; the light reaching our eye has been made red by some property of the petal's surface, but that one should not say the surface is red. Take your choice.

As for what the eye actually does, there is no talk of the retinal image or anything else that Kepler wrote about. In the circles in which a Jesuit professor spent his life, Protestants existed only as part of a monstrous conspiracy against the Church. In a very long discussion Grimaldi shows that species cannot possibly explain vision, but no physiological mechanism is suggested that does explain it. He shows by a diagram equivalent to Figure 6.4 why we see the insect as if it were behind the mirror, but there is no talk of an "intentional" entity formed in the mind, only a remark that the diagram shows the efficient cause of the image we see but not its formal cause. That is, he cannot explain why we actually see the form of the insect. Aristotle still governs the soul. The experiments of Descartes and even of Father Scheiner lay with Kepler's work on the other side of a curtain of silence.

At the end of Grimaldi's Book I are several propositions relating to the rainbow. We will return to this subject in §8; I mention the fact here only because it is clear that though he never mentions Descartes, Grimaldi has read and understood his calculations. And having read what Descartes had to say, he knows it is possible to discuss the physical side of light without mentioning substances, accidents, species, or any other medieval paraphernalia. Perhaps this is why he never really tells us what he thinks about the ancient question. Perhaps he does not consider it is worthwhile to raise people's temperatures over a matter of words. At any rate, when he died of consumption at the age of forty-four his companions in the Society of Jesus wrote a memorial to him in which appear the words HE LIVED AMONG US WITHOUT ANY QUARREL.

Figure 7.4. Mold growing on the leaf of a rose, magnified about fifty times. From Robert Hooke's *Micrographia* [1667], courtesy of the Chapin Library, Williams College.

3. Robert Hooke and the Royal Society

In the same year, 1665, a book was published in England that was widely read though its scientific content was much smaller. The author, Robert Hooke (1635–1703), was a creator of experimental apparatus, scientific demonstrations, and useful devices. He seems to have been the first to put a spiral hairspring into a watch, he built an excellent vacuum pump, he made microscopes of astonishing quality—the list is long, but his education has skipped the mathematics that would have allowed him to go deeply into the new ideas that were boiling around him. In 1662 he was appointed Curator of Experiments for the newly formed Royal Society, a difficult job. The members—gentlemen, lordships, and an occasional bishop—sat in comfortable chairs and discussed one subject after another, sometimes very acutely: customs of a faraway country, an extraordinary circle around the sun, a petrified fish, a monstrous birth, the expansion of solids on heating, the change in volume of a sample of acid when metal is dissolved in it. In these meetings, questions often arose that could be settled by experiment, and one of Hooke's jobs was to design the experiment and perform it in front of the society the next time they met. If none were assigned, he was expected to invent some, "three or four considerable experiments," each time. A glance at Birch's *History of the Royal Society* [1756] for these years shows his prolific ingenuity, but he was unknown to the wider public until his book *Micrographia* appeared, showing how the world looks when viewed through a microscope and portrayed by a master draftsman. The instrument is an improved version of Galileo's and is illustrated in Figure 8.14. Figure 7.4 shows Hooke's drawing of mold growing on the leaf of a rose; other plates, too large to reproduce here, showed the general public for the first time how fleas, mites, and parts of plants really look. From these observations and from his experience with the Royal Society he drew a lesson for those who still sought to catch Nature in the net of Peripatetic words: "There is not so much requir'd towards it, any strength of *Imagination*, or exactness of *Method*, or depth of *Contemplation*, . . . as a *sincere Hand*, and a *faithful Eye*, to examine and record, the things themselves as they appear." *Micrographia* is a landmark in the history of optics, but its main importance for our purpose arises because it contains some speculations on the nature of light.

Hooke's theory, like those of Descartes and Grimaldi, is purely mechanical. He claims that light originates in the movement of particles of matter. For example, burning wood glows because its fabric is being torn apart by the heat of a flame, and this must make its particles move very fast. In general, when a hot object glows without being destroyed, it is because its particles are vibrating. This is clearly the case if the object is hard, for one can heat such a thing for a long time without changing its shape. But the vibrations are not like those of a bell; rather, the particles are imagined as lurching around so that irregular ripples are set up in the surrounding medium. What happens when these ripples encounter a transparent medium such as glass or water? Figure 7.5 shows Hooke's drawing of a narrow ray entering obliquely. Descartes has declared that light moves faster there, and in the medium you can see that the edge of the beam which reached the boundary first has pulled ahead, so that the ripples cross the ray obliquely. This, Hooke says, is what colored light is. Depending on which edge is ahead, its color is either red or blue; all other colors are mixtures. Thus he makes a modest though erroneous contribution to the idea proposed by Descartes and then by Grimaldi, that color is some modification of light and not a separate substance.

Later, in 1680–82, Hooke delivered a series of public lectures on light,

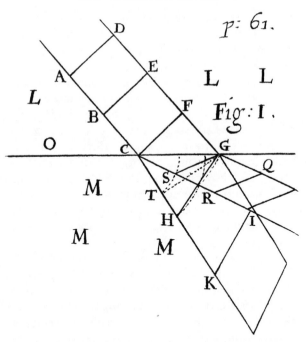

Figure 7.5. Hooke's explanation of the formation of colors when a beam of light enters a dense transparent medium. Ray CK is red light; ray GQ, blue light, bends the wrong way and violates Snel's law. From *Micrographia*, courtesy of the Chapin Library, Williams College.

in which these speculations are not mentioned. Perhaps he thought them too hypothetical for a London audience; perhaps he had decided that they were wrong. Nevertheless, as we now understand it, white light originates in the vibratory motion of the electrons inside atoms. In a heated source like the filament of an electric bulb they vibrate in a random way without changing their position very much, and the light they produce arrives in our eyes as a sequence of irregular ripples. Hooke knew nothing of electrons of course, and we shall see later that the process of radiation cannot really be described in such naive language. Thirty-two years earlier Galileo had written that light emerges from the hearts of atoms; Hooke advances a little farther.

All of us, in our soap-bubble days, have noticed the swirling colors that form on a bubble before it bursts (Plate 7). Water on wet pavement, if a drop of oil gets onto it, produces similar colors. Bubbles and wet surfaces are hard to experiment with, so Hooke studied thin sheets of mica, "Muscovy glasse," he calls it. Its flakes can be split into thinner flakes, and these again, until they are almost invisible. Toward the end of the process, when the flake is not too thick but not too thin, it shows the same colors as bubbles and oil films. Hooke had no way of estimating the thickness, but he could study the surface of a flake under the microscope. He found that if the flake was uniform, the color was uniform; but most flakes vary in thickness, and sometimes they split so as to form a staircase of thin layers, and here the colors ranged through the rainbow: red, orange, yellow, green, blue, indigo, violet, and then started again, repeating as many as ten or twelve times.

It is not just a thin layer of mica that produces color; a thin layer of air can do it as well. Hooke tells how he pressed together two surfaces of polished glass, flat but of course not perfectly flat, and saw "irises," as he calls them, irregular rings of colored light formed where the surfaces did not quite touch. Later he repeated the experiment using two glass plates thin enough so that increasing the pressure would slightly change their shapes.

> I prest them hard together til there began to appear a red-coloured spot in the middle; then continuing to press them closer, I could plainly see several rainbows (as I may so call them,) of colours, encompassing the first plate; and continuing to press the same closer and closer, at last all the colours would disappear out of the middle of the circles, or rainbows, and the middle would appear white; and yet if I continued to press the said plates together, the white would in several places thereof turn into black.

(If you try this experiment, use the thinnest available glass, make sure the surfaces are very flat and clean, press hard, and arrange not to get cut if anything breaks.)

How are the colors produced? Hooke realizes that they originate in the thin layer of air at the interface. Light reflecting from the top surface of this layer somehow combines with that from the bottom, but his account involves new hypotheses and is hard to follow. Before anyone could understand the formation of these colors two questions had to be answered: what is color, and how does colored light differ from white light? The next two sections will describe the steps by which Isaac Newton answered these questions and how he was then encouraged to attempt his own answer to the deeper question "What is light?"

4. Isaac Newton's Theory of Color

When Isaac Newton (1642–1727) arrived in Cambridge from his native Lincolnshire he was eighteen and brought with him an inquisitive mind, skilled hands, and very little money. His early notebooks show that at the university he was exposed to the diluted Scholasticism of those days, which though it leaked from a dozen holes still provided the only available vocabulary for discussing natural philosophy. But his reading notes, in Latin or Greek according to the book, tend to peter out after a few chapters and become collections of astronomical or physical data. They show Newton's interests drifting away from what he was supposed to be learning.

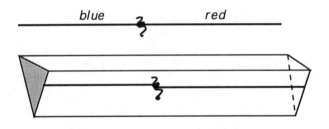

Figure 7.6. Newton's experiment showing that a prism refracts red and blue light differently. The image of two threads knotted together shows the two threads no longer in the same straight line.

First there was mathematics. Calculus was just being invented—formulas had been proved and a few dozen problems had been solved—but there was no system to it. The possibilities were limitless, new avenues of research led in every direction, and between 1664 and 1671 Newton produced about fifteen hundred pages of mathematics, as printed in the oversized volumes of his Mathematical Papers, that he thought worth keeping. Of the sciences that used mathematics, mechanics was in a rather primitive state, and optics had been opened up by Kepler's and Descartes's books on dioptrics with their brilliant calculations and farfetched hypotheses, ready for him to refute or appropriate. In his *Meteorology* Descartes describes how he used a prism to study the formation of color. Newton seems to have acquired a prism early with his meager funds, and Plate 8 shows how a prism spreads a beam of white sunlight into a narrow fan of colors. Prisms available at the time were mostly intended as toys—they contained air bubbles and were not carefully shaped—but with his toy Newton made an observation that pointed the way to all his later work in optics.

Figure 7.6 shows the experiment. He tied a blue thread to a red one and looked at them through the prism. (If you try this, hold the thread against a black background so as to eliminate extraneous colors.) The image of the red thread appeared a little lower than that of the blue one. This means that as Thomas Harriot had already noticed, color determines how much a ray of light is refracted, that is, bent, at a glass surface. Nobody at this time was very clear about where the colors come from when sunlight strikes a prism, but the general view was that the prism creates the colors. Here the colors were in the threads; the prism did nothing except deflect the beams of colored light coming from them to the eye. Newton concluded that the prism refracts blue light more strongly than red.

At this time Newton was working night and day on the creation of modern mathematics. He was the best mathematician in the world, the most creative and the most prolific, yet he was unknown except to a few people at the university.

He received his bachelor's degree in 1665, but soon afterward life in Cambridge was interrupted by two years of plague, which scattered students and teachers to the safety of country villages. Newton spent most of the time at his home in Lincolnshire, creating mathematics and beginning to wonder whether gravity had anything to do with the motions of planets. He probably also played with his prism and thought about light. Back in Cambridge, he was elected a Fellow of Trinity College in 1667 and, after two years, to a very well paid professorship of mathematics. He was then twenty-six years old, and, if he obeyed university regulations, he was set for life. He was supposed to give lectures, and so because nobody there could have followed his mathematics he gave a series on optics. Not many followed those lectures, either. The few who came encountered new experimental facts explained with difficult calculations. His duties required copies of his lectures to be deposited in the library and what he produced was clearly intended to become a book. But as we shall see in a moment, the *Optical Lectures* [1728a] were not published for about sixty years.

Isaac Newton moved into the spotlight as a consequence of an invention he made. The invention was a new kind of astronomical telescope, and it was not entirely his,[2] but the way he explained it interested everyone who owned a telescope or had tried to make one, for none of these earlier instruments were as good as they should have been; the more they magnified, the more the images they produced were blurred and ringed with color. Was it the fault of the lenses? It is much easier to shape a lens into part of a sphere than any other shape, and almost all lenses were made thus, but optical theory showed that they would suffer from spherical aberration. Descartes, and Newton himself, had tried to design lenses with elliptical or other profiles, but nothing seemed to help. Newton announced that, as I have mentioned in §6.3 (and see Figure 6.6), the edge of a lens acts like a prism and produces fringes of colored light, that this *chromatic aberration* is by far the main reason for bad images, and that as long as telescopes use big lenses nothing can be done about it.

He now came forward and presented the Royal Society with a telescope he had made which used a mirror instead of an objective lens and that therefore, except perhaps in the lens of the eyepiece, had no chromatic aberration at all. Figure 7.7 shows how the concave mirror focuses light into the eyepiece via a little flat mirror mounted in the middle of the tube. The instrument magnified about thirty-eight times (comparable with Galileo's best) and was two feet long, stubby compared with Keplerian telescopes of equal power. The weak point was the concave mirror, which had to be shaped and polished with extreme accuracy and which, being metal, tarnished quickly in the air of London. Two weeks later he sent another instrument, twice as long; but further work bogged down under the

[2] Newton explains (*Correspondence*, vol. 1, p. 153) that he got the idea of using a mirror from James Gregory's *Optica Promota* (1663), p. 94, but that Gregory's design was less efficient and much harder to build.

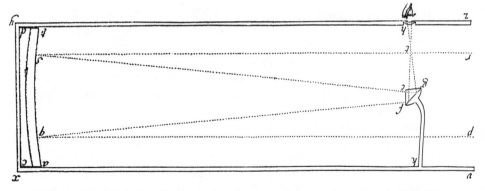

Figure 7.7. Newton's diagram of a reflecting telescope, from *Opticks* [1604].

difficulty of finding a metal that resisted corrosion and was hard enough so that its spherical curvature was not spoiled by occasional polishing.

Along with the telescope came a letter to Henry Oldenburg, Secretary of the Royal Society, announcing "a Philosophicall discovery wch induced mee to the making of the said Telescope, & wch I doubt not but will prove much more gratefull then the communication of that instrument, being in my Judgement the oddest if not the most considerable detection wch has hitherto beene made in the operations of Nature." This from an unknown young man just wading out of the Cambridge fens. Exclude the astronomical discoveries of Kepler and Galileo. Is this claim absurd? Read on; I leave it to the prudent and benevolent reader. The discovery was reported in a paper placed before the gentlemen of the society on 6 February 1671/2, which is so clearly written and so well adapted to its audience that I will quote it rather than the lectures given earlier in Cambridge that covered much the same material. Newton begins:

> To perform my late promise to you, I shall without further ceremony acquaint you, that in the beginning of the year 1666, (at wch time I applyed my self to the grinding of Optick glasses of other figures than *Spherical*,) I procured me a Triangular glass-Prisme, to try therewith the celebrated *Phenomena of Colours*. And in order thereto having darkened my chamber, and made a small hole in my window-shuts, to let in a convenient quantity of the Suns light, I placed my Prisme at its entrance, that it might be thereby refracted to the opposite wall. It was at first a very pleasing divertisement, to view the vivid and intense colours produced thereby; but after a while applying myself to consider them more circumspectly, I became surprised to see them in an *oblong* form; which, according to the received laws of Refraction, I expected should have been Circular.

First enjoyment, then a surprise, something unexpected, something to be explained. Perhaps Aristotle had sometimes felt like this, in the early morning of natural philosophy. Combine Newton's observation (Figure 7.8) with what he had seen in the little experiment with colored threads, which

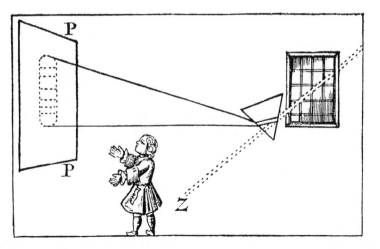

Figure 7.8. Newton's experiment with a ray of sunlight. The prism not only colors the ray but also displays the colors along a line. From Voltaire [1728].

showed that how much the prism deflects a ray of colored light depends on the color of the light. The new experiment again shows blue light refracted more than red, only this time the source of light is not a colored object but the sun. The colored rays diverge gradually. If the screen is put close to the prism, the sun's image is mostly white with a little red and violet at the top and bottom. At twenty feet, where Newton had it, the colors spread out into a beautiful spectrum.

What is going on? Newton considers two possibilities: color is a quality that light receives from the prism; or else white light is a mixture of all the colors of the rainbow, and the prism only serves to separate them. Descartes supports the first possibility: in his model of tennis balls the prism starts them spinning and thereby gives them color they did not have before. Newton has observed that spinning balls curve in flight, and so he does a simple experiment to see whether the colored rays curve after leaving the prism. They don't. Besides, in the experiment with the threads, the light is already colored when it reaches the prism, and the prism adds no more color. To make sure that color is separated, rather than created, by his prism, he performed what he calls a crucial experiment. It uses two prisms and is shown in Figure 7.9. G is a small round hole in the board DE, g is a small round hole in another board de, and the distance Gg is about twelve feet. By turning the first prism in his hand he could make different parts of the spectrum fall on the hole g, from which they passed through a second prism and were refracted onto the chamber wall. In this experiment the second prism received a color that is essentially pure, and the spot on the chamber wall was of the same pure color. Furthermore the spot was round; the second prism spread the beam of light no farther, and if light from the red end of the spectrum dc reached the wall at M, light from the

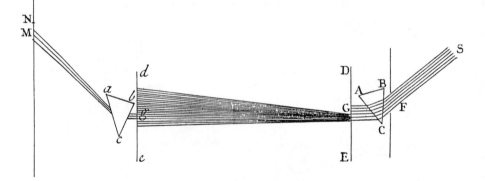

Figure 7.9. Newton's "crucial experiment," described in the text.

blue end was deflected a little more, to N. This is exactly what the experiment with the threads suggested, provided that the colors already exist in the sun's light. The second prism added nothing and changed nothing except the direction of the beam that passed through it. And the change in direction depended on the color; that is all.

> And so the true cause of the length of that Image [the spectrum of white light cast on dc] was detected to be no other, then that *Light* consists of *Rays differently refrangible*,[3] which, without any respect to a difference in their incidence, were, according to their degrees of refrangibility, transmitted toward divers parts of the wall.

There is a mathematical question here with which Newton does not bother his audience: How does he know that the prism does not distort the ray of light from the sun into an elongated spot which then somehow becomes colored? If so, his argument is destroyed. To cope with this he must calculate the shape that the sun's image would have if all its white light were refracted by the same amount. This is the subject of a mathematical analysis in the first of the *Optical Lectures*, and the result is that the image should be very nearly circular. It isn't; and therefore, he concludes, the reason why light makes an oblong image is that it consists of constituents that refract differently. One of the most valuable lessons to be learned from Newton is the importance of mathematically analyzing the results of an experiment in order to find out what it proves and what it does not.

Having thus established that "refrangibility" is the culprit, Newton discusses the relation between this quantity, which can be measured, and color, which by its nature cannot:

> As the Rays of light differ in their degrees of Refrangibility, so they also differ in their disposition to exhibit this or that particular colour. Colours are not *Qualifications of Light*, derived from Reflections, or Refractions of natural

---

[3] I.e., refracted by different amounts.

Bodies (as 'tis generally believed,) but *Original* and *connate*[4] *properties,* which in divers Rays are divers. Some rays are disposed to exhibit a red colour and no other; some a yellow and no other; some a green and no other, and so of the rest. Nor are there only Rays proper and particular to the more eminent colours, but even to all their intermediate gradations.

He has tried in every way to change the color of a ray separated from the rest of the spectrum as in Figure 7.9. He can change its brightness, but the only way he can change its color is to mix it with light of another color. He makes green by mixing blue and yellow light, and it looks much like the green produced by his prism, but the blue and yellow have not really combined, for a prism separates them out again. He calls these homogeneal and compounded colors,[5] and green is of course not the only one that can be compounded. But he does not consider white to be a color. Whiteness is a mixture, "a confused aggregate of Rays, indued with all sorts of Colours." Later we must critically examine this claim.

The microscope and the telescope allow us to see things that unaided sense can never see. Now there is a new instrument, the prism, which distinguishes between colors that look the same. How can the eye be fooled in this way? The ear is much smarter. If we combine two musical tones the ear hears them both and is not fooled. How is the eye different, and how does it distinguish one color from another? It was more than a century before Thomas Young began to answer that question.

So much for colored light; what about colored objects? Newton has found that when he illuminates a colored surface with any "homogeneal" color, the surface is not seen in its own color but takes on the color of the light, "with yet this difference, that [such surfaces] are most brisk and vivid in the light of their own day-light color." That is, a blue object has a surface that reflects blue and absorbs the other colors. If you illuminate it with blue light it reflects the light and looks blue. Under red light it looks nearly black.

> These things being so, it can no longer be disputed, whether there be colours in the dark, nor whether they be the qualities of the objects we see, no nor perhaps, whether Light be a Body. . . . But to determine more absolutely, what Light is, after what manner refracted, and by what modes or actions it produceth in our minds the Phantasms of Colours, is not so easie.

Having claimed that white light is nothing but a mixture of light of different colors, Newton could expect a challenge: well, mix them. He did this in several ways, but one is so simple that we can do it anywhere that a prismatic bit of glass projects a spectrum on a wall or a piece of paper. These

---

[4] We would say innate.

[5] *Homogeneal* is Newton's word, but *pure* rings more sweetly in the ear, and I will generally use it.

surfaces reflect light, which is why they do not look black, but unlike a mirror they reflect it in every direction. Hold a second piece of paper near your spectrum to see if it catches the reflected colors. Unless it is held very close you will find it shows no color at all; they have all recombined into white. Try it.

In a series of communications to the Royal Society, Newton continued to pour out experimental results. Clearly, for him light is substance, not accident. But what kind of substance? This is a hard question to answer, for though it is easy to describe what happens in an experiment, the outward manifestations of a thing rarely tell you what it is made of. In the quest for fundamentals, the searchers of those days carefully distinguished two levels of propositions, Newton called them theories and hypotheses. A *theory* is a proposition whose truth is not at once evident from observations but can be deduced from them. If the observations are correct, the theory is true. A *hypothesis* is a statement that goes farther. It is suggested but not required by experimental facts, and it is intended as a single statement that clarifies many or all of the relevant experiments.[6] In his first letter to the Royal Society Newton lists several propositions, of which one is the relation between color and degree of refraction and another is his conclusion that white light is a mixture of many colors, and he claims that they all follow necessarily from the experiments he has done. Necessarily?

In recent years critics of science have pointed out more and more forcefully that a scientific experiment, by itself, rarely resolves a disputed point; the reason is that to make any sense of an observation one calls upon assumptions that may not have been not stated. Let me give an absurd example. I look out my window at the clock in the church tower across the street. It says nine o'clock, and I conclude that unless something is wrong with the clock that is what time it is. I know that light takes an instant to go from the clock to me, but its speed is so great that the little delay makes no difference. I know it because I am familiar with the behavior of light, as experienced every day and as represented in the theory of electromagnetic radiation (Chapter 9). But what if I am mistaken? What if light, here and on this particular morning, takes half an hour to get from the church clock to my eye? I dismiss the possibility, but it illustrates the fact that every observation staggers into our consciousness loaded with the hypotheses that tell us what it means. Newton's conclusions from his marvelous experiments depend on assumptions he never discusses: for example, that light travels in the form of rays. And because a ray is by its nature a thing rather than a quality, he has already assumed that light is a substance with quali-

[6] On the last page of Newton's *Principia* occurs his most often quoted remark, *Hypotheses non fingo*, "I do not invent hypotheses"—more exactly, "I do not snatch hypotheses out of the air." This echoes another well-known dictum of Francis Bacon (Bacon, *Novum organum* II.10), *[Non] fingendum, aut excogitandum, sed inveniendum, quod Natura faciat, aut ferat*, "We are not to invent what Nature does or may be made to do, or deduce it from arguments, but instead we must find it out."

ties when he concludes that "it can no longer be disputed . . . whether Light be a Body." This theory is the most economical one that can be drawn from the observations and the assumptions that Newton might reasonably have made. For more than two centuries it was accepted by the experts and challenged only by a few outsiders such as Goethe (§8.5); yet for the last hundred years we have believed that to the extent that it has been clearly stated, it is wrong.

But what light is, what color is, these are not so easie.

5. Particles of Light

What Newton had told the members of the society was news to them, and so the secretary asked Robert Hooke, older and better known, to give them his reactions. What he said about Newton's color theory can be summarized quickly: the experiments are beautiful, but when he claims they prove that white is "a confused aggregate of rays," he goes too far. Because Hooke understands Newton as saying that a ray of white light consists of particles of different colors, he challenges him to produce a mixture of colored powders that comes out white. Hooke argues that the experiments are explained just as well if one assumes, as in his *Micrographia*, that "light is nothing but a pulse or motion. . . . And that Colour is nothing but the Disturbance of that light . . . , that whiteness and blackness are nothing but the plenty or scarcity of the undisturbd Rayes of light, and that the two colours [red and blue] . . . are nothing but the effects of a compounded pulse or disturbed propagation of motion caused by Refraction." Hooke does not see why the colors should all lie hidden in white light, any more than "all those sounds must be in the air of the bellows which are afterwards heard to issue from the organ-pipes."

Newton has not told the society what light actually is, but in his reply he moves closer to Hooke as he sets out to produce a theory that covers the facts. The first fact, the most obvious one, known from antiquity, is that light moves in straight lines. I can hear you from around the corner of a building but I cannot see you. This is what Descartes's theory of pressure and Hooke's theory of pulses that are just brief fluctuations of pressure cannot explain. (Later, in his *Principia* [1687], Newton proves this with considerable pomp, but it is obvious anyhow.) Therefore he conjectures that light consists of particles that move through space in straight lines. Second, when these particles encounter a refracting or reflecting surface they set up vibrations in the particles that compose it. These in turn

excite vibrations in the Æther of various depths or bignesses, wch being promiscuously propagated through that Medium to our eyes, effect in us a sensation of light of a white colour; but if by any means those [vibrations] of unequall bignesses be separated from one another, the largest beget a sensation of a Red colour, the least or shortest of a deep Violet, & the intermediate ones of intermediate colours.

It is clear from the text surrounding this passage that "vibrations" denotes what we would call a wave and "bigness" refers to the length of the wave, the distance between its successive crests, and here Newton has snatched a great discovery from under Hooke's nose, the physical property that distinguishes lights that seem to us of different colors. Recall what Democritus said in §1.3, that color and other sensations exist by convention, that in reality there are only atoms and emptiness. If this wave is a wave in ether one might say it is a property of emptiness, though that distorts Democritus's meaning a little. But the point is that whereas red and green look completely different to an ordinary eye, Newton is saying, here and in the quotation that begins this chapter, that the wave we call red is exactly the same kind of thing as the one we call green, except that its "bigness" is about 15 percent greater. The great difference resides in our sensory apparatus, not the light itself.

Hooke might have guessed the connection of color to wavelength several years before Newton, had he talked about waves instead of pulses.[7] And not only has Newton understood that there is a relation between color and wavelength; he has got the relation right: the longest waves produce the red end of the spectrum as we see it. It is clear to Newton that some sort of wave theory will explain Hooke's observation that the colors of thin films range repeatedly through the spectrum as the films are made thinner and thinner: the thickness of the film is a whole number of waves of whatever color it is, but because waves travel around corners and light does not, he decides that such a simple explanation cannot be right.

The exchange of views continued as Newton's radical theory of color was criticized by scholars from England and the Continent. The tone of these letters, especially those from Newton and Hooke, sometimes shows an impatience that goes a little beyond scholarly enthusiasm. But Hooke was burdened with his job, and Newton, while he was pushing back the frontiers of mathematics at incredible speed, had to produce more than twenty-four thousand words during the next fifteen months to defend and explain his ideas on light. It is not surprising, when each word provoked fresh objections, that he gave up preparing his *Optical Lectures* for publication.

What, then, is light? Is it the particle, or is it the wave? The methodology expounded by Francis Bacon and promoted by the Royal Society was designed to liberate scientific discussion from plausible guesses and arguments by analogy. This was necessary in that period of transition, but in

[7] I think Hooke was impeded by lack of a word. "Wave," in the seventeenth century, referred only to ocean waves or something that looked just like them, such as waves when one shakes a long piece of cloth. Newton's word "vibrations" is clear only if one knows what he means. In later writings he sometimes says "wave," but the word's general use in the sense applicable here begins in the nineteenth century.

fact it is often by guess and analogy that creative scientists find new ideas. A guess is made, and experiment shows the guess is right, wrong, or maybe; the argument goes forward or back. Newton's head was full of ideas, but his optical writings explain them in a minimal way that is ostentatiously free from hypotheses. In 1672, responding to critical letters from Father Ignace Gaston Pardies of Paris, he tries to define light without committing himself:

> I understand light to be any entity or power of an entity (whether substance or some force, action, or quality of it) which proceeds directly from a bright body and is able to excite vision: and I understand rays of light to be the smallest parts, or at least the indefinitely small parts of it, which are mutually independent, as are all rays which luminous bodies emit along straight lines either successively or at the same time.

Lines do not move, but Römer has shown that light moves. Then what is this ray? Newton's carefully chosen words are not easy to paraphrase, but they came to be interpreted as meaning that a beam of light is a bundle of discrete paths with discrete particles moving along them in the manner of cars keeping to their lanes on a highway. A beam of light may be made smaller and smaller in diameter and, if one imagines chopping it by opening and then closing a shutter, also in length. Can this process be continued indefinitely? Newton says no, there is a smallest unit of light, in length, breadth, and thickness (one car on one lane), and this is what he calls a ray. I think most people would call it a particle. A hypothesis lurks in Newton's definition: there exist particles of light. Why not? Matter is made of particles, and Nature, as Newton says many times, is very conformable to herself. But whatever these rays are, we do not see them; we see colors, and these are communicated by waves, not rays. They "cause a sensation of light by beating & dashing against the bottom of the eye," just as sound waves affect the ear. Where does he get this idea? I just said that creative people argue by analogy all the time, but he also knew about Hooke's experiment with thin films, how as they are made thinner and thinner the color they show goes down through the spectrum from red to violet and then starts again, a sign of spatial periodicity. What, then, is the connection between waves and rays? They are very different. Waves spread out in every direction from a disturbance (think of what happens when a stone is tossed into a pool). Particles move in a straight line and arrive at some definite point. As I understand Newton, he supposes that though the wave may go everywhere, we perceive it only at the point where a ray arrives. Thus light communicates itself to our view in straight lines, even if perhaps it does not move that way.

In 1675 Newton wrote a long letter to the Royal Society headed "An Hypothesis Explaining the Properties of Light." It tells of the experimental work he has done in the last three years and proposes a hypothesis to explain what he has found. He does this unwillingly, for he knows better than anyone how wrong he may be, but, he says, "I have observed the heads of some great virtuoso's to run much upon hypotheses, . . . & found, that some when I could not make them take my meaning, when I spake of the nature of light & colours abstractedly, have readily apprehended it when I illustrated my Discourse by an Hypothesis." There follows a long fantasy in which light consists of particles whose difference in some respect, perhaps mass or size, corresponds to differences in perceived color. The impact of these particles produces waves in material substances as well as in ether, both of which react back on the particles to change their motion. With this imaginative picture he is able to qualitatively explain why light is refracted at the surface of a transparent medium, why thin films are colored, why light diffracts into a shadow, in fact most of what was known about light. Qualitatively—there are few numbers, and he must have been keenly aware that numbers were the rocks that could sink his boat.

6. *Opticks*

It was many years later, in 1704, that Newton published the book *Opticks*, which contains the final account of his experimental labors and the conclusions he drew from them. It is written in English, much of it in the form of a narrative of experiments he has done. Later editions appeared in 1717, 1721, and a last one in 1730, "corrected by the Author's own Hand, and left before his Death with the Bookseller." They are almost identical except for the last part of Book III. Newton's head was full of ideas for research that he would never do, hypotheses that he could not prove or disprove but that should not die with him. Starting with the second edition he begins to list them in the form of queries, so no one could accuse him of making statements he could not defend. By the fourth edition there are thirty-one of them. I will mention a few in the following paragraphs, as counterpoint to the recital of experimental facts.

Opticks begins:

> My Design in this Book is not to explain the Properties of Light by Hypotheses, but to propose and prove them by Reason and Experiments: In order to which I shall premise the following Definitions and Axioms.
>
> DEFIN. I
>
> By the Rays of Light I understand its least Parts, and those as well successive in the same Lines, as Contemporary in several lines.

This is the same hypothesis made in his letter to Pardies more than thirty years earlier. It is more than a definition, it is a hypothesis; and clearly it arises from his belief that even if light is not actually a material substance, its structure must somehow be analogous to the atomic structure of matter. Nature is conformable to herself, and fertile minds abound in analogies. . . . The hypothesis plays little part in the rest of his explanations, and anyone may guess why he began in this fashion. In Book III he molds the definition into something more specific: *Query 29. Are not the Rays of Light very small Bodies emitted from shining Substances? For such Bodies will pass through uniform Mediums in right Lines without bending into the Shadow, which is the Nature of Rays of Light.* In Newton's heart, light is a material substance.

Opticks continues with definitions and some axioms that include Snel's law of refraction. With this he can discuss simple convex lenses, and he gives the formula, which he could have taken from Kepler but probably worked out for himself, that tells where the lens focuses its image. He explains what Alkindi told us long ago, that the image is formed point by point, corresponding to the visible points of the object. He describes the experiments which persuaded him that white is a composite of colors corresponding to rays that refract by different amounts, and then he explains refraction. The idea is essentially Descartes's, only Newton doesn't use a tennis racket to deflect the particles of light. Instead, he assumes that at the surface of a transparent medium there is a very thin region where a force acts to pull the rays of light down into the medium. He can then derive Snel's law, concluding what Descartes assumed, that light travels faster in the medium than in air or vacuum. On this model he can easily explain why different colors refract differently and why the same color refracts differently in different media. Figure 7.10a (taken from *Principia*) shows a ray coming down out of a piece of glass into the air below it. The force acts upward, and the ray is deflected away from the perpendicular. Figure 7.10b shows what happens if the ray arrives too obliquely or if the upward force is too strong: the ray is reflected. *Query 1. Do not Bodies act upon Light at a distance and by their action bend its Rays; and is not this action (*cæteris paribus*) strongest at the least distance?*

The same hypothesis explains diffraction. While he was preparing *Principia*, published in 1687, Newton repeated some of Grimaldi's experiments, and Figure 7.10c, in which AsB represents the edge of a knife, shows how the rays are deflected. But the situation is more complicated than this, for Newton, like Grimaldi, had found three colored fringes at the edge of the shadow. *Query 3. Are not the rays of Light in passing by the edges and sides of Bodies, bent several times backward and forwards, with a motion like that of an Eel? And do not the three Fringes of colour'd Light above-mention'd arise from three such bendings?*

But things are still more complicated, for the shadow cast by a small object (Figure 7.1) is larger than the geometric shadow would be. This implies that the force exerted by the material object is not attractive but repulsive, and Figure 7.11, from *Opticks*, shows what happens. Then is the force sometimes one, sometimes the other? Newton doesn't know and at this point in his life hasn't time for more experiments. He lets the matter

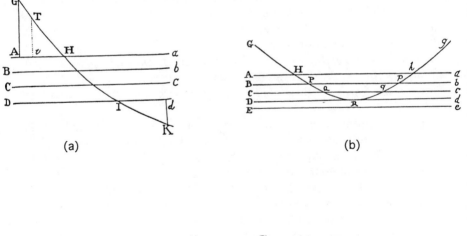

(a)   (b)

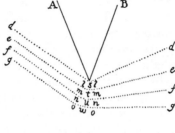

(c)

Figure 7.10. Three figures from Newton's *Principia* [1687], showing how a ray passing downward from glass into air experiences forces that (a) refract or (b) reflect it. Figure (c) explains diffraction by a sharp edge.

drop with a guess: *Query 4. Do not the Rays of Light which fall upon Bodies, and are reflected or refracted, begin to bend before they arrive at the Bodies; and are they not reflected, refracted, and inflected [i.e., diffracted] by one and the same Principle, acting variously in various Circumstances?*

Part II of Book I discusses color, and Newton repeats his conviction that once a ray of a single color has started out, nothing can be done to change the color. "If the Sun's Light consisted of but one sort of Rays, there would be only one Colour in the whole World."

But there are many colors in the world; how are they produced? The simplest case to study is the color of thin films of oil or mica, first investigated by Hooke. To explain these colors Newton has first to quantify what happens. By a diabolically clever experiment known as Newton's rings in which the film is a very thin layer of air between two suitably shaped pieces of glass, he can calculate the exact thickness of these infinitesimal layers (a typical value is $1/88{,}000$ of an inch) and finds perfect periodicity: if the film

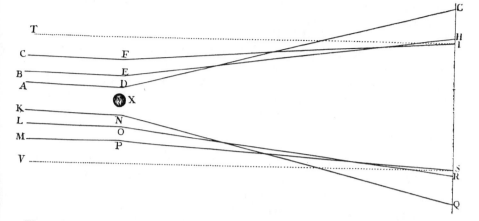

Figure 7.11. Another attempt, from *Opticks*, to explain diffraction by a needle (shown in cross section).

glows yellow at a certain thickness, then it will be yellow at twice that thickness, three times, and so on.

To explain this periodicity Newton gives us his strangest hypothesis, that when a ray of light (remember that it is essentially a stream of particles) arrives at a surface it is put into a sort of vibration in which it oscillates between two states to produce what he called "fits of easy transmission and reflection." In one state it will pass through the next surface it encounters, in the other it will not. If a film is thick enough to allow one oscillation of a ray of yellow light, or two, or more, before it reaches the next surface, the ray will go through; otherwise it will reflect. Similarly, if the film is $1/2$ this thick, $3/2$, $5/2$, and so on, it will be reflected from the second surface but transmitted when it reaches the first one again and so return in the direction from which it came. Newton showed that if this is what happens, he can explain not only the colors of these films but all colors. He assumes that "the least parts of almost all Natural Bodies are in some measure transparent" and that "the transparent parts of Bodies, according to their several sizes, reflect Rays of one Colour, and transmit those of another, on the same grounds that thin Plates or Bubbles do reflect or transmit those rays. And this I take to be the ground of all their Colours." If the petal of a rose is red, its "least parts" near the surface reflect that color just as Hooke's oil film does; the other colors pass through and get lost in the mass of "least parts" beneath.

When Newton died he left behind his *Principia* and *Opticks,* and also several books he had written and not thought fit to publish: *Optical Lectures* [1728a], the *Chronology of Ancient Kingdoms* [1728b], a treatise on astronomy called *The System of the World* [1728c], and *Observations on the Prophesies of Daniel* [1733]. *Principia* [1687], almost from the beginning, has been more praised than read, for within twenty years Newton's cumbersome geometrical constructions had been largely replaced by the supple and easy methods of calculus to which he had contributed so much. *Prophesies* appealed to a limited audience, and the *Chronology*, which depended on accepting ancient legends as literal fact, was taken seriously by only a few. But *Opticks*, with its classic experiments interspersed with daring hypotheses, lives on. For over a century, especially in England, investigators believed it in the way their forbears had believed Aristotle, and this belief had to be fought, or accommodated, by anyone seeking to advance the study of light.

The speculations in *Opticks*, as well as others in *Principia* and the posthumous works I have mentioned, show an imagination that jumps the walls of reasoned discourse and forages in the unexplored territory beyond. As an example of the thoughts that ran in Newton's mind while he was carrying out his experimental program there is an interesting manuscript in the Library of Congress, known by the words of its first line, "Of Natures Obvious Laws and Processes in Vegetation"[8] It is a private memorandum of twelve pages setting out some assumptions about the nature of the world and has been dated at about 1670, the same time that Newton was making his reflecting telescope and preparing his first paper on color for the Royal Society.

The manuscript seeks to explain changes of inorganic matter using concepts and terms that describe processes of life. Here are a few lines from it.

Thus this Earth resembles a great animall, or rather inanimate vegetable, draws in æthereall breath for its dayly refreshment & vitall ferment & transpires again wth grosse exhalations. And according to the condition of all other things living ought to have its times of beginning youth old age and perishing.[9] (This [breath] is the subtil spirit wch searches the most hidden recesses of all grosser matter which enters their smallest pores & divides them more subtly then any other natural power w^t ever. (not after the way of common menstruums [solvents[10]] by rending them violently asunder &c) this is Natures universall agent, her secret fire, the onely ferment & principle of all vegetation. . . . This spt perhaps is the body of light 1 becaus both have a prodigious

---

[8] I.e., of the natural laws and processes that can be seen operating in living matter. I hope I have correctly transcribed Newton's hasty scrawl.

[9] The living Earth was not a new idea. For Newton's older contemporary Otto von Guericke, whom we will meet in Chapter 9, the hypothesis explains facts that are otherwise mysterious: that the Earth keeps spinning and moving in its orbit and never slows down, that it stays warm when there are no heat sources outside it, whereas the fact that like any organism it grows a little introduces an irregularity into its motion that explains the observed precession of the equinoxes.

[10] The word shows the assumptions behind it. In the medieval physiology of reproduction, the male contributes semen, which is dissolved on contact with the female's menstrual fluid. Fertilization cannot take place until the seed has been destroyed, and the same process is imagined here: the properties of a substance must be destroyed before it can be born again as something else.

active principle, both are perpetuall workers 2 becaus all things may be made to emit light by heat, 3 the same cause (heat) banishes also the vitall principle. 4 tis suitable wth infinite wisdom not to multiply causes without necessity. 6 Noe heat is so pleasant & beamish[11] as the suns. . . . Noe substance soe indifferently, subtily, and swiftly pervades all things as light & noe spirit searches bodys so subtily percingly & quickly as the vegetable spirit.

The proposal "This spirit is perhaps the body of light," equating spirit in one conceptual domain with body in another, fills the gap in Newton's reasoning. The spirit that vivifies Nature in the philosophy the young Newton had imbibed in Cambridge is not enough, in itself, to make anything happen. Physical action needs body to cause it, and light must have body, for sunlight warms our hands and excites our organs of vision. Once more, in the tradition of Robert Grosseteste, light provides the force that runs the world.

Newton sought the link between the living world and the world of inanimate matter over many years in the study of alchemy, and even in the *Opticks* there is a trace of it. Here is the beginning of one of the last queries: *Query 30. Are not gross Bodies and Light convertible into one another, and may not Bodies receive much of their Activity from the Particles of Light which enter their Composition?*

A very striking thing: all Newton's particles of light move at the same speed. (We know this because if they did not, the light from the planets would look like colored streaks spread out across the sky.) That is not the case with material bodies; we can throw them hard or gently, but no one has ever seen slow light. In the letter of 1675 that I have mentioned, Newton suggests that light rays are "continually urged forward by a Principle of motion, wch in the beginning accelerates them till the resistance of the Æthereall Medium equal the force of that principle," at which point they attain constant velocity. At any rate their dynamics is obviously not that of ordinary material bodies, even though they enter into the composition of matter. Light is a special substance, created at the beginning of the world, and it continues to play a special role in the operations of Nature.

Newton's efforts to understand the physical principles of life came to nothing, but I think the study of light was an important part of this effort. The queries ponder what ether is, how it acts, what it does: *Query 23. Is not Vision perform'd chiefly by the Vibrations of this Medium, extended to the bottom of the Eye by the Rays of Light, and propagated through the solid, pellucid and uniform Capillamenta of the optick Nerves into the place of Sensation? . . .*

Query 24. Is not Animal Motion perform'd by the Vibrations of this Medium, excited in the Brain by the power of the Will, and propagated from thence through the solid, pellucid and uniform Capillamenta of the Nerves into the Muscles, for contracting and dilating them? I suppose that the Capil-

---

[11] This jabberwockian word is in the dictionary and means what it seems to mean.

lamenta of the Nerves are each of them solid and uniform, that the vibrating Motion of the Æthereall Medium may be propagated along them from one end to the other uniformly, without interruption. . . .

Ether is what was called spirit in the old philosophy.[12] Its vibrations transmit light signals from the outside world into the consciousness, and also from the brain to the muscles. Thus our experiments with light and vision should help unlock the mysteries of animal motion and other actions of the nervous system which occur entirely within the body and cannot be studied directly. There was much to be learned from the study of optics, and he must have regretted finally having to publish his book and let it go.

Newton died in 1727, full of honor, and is buried in Westminster Abbey. His talents were extraordinary, perhaps unique in history. As Einstein said of him, "In one person he combined the experimenter, the theorist, the mechanic, and, not least, the artist in exposition. He stands before us strong, certain, and alone: his joy in creation and his minute precision are evident in every word and every figure." As a mathematician he has few equals. In formulating the law of gravity and the principles that govern the motion of material bodies, he used mathematics to express profound insights into the nature of the physical world. Working in alchemy and physiology where he had neither experiment nor mathematics to tell him what was happening, hypotheses boiled up out of his imagination. They tended to take control of him. It did not matter that in his writings he was scrupulous to label them as hypotheses and thus only tentative; they possessed his mind and kept him from considering other people's thoughts. It is amazing to see how little his idea of what constituted a light ray changed in the more than fifty years between his letter to Father Pardies in 1672 and the last edition of *Opticks*. And though a wave theory of light (to be discussed in the next section), which ultimately triumphed over Newton's rays, was available for his study and comment for fourteen years before *Opticks*, he never mentioned it.

Loosely translated, Newton's epitaph in Westminster Abbey reads, "Let all men rejoice that so great a glory to the human race has now appeared." It is hard to imagine what the world would be like today if he had not lived. As Isaiah Berlin has written,

> The entire program of the Enlightenment, especially in France, was consciously founded on Newton's principles and methods, and derived its confidence and its vast influence from his spectacular achievements. And this, in due course, transformed—indeed largely created—some of the central con-

---

[12] The last paragraph of *Principia* is a hymn to the powers of spirit, and a few pages earlier we find a hypothesis as to where it comes from: comets. Newton had been troubled by the question "Why do comets exist, and what part do they play in the great design?" He suggests that comets' tails shower Earth with moisture to replace what is being lost all the time by evaporation, and "I suspect moreover that it is chiefly from comets that spirit comes."

cepts and directions of modern culture in the west, moral, political, techno-
logical, historical, social—no sphere of thought or life escaped the consequences of this cultural mutation.

7. Christiaan Huygens, the Dutch Master

The gospel of Newton was that light involves a wave and a particle, but that
the particle governs the wave. The connection between them is obscure.
Newton never tried to make it clear, and imagination offers no clue. The
particle was necessary because light moves in straight lines—but it doesn't
always, for Grimaldi had made it bend a little. In the next few pages we will
return to a man older than Hooke or Newton who toward the end of his
life published a small book proposing a solution to Newton's mystery:
there is no particle.

When you read Christiaan Huygens (1629–95), even if the subject is
something like ether and you don't necessarily agree with what he says, you
still feel that you are reading someone who thinks the way modern people
do. His work in any one of the fields of mathematics, dynamics, optics, and
astronomy would have made him great, and hardly a word of it needs to be
excused or set aside as contaminated by the seepage of old ways of thinking.

He came from a family of eminent public servants; his father Constantijn
was a poet and composer as well as a diplomat, a friend of Descartes, and
often his host when he came to Holland. Christiaan and his brother Constantijn, who helped in many of his studies, were educated in languages and
music, and concerning the natural world they had the immense advantage
of being largely self-taught. Christiaan had a brief period of university life,
but from the age of twenty-one to thirty-seven he lived and worked and
wrote at home. He learned from Descartes, and though he ended up differing from him on almost every important point, he adopted the essential
Cartesian program as his own.

Huygens's first work was in mathematics. Fermat, a generation older,
had laid the foundations of differential and integral calculus, but because of
his reluctance to publish few people knew what he had done. Huygens did
much of it again, and then turned to the theory of probability, which in
that age was the urgent concern of insurers and gamblers. Next, he and his
brother started making microscopes and telescopes, and succeeded so well
that Christiaan discovered Saturn's rings as well as one of its satellites,
Titan. (Galileo and others had seen that Saturn had some sort of ears sticking out from it, but had not been able to see what they were.) He was of
course a strong Copernican, and living in a Protestant country outside university circles he did not have to waste time defending his views. In the late
1650s, when Newton was still a schoolboy, Huygens became interested in
making clocks, and in the course of his studies of pendulums he found the

formula that gives centrifugal force. Newton rediscovered it some years later and used it in his theory of planetary motion. (Huygens's argument anticipates Einstein rather than Newton, but we cannot go into that here.) Presently he was invited to Paris with a stipend from Louis XIV, and when the Académie des Sciences was founded he became its most prominent member. Hearing swords being sharpened he left there in 1681, the same year as Römer, and passed his remaining years in the Hague.

The work that concerns us here is mostly to be found in his *Treatise on Light* [1912], written in 1678 and first published in 1690. It is a particularly important step in the program that sought to explain Nature in terms of simple mechanical interactions between structurally simple pieces of matter. We have already seen examples of this program in Descartes's vortex theory of planetary motions and in the three analogies with which he sought to explain light, but Huygens's superior physical insight enabled him to do much better.

The *Treatise* states its mechanistic principle at the beginning: light is matter in motion. Yet it is not a beam of material particles, any more than sound is. Sound is air in motion, and air is matter, but air does not travel from the mouth to the ear. If light were a beam of particles, Huygens argues, there would be collisions between one beam of light and another, and this never happens. Sounds do not collide; the motion of light must be similar to that of sound but not identical, for we must think of the processes in which light originates. You will remember that Galileo and Hooke thought of light as originating with atoms; Huygens is less specific, but his theory makes the same assumption. I do not know whether he derived this idea from anyone else, but he must have seen Hooke's *Micrographia*.

The other premise is that space is full. This is Aristotle's premise, but the philosophers of the seventeenth century adopted it for a different reason: they were mechanizing the universe, and mechanisms are easiest to understand when one part touches another. Huygens distinguishes between air and ether. If you put a bell into a closed jar and pump out the air you can no longer hear the bell, but you can still see it; therefore the ether that carries light is not removed by the pump. This is because its particles are so small that they move easily among the particles of solid matter. Sound is a wave of pressure and motion in air; light is the same in ether. Why does light travel so much faster than sound? It is because the particles of ether are so very hard that when there is a little tap on one side of a particle of ether the motion is transmitted at once through it to the particle on the other side. Particles of air are much softer and transmit the impact more slowly.

Now we must consider how light starts out from its source. The source consists of particles, each of which, when the source is hot, rattles around, communicating its motion to nearby particles of ether. There is no order in

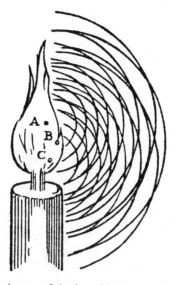

Figure 7.12. Huygens's picture of ripples of light spreading from three arbitrarily chosen points of a candle flame. You can see how the ripples begin to combine to form a smooth front. From *Treatise on Light* [1790].

the motion, and yet from the entire source something like a smooth wave of light comes out. How does this happen? Figure 7.12 shows ripples of motion spreading out from three of the innumerable luminous points in the flame of a candle. Huygens cautions us not to misread this diagram. The waves from a given particle are not a steady and uniform series like sound waves from an organ pipe, for the particle that produced them is not in any sort of steady motion; it is, as I said, just rattling around. Instead, the lines that suggest waves in this and subsequent pictures represent a series of snapshots of the same ripple as it spreads out in space. But notice what happens as the waves move farther from their source. The chaos of ripples smoothes out, and if one imagines not three particles of the flame but millions, it becomes smoother still. By the time they reach us from a distant star they will have formed a single, simple ripple.

Two waves can cross, as one sees in shallow water on a beach. In the same way, two people can look into each other's eyes without interference, but this is not easy to explain "according to the views of Mr. Des Cartes, who makes Light to consist in a continuous pressure merely tending to movement. For this pressure not being able to act from two opposite sides at the same time, against bodies which have no inclination to approach one another, it is impossible so to understand what I have been saying about two persons mutually seeing one another's eyes, or how two torches can illuminate one another."

Huygens is not quite sure how to explain the transparency of glass— whether the particles of glass vibrate and produce a new wave on the other

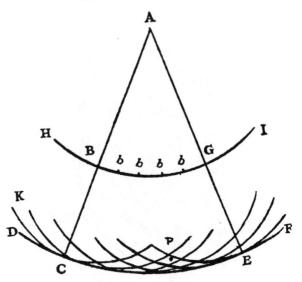

Figure 7.13. Huygens's construction, showing how waves along a smooth front BG generate the next front DF: each point of BG acts as the source of a new wave. From *Treatise on Light*, point P added.

side, whether the ether inside continues to transmit its vibrations uninterrupted, or whether ether and glass vibrate together—but he prefers the third alternative, since it explains why some substances, particularly metals, are not transparent. These, he proposes, consist of a mixture of very hard particles from which light bounces, together with soft, spongy ones which absorb the activity of whatever light penetrates the surface and eventually damp it out.

We have seen how Descartes explains the refraction of light entering a transparent medium from air by modeling it as a tennis ball that speeds up as it enters the medium. Newton assumed a force exerted by the medium that has the same effect. Can Huygens explain it in some way that does not need these aids? For this we must think more carefully about how light actually moves.

Let me recall how Kepler described the motion of light. As was shown in Figure 6.3, he asked us to think of a surface moving outward along the rays that issue from the source, expanding as it goes. Huygens now explains the nature of this surface: it is a wave, in three dimensions, analogous to the long straight swells that one sometimes sees on the ocean's two-dimensional surface. And furthermore, he can explain how it moves.

Figure 7.13 shows a wave that starts from A. A moment later, it has arrived to form the wave front HI, which is a surface of the kind shown in Figure 6.3. Another moment and it is at DF. How did it move from HI to DF? To answer, consider some particle of ether at one of the points marked b. It receives a jiggle from the particles immediately behind it, which it

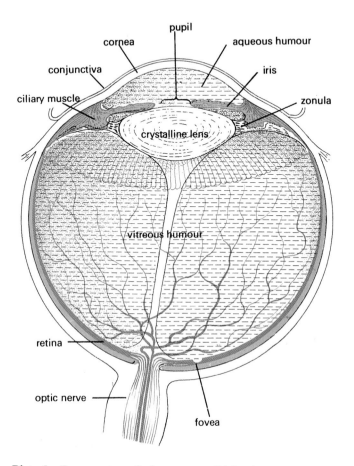

Plate 1. Cross section of a human eye. (Richard L. Gregory, *Eye and Brain*, 4th ed. [Princeton, N.J.: Princeton University Press]. Copyright © 1990 by Richard L. Gregory. Reproduced by permission of R. Gregory and Princeton University Press.)

Plate 2. Mosaic, c. 1150, in the Cathedral of Cefalù, Sicily. The Bible is open to the words, in Greek on one side and Latin on the other, "I am the light of the World." (Scala/Art Resource, New York)

Plate 3. Refraction breaks a pencil.

ASOTVM.

Plate 4. The city of Ashdod, fresco in the cathedral of Anagni, Italy, c. 1255. (Juan Ainaud, *Romanesque Painting* [New York: Viking Press, 1963]. Copyright Andre Held.)

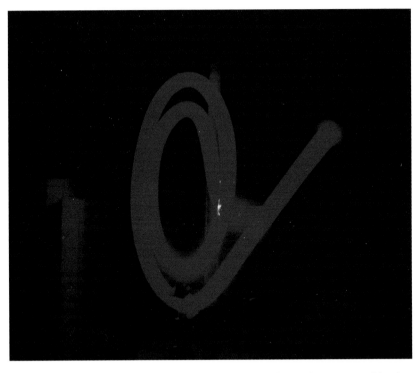

Plate 5. Light enters a coil of transparent plastic, starting at the center and issuing from the end.

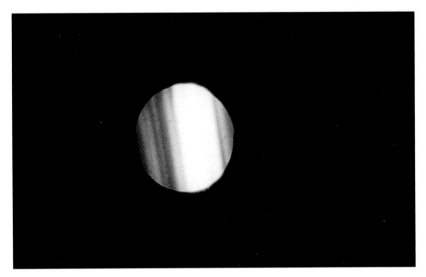

Plate 6. Grimaldi's experiment: colored fringes of light behind a fine wire in a narrow beam of sunlight.

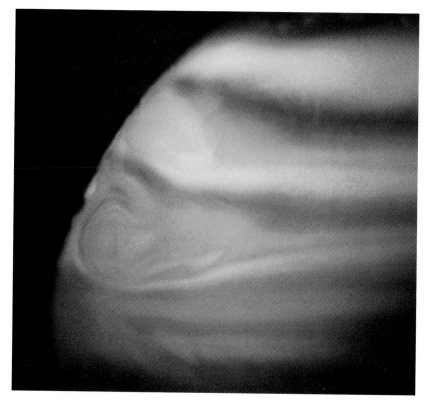

Plate 7. Formation of colored fringes in a soap film.

Plate 8. A ray of white light enters a prism and is spread out into its component colors. Note that the spreading is not very great; the distance to the wall is about a meter.

Plate 9. Looking through a prism at a black line drawn on white paper. The black line is at the top. The blurred line is an edge of the prism. The colored line at the bottom is the black line as seen through the prism. Note that the white paper as seen through the prism is still white.

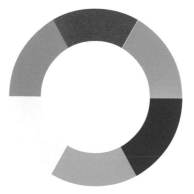

Plate 10. Goethe's color wheel. The primary colors are yellow, blue, and "purple," a combination of pure red and violet.

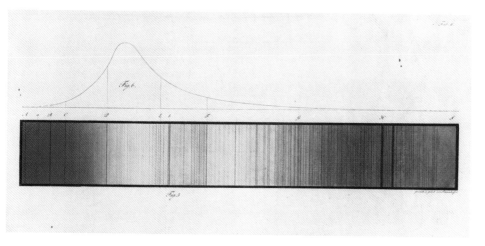

Plate 11. Joseph Fraunhofer's drawing of the solar spectrum, 1814, showing a large number of dark lines. (Deutsches Museum Munich, by permission)

Plate 12. Spectrum of neon gas. Mix the colors together and you have the familiar color of a neon sign.

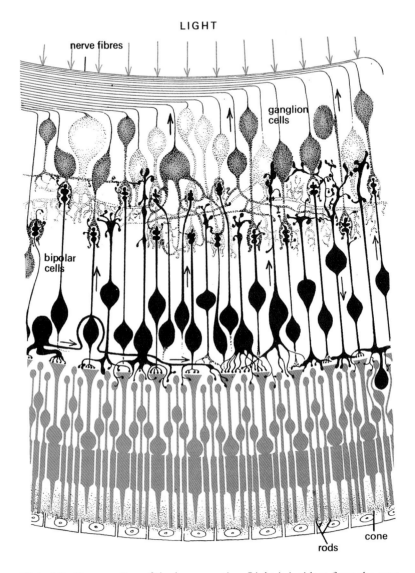

Plate 13. Cross section of the human retina. Light is incident from the top; rods and cones are at the bottom. (Richard L. Gregory, *Eye and Brain*, 4th ed. [Princeton, N.J.: Princeton University Press]. Copyright © 1990 by Richard L. Gregory. Reproduced by permission of R. Gregory and Princeton University Press.)

transmits to some of those immediately in front. But the latter have no way of knowing whether that jiggle came from a particle of the source or a particle of the ether; they respond the same in either case. Therefore each point b serves as the source of a new wave, but as the figure shows, what results is not chaos but the new smooth surface DF, for the wavelets from the points b tend to cancel one another, except on the surface DF, where they reinforce one another. This way of representing the spread of a wave is known as Huygens's principle, and every student of optics uses it even if its explanation in terms of particles of ether is no longer in fashion. It is, instead, a mathematical fact that can be proved as an approximation from modern theories of light.

I can imagine a philosopher in one of the universities of those days shaking his head as he works his way through the proof. Yes, of course, each ripple gives rise to many others, and the wave moves forward that way without any transfer of matter. It is the multiplication of species, except where are the species? How does one see anything? He puts the book down with a sigh. Philosophy gets more and more mathematical, he laments softly, and it explains less and less.

In stating Huygens's principle I wrote, "The wavelets from the points b tend to cancel one another, except on the surface DF, where they reinforce one another," as if it were obvious. It shouldn't be. Look first at the point P, which I have added to the original diagram. If the wave front HI has moved on to DF, this implies that at P the ether is once more at rest. Why would anyone say that? There are all these waves running around. The statement needs mathematical proof, but I will try to show what is important without mathematics. First, what is a wave? In water it has crests and troughs, lines along which it is higher and lower than the average height of the water. In a wave of light as Huygens imagined it, we can think of it as a motion not up and down but back and forth along the direction in which the wave is traveling. As the ether moves back and forth it produces increases and decreases in pressure, so that the wave is also a wave of pressure, just like a sound wave in air. Give an increase in pressure a plus sign, and a decrease a minus sign. Suppose the arcs in the figure represent the crests of the wavelets. Then midway between arcs is a trough. The point P is close to the crest of one wave and the trough of another. The principle here is that at any point you can add the pressures in all the waves, with their appropriate signs, and that if you add a lot of them, many more than the six that are drawn, there will be about equal numbers of pluses and minuses, and you will get something close to zero except along a line where the crests coincide. The idea that you can add pressures (or velocities, or displacements, or densities) in a wave is one of the crucial principles of every theory of wave motion after Huygens. It is called the Principle of Interference or the Principle of Superposition. One says that the waves at P interfere destructively, and those along DE, where the crests coincide, interfere

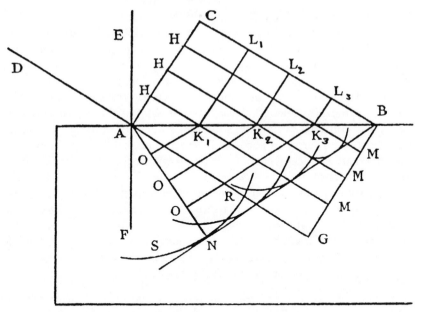

Figure 7.14. Assuming (contrary to Descartes) that light travels more slowly in a denser medium, Huygens explains why a beam of light is refracted at the boundary.

constructively. Not all kinds of wave motion obey this principle. It is a rather poor approximation for ocean waves, but it describes light waves and all but the loudest sound waves very well. The intuition of Huygens produced it when it was needed; he used it without really saying what he was doing; then he went on to discuss something else.

With Huygens's principle we can now explain the refraction as light passes from air into glass. We are going to make Fermat's common-sense assumption that light travels more slowly in glass than in air. In Figure 7.14 the lines AC, K_1L_1, K_2L_2, K_3L_3 represent successive positions of the crest of a wave arriving along the direction HM. According to Huygens, each point of these wave fronts serves as the source of new spherical wavelets. Look at the ones starting from A, K_1, K_2, and K_3. The crest arrived first at A; therefore the radius AN is the longest, while that from K_3 is the shortest. Draw the wavelets from K_1, K_2, K_3 and, in imagination, the millions of points in between, and you can see that just as in Figure 7.13, there is constructive interference along the line BN and destructive elsewhere, so that BN is the crest of the wave after it has entered the glass. The line AN is steeper than the line DA, and a few steps of simple geometry show that their directions agree with Snel's law.

We now have two different proofs of that law which assume that light travels more slowly in a dense medium. One uses Fermat's principle of least time, and the other uses Huygens's principle of wavelets. Both give the same result, but which assumption is the right one? At the end of his discussion of refraction Huygens shows by a mathematical tour de force that Fermat's principle follows from the construction of wavelets. We have one explanation, not two. The little book that challenged the orthodoxies of Descartes and Newton was soon almost forgotten, but he was right about the waves.

Finally, is there really any excuse for taking seriously a theory whose fundamental premise is that space is packed with little marbles? Don't we now know that empty space is really empty? Let us be very careful before we speak in such unqualified terms. No physicist, even today, thinks that empty space is really empty, but just how to say what it contains is a delicate scientific problem. We shall have to discuss it later.

8. The Rainbow Is Caught

We dropped the subject of the rainbow where Aristotle dropped it, in 2.6. It is time to pick it up again and settle the whole thing. There are several facts to be explained. Sometimes there are two rainbows. If the sun is on the horizon, the top of the primary rainbow is about 42 degrees above the opposite horizon, and that of the secondary one, if you can see it, is at about 52 degrees. The space between the two rainbows is noticeably darker than the space below the primary and above the secondary. The top of the primary rainbow is red; that of the secondary is blue or violet. To clarify, let us say that there are four questions to be answered:

1. Is a mathematical discussion possible?
2. Why are there two bows, how are they formed, and why are the two bows colored in opposite order?
3. What determines their angles above the ground?
4. Exactly why are the bows colored?

In the following short discussion I will pretend that each of these questions was answered by one person, emphasizing what he said that is right and ignoring misconceptions. This will give a grossly oversimplified view of what actually happened, but at the end you will know the answers to the questions, if you do not know them already.

As to the first question, let us credit Aristotle with the imagination and optimism that led him to attempt a mathematical theory, even if it is completely wrong.

The second idea, the crucial one, was contributed by a Dominican monk, Dietrich (in Latin, Theodoric) of Freiberg (c. 1250–1310). You will remember that Aristotle explained the rainbow in terms of reflection from the surface of a cloud. Dietrich (following Albert the Great) said, "We will understand it when we have understood what happens in a single drop of rain

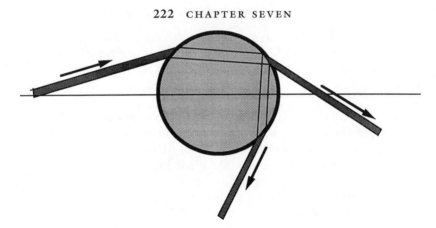

Figure 7.15. Theodoric of Freiburg explains how the primary rainbow is formed by the reflection of light in each droplet of a cloud (redrawn from a manuscript of his *On the Rainbow*).

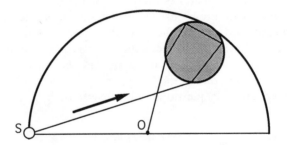

Figure 7.16. Theodoric's diagram explaining the secondary rainbow (redrawn).

or mist." He has seen rainbows near fountains and waterfalls and has studied them in the dewdrops on a spider web in the early morning. Clouds are not needed. But why struggle with dewdrops when we can make a model of one with a glass sphere or a spherical flask of water and study it on a conveniently large scale? This is what Dietrich did. (You can also do it with a cylindrical bottle of water, but the glass must be uniform and thin.) Figure 7.15, redrawn from one in a manuscript of Dietrich's, shows two rays, A and B, leaving the sun, which is on the horizon, and reaching the drop. (He might better have drawn the rays parallel, since the sun is so far away.) Ray A goes straight through. Ray B, encountering the drop at *k* near its edge, passes inside it. When it reaches *l* at the back of the drop most of it goes through, but as I have mentioned, a little light is always reflected at the boundary between two media. The reflected beam goes on to *m* where it is refracted and leaves as ray C in the direction of the eye. The angle between rays B and C in the manuscript is about 48 degrees. In a rainbow it is about 42. If you try it with a bottle of water you can see ray C coming to you from the lower edge of the bottle, and the ray is colored. "It is found," writes

Dietrich, "that if the sun is shining and little spheres are mounted in a fixed position, then when the eye is raised and lowered it sees a variety of colors." Red is at the top, blue is at the bottom as they should be.

The second rainbow is explained in a similar way. Figure 7.16 shows a ray arriving near the bottom of a raindrop and reflecting twice before it is refracted toward the eye. At each reflection most of the light is lost, so that the second rainbow is fainter than the first. And since the second ray has toured the raindrop in the opposite direction, the colors also are spread out in the opposite order, with red at the bottom. The figure shows that Dietrich is using Aristotle's geometry (Figure 2.1), which assumes that the sun is no farther from the observer O than the raindrop is. That is absurd, but luckily for his discussion it doesn't matter.

A question leaps to the mind: in Figure 7.15, what is so special about the ray B? Rays from the sun strike the whole left side of the raindrop, and a few casual sketches will show that they leave it in many different directions. So what picks out this particular ray, and what happens to the rest of them? For that we will have to wait three hundred years and then ask René Descartes.

Addressing the third question, Descartes in his *Meteorology* devotes several pages to the rainbow. Dietrich's theory was available in the works of several commentators and Descartes repeated the experiment with a spherical flask of water. He claims to have seen both primary and secondary rays coming from it at the appropriate angles. His Figure 7.17 shows the whole idea in one picture. The flask is drawn so large that ray ABCDE, reflecting once and arriving from the bottom of the flask, arrives at 42 degrees, while FGHIKE, going around the other way, arrives at 52 degrees. He puts the flask up in the sky where the raindrops are, but the picture is a little deceptive, for no one drop contributes to both primary and secondary bows for the observer at E. Figure 7.18, from Newton's *Opticks*, shows how things really are.

To find out what happens to the other rays that are not seen, Descartes uses Snel's law to calculate how they refract at the boundary of a drop. He does this for rays that aim at the center of the drop, its edge, and many points in between, nineteen in all. Figure 7.19 shows what happens to four of them. Note that the drawing requires no scale of sizes; it shows either a flask or a raindrop. Say that the radius of the drop is one unit. You will notice that the ray that arrives toward the edge, at $h = .86$, is refracted less than the others, and in fact the rays arriving with h close to .86 leave in a bundle in almost the same direction, which turns out to be 42 degrees below the horizontal. Rays arriving with other values of h go off in a fan of directions, but no ray forming the primary bow reaches the ground at a steeper angle than this. Because all values of h are equally probable,[13] more rays reach the ground near 42 degrees than near any other angle, and therefore the light in that direction is brighter.

[13] Mathematical sophisticates will see that this statement must be taken *cum grano*, but the calculation shows what it is meant to show.

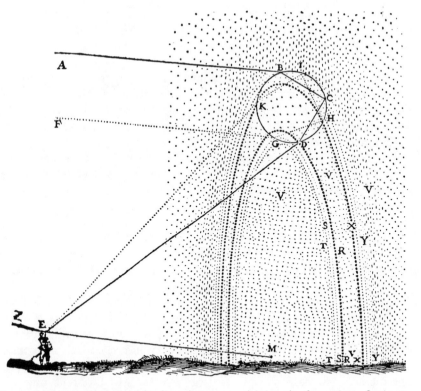

Figure 7.17. Graphic but misleading diagram explaining the formation of primary and secondary rainbows. (A raindrop isn't really that big.) From Descartes's *Météorologie*.

Descartes calculates the secondary rainbow in the same way, and now it turns out that when *h* = .95 the critical angle is a *minimum*, at just under 52 degrees, and again there is a bundle of rays near that one. No rays from either bow arrive at an angle that would put them between the two bows, and that is why the space between is dark.[14]

Thus far I have not mentioned the rainbow's color, my fourth question. Dietrich gives an explanation based on Aristotelian ideas that need not detain us, and Descartes applies his theory of spinning tennis balls to explain why the top of the bow is red and the bottom is blue. With Newton's theory of colored light the whole thing is plain. The angle of 42 degrees was calculated using Snel's law for the refractions that occur when a ray enters and leaves the drop. But Harriot and Newton have shown us that the different colors in the sun's white light are differently refracted, so that they do not all leave at exactly the same angle. And since Newton knows how

[14] Carl Boyer [1987, p. 216] points out that this explanation was within the reach of Dietrich of Freiberg. Snel's law is not required; the same results could have been calculated almost as accurately from the refraction tables of Ptolemy and Witelo.

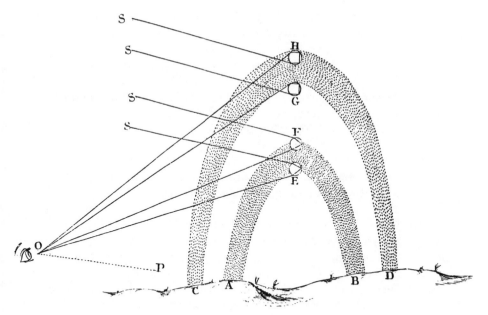

Figure 7.18. More accurate rainbow diagram, from Newton's *Opticks*.

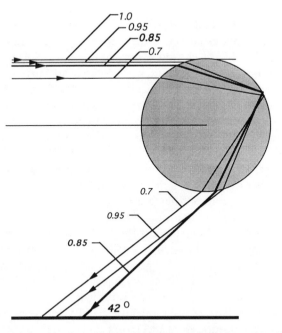

Figure 7.19. Diagram showing why the sun's rays reflected once inside a raindrop will cluster around a direction 42° away from the direction of the rays. The labels 1.0, 0.95, 0.85, etc. indicate a ray's distance from a line to the center of the drop.

great the difference is, he easily calculates the widths of the two bows, almost 2 degrees for the primary and a little over 3 for the secondary. Figure 7.18, taken from *Opticks*, shows all the trajectories.

Finally, what happens to the rays that do not fall into the bows we see? In the primary bow they are refracted less, and will appear to come from below the bow. Here there will be a rather dim mixture of rays of different colors that reached the drops at other than the critical value of h. But such a mixture of colors will appear white, or rather a pale gray, and we can see it when there is a black cloud in the background.

Newton's death ended the period during which Europeans were bent on demolishing the forms of thought in which they had been brought up. The task was difficult because the tools they used in the demolition were the tools they had, mostly forged in the earlier age whose remains they were trying to destroy. Those who followed were not burdened in this way. Up they climbed onto the shoulders of Fermat, Leibniz, Huygens, Newton, to survey new territory that their forbears had not dreamt of. But before we enter the age called the Enlightenment, I will close this chapter as it began, with a measurement of the speed of light.

9. James Bradley Goes for a Sail

In the days of James Bradley (1693–1762) most telescopes belonged to families that were interested and could afford them, and Bradley was born into such a family. He was elected Fellow of the Royal Society at the age of twenty-five, and three years later he was professor of astronomy at Oxford, but he seems to have spent most of his time in the family observatory. By then the Copernican system was generally accepted in intellectual circles, but there was some discomfort, for no piece of clear observational evidence that would have established it without question had ever been found. One, in particular, had been looked for. Figure 7.20 shows the Earth sailing around the sun with a telescope pointed toward a nearby star S located close to the axis of the Earth's orbit. It also shows the appearance of the background stars beyond S (they are of course at different distances, but one does not see that). At the moment when Earth is at P, one would expect to see S at the position p against the background, and in the course of a year p should describe a little circle in the sky. This apparent motion of a nearby star against the distant background is known as *parallax*. Again and again the critics of the new astronomy had pointed out that if the Earth moves and if the stars are not all stuck on a great Ptolemaic sphere, we should see parallax. The answer, that the parallax must be there but too small to see, struck disinterested spectators as singularly feeble and satisfied only those already convinced. A reasonable person could still be skeptical. Detection of the little circle would be the first incontrovertible proof of the

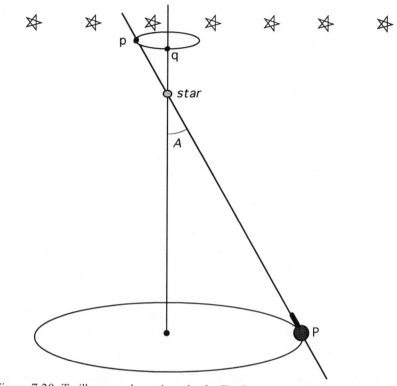

Figure 7.20. To illustrate aberration. As the Earth goes around the sun, a nearby star's position seems to move in a little circle against the background of more distant stars. Parallax, when the Earth is at P, would place the star at p, but aberration, which is much larger, places it at q, and this is where it is seen. Aberration is due to the Earth's orbital velocity and not to its displacement from the sun.

Earth's annual motion around the sun, and it would pay a dividend: by measuring the circle's apparent size one could then calculate the distance to the nearby star.

James Bradley chose a star called Gamma Draconis to study, largely because it was near the orbital axis, and he was delighted to find that in the course of a year the star's image did move in a tiny circle. The measurement was delicate, for the angle A in the figure is very small; it was as if he had measured the size of a book a mile away and determined (correctly) that it was between 6.2 and 6.3 inches wide. But there was something paradoxical about the observation. At the moment the Earth was at some point P on its orbit he expected to find the star at p in its little circle. Instead, it was at q, and remained 90 degrees away from where it was expected to be as the year went on and the point P moved around its orbit.

There is a story that one afternoon in 1728 the professor went sailing on the Thames and noticed that when his boat changed its course a little pennant on the masthead changed its direction. He asked the boatman why this happened and was told that it was not because the wind had changed but because the boat was moving in a new direction. People who sail are familiar with this effect. It occurred to Bradley that the light from the star is analogous to the wind, and that as the Earth carrying its telescope traverses its orbit, the direction from which the light comes should appear to change. The phenomenon Bradley discovered is called *aberration,* and you do not need a boat to experience it. All you need is vertical rain and an umbrella. Stand in the rain and you keep dry by holding the umbrella directly over your head. Walk, and you must hold it a little in front of you. The reason is obvious. The telescope acts like the umbrella, and since it is mounted on the moving Earth it must always be pointed a little in front of the straight line from it to the star. Bradley measured several more stars in the course of the year and found circles of exactly the same size. He had succeeded in showing that the Earth moves in its orbit, but not the way he intended to; he had failed to measure the distance to Gamma Draconis but on the other hand he had measured the speed of light, for there is a simple relation between this speed and that of the Earth. He calculated that light travels from the sun to the Earth in eight minutes, twelve seconds. Römer's figure, fifty-seven years earlier, had been eleven minutes; the modern value is eight minutes, nineteen seconds. From his failure to observe the effect he was looking for, Bradley concluded that the stars are farther away than anyone had thought, and he calculated that light would take more than six years to reach us from Gamma Draconis. He ends his account with a sarcasm which reveals that Aristotle's teachings still survived in the dark corners of the academy:

> There appearing, after all, no sensible parallax in the fixed stars, the anti-copernicans have still room on that account, to object against the motion of the Earth; and they may have, if they please, a much greater objection against the hypothesis, by which Mr. B. has endeavoured to solve the forementioned phænomena; by denying the progressive motion of light, as well as that of the Earth.

In 1742, after the death of Edmund Halley, of cometary fame, Bradley was named Astronomer Royal and devoted himself to developing instruments and techniques for improving observational accuracy. When he died he left sixty thousand very exact observations to be corrected for atmospheric refraction and expressed in the coordinates with which astronomers locate stars and planets. These measurements are used today by people studying the slow drift of certain stars across the sky.

Finally, a century later, Friedrich Wilhelm Bessel found the parallax that Bradley was looking for, but only after he had disentangled the parallactic angle A from the effect of aberration, which is sixty times larger. The method gives the distance to the nearest stars, but the rest are all too far away.

The explanation of rainbows showed the importance of mathematical thinking, and the discovery of aberration showed what can be done with a good telescope. Neither of these provided any insight into what light actually is, but both are easily explained if one thinks of light as a rain of particles; in fact it is harder to explain them by a theory of waves. Almost everyone was happy with the Newtonian presentation of light. It fitted with the general belief that matter is composed of atoms, and although scientists were aware that atomic theory was 10 percent evidence and 90 percent plausibility, they were at least satisfied that they had a harmonious theory of light and matter. The next chapter will describe the gradual erosion of Newton's theory of light as facts piled up which it was unable to explain, but the dam did not break for another hundred years.

8

ENLIGHTENMENT

The unpursued speculations of Newton, and the opinions
of Hooke, however distinct, must not be put in competition,
and, indeed, ought scarcely to be mentioned, with the
elegant, simple, and comprehensive theory of Young,—
a theory which, if not founded in Nature, is certainly one
of the happiest fictions that the genius of man ever invented
to grasp together natural phenomena, which, at their first
discovery, seemed in irreconcilable opposition to it. It is,
in fact, in all its applications and details, one succession of
felicities, insomuch, that we may almost be induced to say,
if it be not true, it deserves to be so.
(John Herschel, 1827)

1. The Eighteenth Century

THE BELLS that tolled the death of Isaac Newton in 1727 rang
long and loud. Alexander Pope expressed the sentiment of the day
in a couplet intended for Newton's tomb in the Abbey, though it
was not used,

Nature and Nature's Laws lay hid in Night.
God said, *Let Newton be!* and all was *Light.*

Nature shed much of its mystery once people understood that Kepler's ellipses are required by Newton's theory, that eclipses happen when they
must happen and at no other time, and that one can calculate the length of
a lunar month on the back of an envelope. The *Opticks* revealed to ordinary
readers the simple principles that explain the miracle of color, principles
that were soon part of the body of natural knowledge and metaphor available to poets and essayists.

It took longer for the new science to remodel scientific and philosophical discourse, for it pulled them in two directions. The more readers admired the abstract Newtonian principles and understood that they control
the heavens with impersonal precision, the less they felt enfolded in a loving Nature created specially for humankind, or sensed the everyday providence they had been taught to find in small events. For many believers,
what Newton took away was more than what he had given, and the mysteries he had erased made even darker those he had not.

But the course of events set going by the new science was not slowed by anybody's second thoughts. Once Newton had shown that God's creation is governed by exact mathematical law—in dynamics, in astronomy, in optics and other less noticed areas of science—analogy slipped its leash and roamed unchecked. When Thomas Jefferson, in the Declaration of Independence, speaks of the right of the former Colonies "to assume among the powers of the Earth the separate and equal station to which the Laws of Nature and of Nature's God entitle them," he imagines a political universe that is as law-governed as the motions of the planets. An eloquent expression of the same faith appears in the words of James Wilson (1742–98), a signer of the Declaration, one of the principal theorists and drafters of the Constitution, and a member of the first Supreme Court. Early in 1790 he lectured to law students in the College of Philadelphia: "Order, proportion, and fitness pervade the universe. Around us we see; within us we feel; above us, we admire a rule, from which deviation cannot, or should not, or will not be made."

The American theorists owed much to the French *philosophes* and the great *Encyclopedia* (1751–72) in which they expressed their views, but in fact the *philosophes* were not thinking in the same way. If God's legislation is written in mathematical symbols, they argued, it can apply only to the motions of planets and other machinery, for it does not distinguish right from wrong, or stipulate the quality of anything. At best, mathematics tells what happens, not what ought to happen. The law of human beings must be a law of freedom, of choice, and its source must be man, not God. In his article on natural law, central to what he conceived as the social program of the Enlightenment, the *Encyclopedia*'s editor Denis Diderot denies this law's divine origin and declares to his readers that it is the product of human reason as manifest in the general will of humankind. "You have the sacred natural right to whatever is not disputed by the race as a whole. . . . Say often to yourself: I am a man, and I hold no inalienable natural rights beyond those of humanity." This was the spirit that inspired the French Revolution, a few years after the American one, to take over the Cathedral of Notre Dame, screen its main altar with a great pile of earth, and call it the Temple of Reason.

While philosophers tried to define the source and quality of a human law that exists alongside the divine but impersonal principles that Newton had discovered, the truth of those principles was generally not contested. We must now look at them more closely, especially those related to optics. What do they claim, what do they not claim, and are they true?

Newton ornamented the world with two great books, *Principia* and *Opticks*, and the authority they carried was enough to almost immobilize scientific progress along those lines for a century. It is hard to fight with *Principia*, since most of its mathematical results are rigorously valid, and the experiments reported in the *Opticks*, repeated today, still give the same results. But the conclusions Newton drew from his studies in optics were misrepresented by his supporters and at last were largely discredited by appeal to the principles that he had labored so hard to establish. His mistake was to appease his friends the "great virtuoso's" by throwing them hypotheses, mental pictures to make his discoveries easier to understand and remember; for in time these hypotheses were received as doctrine and considered integral parts of the system. Newton certainly took some of them more seriously than others, but he never failed to distinguish between conclusions that he thought necessarily followed from his experiments ("theories") and hypotheses which experiment suggested but did not require.

In England, it became a fixed belief that a light ray is a stream of particles (see Newton's words in §7.5). On the Continent, Cartesian doctrine was less explicit, but Newtonians and Cartesians agreed on one point: light is not a wave. Still, there were difficulties. Cartesians made no attempt to explain Grimaldi's diffraction fringes, while Newtonians had to suppose (as in Query 1, quoted in Chapter 7) that the edges of the diffracting body exert a force that pulls passing particles into the shadow. To explain the colored fringes (Query 3), the orthodox had to assume further that "the rays of Light . . . [are] bent several times backward and forwards, with a motion like that of an Eel." These are vivid images, but considered as science they are vulnerable to the challenge that Newton had thrown at Hooke and other critics: "If you know so much, why don't you calculate what will happen in some particular experiment and then we will see if it does." Nobody had a formula for the force required in Query 1, and science still awaits a mathematical theory of the eel. Often during the century following Newton's death the edifice he had created tottered and shook, but there was no one with the scientific authority to condemn it and begin the work of demolition. Instead, scientists moved away from the dangerous shadow of those great walls to explore two new sciences that Newton mentioned only in his queries, chemistry and electricity. As people began to think atomically, chemistry was ready for great advances; and electricity, in the following century, resolved one of the mysteries of light.

The great chemist of the period was Antoine Laurent Lavoisier (1743–94). He led the effort to identify the substances called elements, the basic materials of which all matter is composed, and in 1789, five years before he was extinguished during the Terror by a tribunal which declared that the Republic had no need of learned men, he published a tentative list of twenty-eight candidates. The first is light. The second is caloric, otherwise known as heat, which for a few more decades would be regarded as a substance added to or subtracted from a thing by flame or ice. The rest, beginning with oxygen, nitrogen, and hydrogen, are all correctly identified as chemical elements. By now, caloric is gone without a trace. Light does not belong in Lavoisier's table because considered as a substance it is more

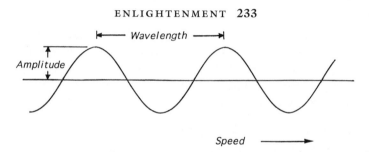

Figure 8.1. Defining amplitude, wavelength, and speed of a sine wave.

simple and elementary than oxygen, nitrogen, and the rest. In fact, as we shall see, it is a constituent of the elementary atoms, though in a way that will need explanation.

One idea that had disappeared in the shadow of Newtonian authority was the possibility that colored light is a regular periodic wave, and that the color we perceive depends on its wavelength. This idea will shortly reappear, and to continue the discussion we will need some terms. Imagine a wave produced by moving the end of a clothesline rhythmically up and down a few times. A pattern (Figure 8.1) travels along the line, reflects at the end, and comes back. If the pattern is shaped like that in the figure it is called a *sine wave*. Such a wave has four measurable characteristics:

its amplitude, measuring how far from its midline the wave extends;
its wavelength, measuring the distance crest to crest;
its frequency, measuring how many waves pass a given point per second;
and its speed, the rate at which the pattern travels down the line.

The last three are not independent, for a little thought reveals that the speed is equal to the product of the wavelength and the frequency. A sound wave of a certain wavelength produces in our ears the sensation of a certain tone: the shorter the wavelength the higher the pitch. Unlike the waves in a string that I mentioned just now to make the picture more vivid, sound waves are waves of pressure: the air moves back and forth in the direction in which the wave travels, rather than up and down; but the same terms are used. When Newton, or anyone else before the early 1800s, thought of light as a wave, it was as a wave of pressure in the ether. This kind of wave is called *longitudinal*; the wave in the string is *transverse*.

We have seen in §7.5 that Newton proposed a wave theory of light and at once discarded it because it could not explain to him why light travels in straight lines, but the idea remained in his mind, and many years later, in Query 14 of the *Opticks*, he carried it further. He suggests that the sensation of color is produced by a periodic motion: "May not the harmony and discord of Colours arise from the proportions of the Vibrations propagated through the Fibres of the optick Nerves into the brain, as the harmony and discord of Sounds arise from the proportions of the Vibrations of the Air?"

Specifically, are not the colors of gold and indigo "agreeable to one another" for the same reason that C and E sound sweetly together on a piano? The possibility of an analogy with sound was obvious and appealing, but perhaps Newton's reading of Descartes and many others had sensitized him to the perils of analogy. At any rate, he refused the bait.

In England Newton's views were quickly absorbed and accepted, but it was not so on the Continent. There the opinion was that though he was a brilliant mathematician he was really no physicist at all, for he left the important questions unanswered, including the cause of the gravitational attraction, the reason why his proposed laws of motion are true, and the reason why refraction varies with color. In 1681 the Abbé Edme Mariotte, a respected experimenter, published a little book *On the Nature of Colors*, in which he says that when Newton's "crucial experiment" (Figure 7.9) is done right, the second prism produces further colors in the supposedly monochromatic beam that issues from the first. This, plus doctrinary opposition from the Cartesians, greatly delayed the acceptance of Newton's ideas. In 1706 the Latin translation of *Opticks* appeared, and the influential Father Nicolas Malebranche wrote, "Though Mr. Newton is no physicist, his book is very interesting and useful for anyone with good training in physics; he is also an excellent mathematician." The Reverend Father himself was not yet convinced by anything he read, and it was not until 1712 that he accepted the Newtonian doctrine. A few years later the "crucial experiment" was repeated correctly and, forty years after Mariotte's mistaken attempt, the tide began to turn. By 1730 Newtonian philosophy was trampling what remained of Descartes, and in 1738 there finally appeared a book that explained it to the French public.

The playwright and satirist Voltaire (1694–1778) spent several years as an exile in England. He attended Newton's funeral in Westminster Abbey, discussed science with members of the Royal Society, and on his return to France immersed himself for a while in Newtonian experiment and theory. Since his countrymen were so slow to adopt the new ideas, he pushed them along by writing a book, *Elémens de la philosophie de Neuton* [1738]. It is one of the best popular accounts of new developments in science: the language is simple and clear, conclusions follow from premises, and opposing views are cleverly demolished. I mention it here because it is the bridge over which Newton crossed the Channel and reached large numbers of intelligent readers in France. The first half, on optics, details Newton's ideas with a few additions from the author, some of which show how hard it was to set aside the Scholastic categories: "Experiments by the most skilful hands teach us that some materials gain weight after long exposure to light. The particles of fire have penetrated their substance and increased it. But even if one doubts these experiments, fire is a substance; therefore it has weight, and light is nothing other than fire." The second half expounds *Principia*. Everyone with liberal opinions read Voltaire. His book did much to dissipate the silence and confusion that had greeted Newton's work, but its persuasive discourse tended to replace one dogma with another. I turn now to another respected author whose popular book, a generation later, tried and failed to free his readers from ideas that Voltaire had done so much to establish.

2. The Musical Atoms of Leonhard Euler

In the eighteenth century there was only one prominent scientist who publicly advocated a wave theory of light. As a mathematician, Leonhard Euler (1707–83) occupies a place in history roughly analogous to that of Bach among musicians. He was born in Basel and educated in theology, but mathematics took hold of him early and he spent his life as a member of learned academies, generously and sometimes lavishly supported by monarchs who wanted their courts to be bright as well as beautiful. In the eighteenth century, science and literature poured out of these institutions much more than from universities, which were generally the comfortable home of sleepy old opinions. Aged twenty, Euler arrived at the Academy of Saint Petersburg newly founded by the Empress Catherine, widow of Peter the Great. Thirteen years later Frederick the Great lured him to Berlin, but after twenty-six years there he returned to the comparative quiet and civility of Petersburg. His mathematical powers were legendary, as was his memory (he knew the Aeneid by heart) so that when in the latter part of his life he slowly went blind, his mathematical output hardly faltered. The publication of his collected works started in 1911 and still continues after seventy-eight quarto volumes. No other mathematician has ever written so much, and few have wrought so well.

Euler's interests outside mathematics included agriculture, shipbuilding, and a dozen other topics; there are five volumes on optics. Most of this work explores what can be done with combinations of lenses; we shall encounter some of it in 6 of this chapter, but the first long essay, "A New Theory of Light and Colors," written in Berlin and published in 1746, goes deeper. It develops the idea of light as a wave, making full use of the analogy with sound. In sound, a pure tone originates in a string or a column of air that vibrates at a certain frequency; the vibration is communicated to the air (or some other medium) in the form of a sine wave, and when that reaches our ear it excites vibrations in the ear's inner structure and is heard. In exactly the same way, said Euler, light of a certain color starts with particles in the glowing source that vibrate with a definite frequency and produce sine waves in the ether, a very thin and elastic fluid that transmits its waves much faster than air does. Ether mingles with the particles of all kinds of matter, so that light waves can pass through them unless hindered by opaque particles. The waves are perceived just as Newton proposed in Query 23 (see §7.6). It is clear that the whole argument depends more on analogy than on observation (for instance, Euler misremembers Newton's measurements and claims that red light has a shorter wavelength than violet), but the very simplicity of his vision, unencumbered by the cautions and caveats that arise from much knowledge, led him to say things that turned out to be right where others were wrong. He proposed that a pure color, like a pure sound, corresponds to a sine wave with a definite wavelength and frequency. Nobody had ever said this before. He saw the sun as composed of particles of different kinds, each of which vibrates at some particular frequency, the ensemble producing the effect of white; and here is how he explains the color of a colored object:

> The nature of the radiation by which we see an opaque object does not depend on the source of light but on the vibratory motion of the very small particles of the object's surface. These little particles are like stretched strings, tuned to a certain frequency, which vibrate in response to a similar vibration of the air even if no one plucks them.[1] Just as the stretched string is excited by the same sound that it emits, the particles of the surface begin to vibrate in tune with the incident radiation and to emit their own waves in every direction.

Euler is saying that if a thing looks red this is because the particles in its surface are made so that they vibrate at the frequency of red; if placed in white light they will respond to the red vibrations in it and reemit red light. Thus the reflection of light is not like the bounce of a tennis ball; the light that comes back is newly produced by the vibrating atoms, and Euler goes on to explain the direction of its ray exactly as Huygens explained it. Throughout his discussion Euler uses only the simplest mathematics, not realizing that Huygens's principle can be expressed in mathematical terms and used to perform calculations. The mathematical theory of waves rests on d'Alembert's equation, which Euler had never seen, for it was published only in the following year. But even with the equation in hand, it was a long time before anyone learned to solve it so as to understand why light seems to travel in straight lines while sound, obeying exactly the same equation, does not.

Euler's ideas reached the public through a book called *Letters to a German Princess* [1768], in which he explains science in language anyone could read. He begins the first volume with a discussion of light that sets out the theories of Descartes and Newton, shows what is wrong with them, and puts forth his own ideas; then he goes on to explain Newton's theory of planetary motion. The second volume concerns electricity, the great topic of eighteenth-century physics, and so is post-Newton. In spite of Euler's authority as a mathematician and the immense circulation of his book, the scientific community was embarrassed by his theory of light and ignored it, to the extent that the translator of the German edition (1792)

---

[1] This phenomenon, known as resonance or sympathetic vibration, can be illustrated with a piano. Press down the damper pedal so as to release the strings, and sing a single note—*ohhhh* or *ahhhh* or *eeeee*—loudly. You will hear it faintly returned by the strings of the piano.

thought he had better insert a note warning readers that this theory was not supported "by a single person of prominence."

Because I hope to persuade you later that visible light of a pure color comes out of individual atoms, this is a good place to review what has been said on light and atoms thus far. Galileo began it, saying in §5.7 that when the atoms of a substance are released by friction, light bursts forth. Christiaan Huygens and Robert Hooke were led astray by the fact that light is associated with heat and therefore with random vibrations of the particles of the hot object. The particles bump against each other and also against the adjacent particles of ether and start irregular pulses moving through it. This rattling around could not produce a sine wave. Huygens has no ideas about color, and his book never mentions it, while Hooke looks for a different explanation. For Newton, light of a certain color is simply a ray that a transparent medium refracts in a certain way; he does not know how the ray begins or why we perceive it as colored. Euler answers the questions that his predecessors did not. He goes further, remarking that a stretched string can be made to vibrate at frequencies twice or three times its fundamental frequency, or faster, and he expects that something similar will happen with his vibrating atoms. When I was a child we played a game in which we hid a cookie or thought of a word, and while somebody tried to find it, the others gave clues: "You're getting warmer—no, now you're getting cold." Leonhard Euler was getting hot.

I think this was the first time anyone who believed in atoms ever suggested that they have a vibrating internal structure. The atoms of Newton and Boyle are clusters of hard little balls; Euler's atoms are like musical instruments. His clairvoyant insight was rediscovered much later, and when it was, nobody remembered who had it first.

By the time we are speaking of, many people had begun to think that though Newton's universe of mechanism and mathematical law might be an interesting place to visit, they would not care to live there. To them it felt like a prison, and considered as an intellectual structure it raised more questions than it answered. In this universe humanity did not sit at God's feet but spun around on one of several circling planets; if our own star has such planets, why not others, by the thousands—why not living creatures, by the billions? And if so, does God cherish every one of them equally? Has his Son died a thousand times? And what becomes of our relation with Nature when "she" is revealed as an automatic machine? By 1734 Alexander Pope's enthusiasm had yielded to second thoughts:

> Know then thyself, presume not God to scan;
> The proper study of mankind is man. . . .
>   Superior beings, when of late they saw
> A mortal man unfold all nature's law,

> Admired such wisdom in an earthly shape,
> And showed a Newton as we show an ape.
>   Could he whose rules the rapid comet bind
> Describe or fix one movement of his mind?
> Who saw its fires here rise, and there descend,
> Explain his own beginning, or his end?

(*Essay on Man*, Epistle II)

Literary and philosophical movements sprang up in Europe determined to assert the primacy of feeling over reason, to restore Earth and its inhabitants, mentally at least, to the center of the universe, to put science in its proper place, to mend the broken link between humanity and Nature. At the end of the century William Blake challenges the science of Descartes and Newton, which situates humanity as spectators and analysts of a dead world: " 'What,' it will be Question'd, 'When the Sun rises, do you not see a round disc of fire somewhat like a Guinea?' O no, no, I see an innumerable company of the Heavenly host crying 'Holy, Holy, Holy is the Lord God Almighty'" (*A Vision of the Last Judgement*). And he expresses his contempt for science in derisive verses:

> Reason and Newton are quite two things
> For so the Swallow & the Sparrow sings
> Reason says Miracle. Newton says Doubt
> Aye thats the way to make all Nature out.

Blake's vision of the universe pours out of the Bible and his own imagination. The next two sections will describe the work of another great man who rose up against Newton, who tried to make a science of light that sprang from human perceptions, not prisms, and who explained what we see without treating the universe as a machine.

3. Goethe Does Not Believe What He Reads

The success of the Newtonian program, the philosophical revolution it began, the verifiable correctness of its conclusions on light and planetary motion (not to mention other sciences less known to the public)—all these affected the progress of science in the way one might have predicted. As I have already said, theory became dogma. Newton's mathematical principles, once they had been cleaned up and expressed in simpler language, are used today, in all parts of the world, all the time. His hypotheses as to the nature of things, even the ones he fancied he had deduced rigorously from experiments, have fared less well, but during the eighteenth century and well into the nineteenth, these too had the force of law, and people with other ideas found themselves condemned as crackpots. Among them was Johann Wolfgang von Goethe (1749–1832), poet, dramatist, and statesman, and it may help to understand the strengths and weaknesses of Newton's theory of light if we contrast it with Goethe's opposing theory, both of them based on experiments but interpreted in terms of different assumptions.

Goethe was born in Frankfurt, studied law at Strasbourg, and returned home to enter the profession. For a few years he was active in the city's literary life, and in 1775 he was invited to serve the ducal court at Weimar. Here, except for a few short trips and 18 months in Italy, he spent the rest of his life, serving in the highest offices of the state government and turning out poems, novels, and plays that culminated in his *Faust*, finished just before he died.

Goethe was a connoisseur of painting and sculpture, and he seems to have become interested in color as he studied how painters represent light and shade in a landscape. For this, Newton's theory of spectral colors is not very helpful. He associates color with "refrangibility," which is represented by a continuous range of numbers, but this association suggests that color varies continuously through the spectrum in a way that experience does not really support. Blue and green are large areas of the spectrum that look fairly uniform, with a sudden transition between them; spectral red looks rather orange; yellow is a narrow band between orange and green; some people say they see indigo and some don't. And when Newton points out that the same green can be produced either by extracting it from white light with a prism or by mixing blue and yellow, he seems to destroy the correspondence between color and refrangibility. Newton, following the ancients, bases his mathematical treatment on the idea of a ray without ever exactly defining it or telling how it can be seen, and he has little to say about what happens when we see something.

Goethe had studied Newtonian optics in school, from texts that were both shallow and dogmatic, and he found little that fitted with his everyday perceptions or with the practice of painters. In Weimar he borrowed a prism from a friend but was too busy to make any observations with it until the moment came when he had to return it; at that moment he looked through the prism at a white wall and was surprised to see that it still looked white. His education had led him to suppose that white light seen through a prism would be broken up into colors, and he concluded at once that Newton was a charlatan, a buffoon, a "Cossack." Goethe's mistake is a simple one: he had forgotten, or else never knew, that Newton's white light originated in a ray of sunlight narrowed by passing through a small hole, and not in a broadly illuminated surface. If one looks at a surface through a prism, the prism picks up colors from various points of it and sends them into the eye, but if the whole surface is evenly illuminated the eye receives equal amounts of all the colors from these different points and registers white. Friends hastened to explain Goethe's error to him, but with iron determination he refused to believe them and went on to develop his

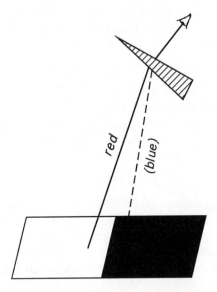

Figure 8.2. Showing why one edge of a black stripe on white paper looks red when seen through a prism or the edge of a magnifier. When one looks at the edge the blue light is missing; red remains. The other edge of the stripe will look blue; see Plate 8.

ideas along different lines. In 1791 and 1792 he published a pair of pamphlets under the title of *Beiträge zur Optik*, Contributions to Optics, in which he describes his experiments.

If you look through a prism at a surface that is white or of any uniform color, the prism changes nothing except at the surface's boundaries. But the visual world in front of us is composed of many surfaces with many boundaries, and it is here, Goethe decided, that the prism delivers its colors. If you want to try it and have no prism, the edge of an ordinary magnifying glass is prismatic enough (Figure 6.6) to show you what Goethe saw in his first experiments. Take a piece of white paper and paint on it a very black strip about a quarter-inch wide. Lay it on a table in a good white light with the strip directed across your field of view, focus on the strip, and look at it through the prism or else the very top of the lens, next to its edge. The top of the strip will look yellowish, the bottom will look blue. Figure 8.2 gives the Newtonian explanation. I have drawn two rays; the prism bends the red one less than the blue. The red ray comes from the white part of the paper and there isn't any blue, for the place from which blue would come is black and reflects no light. Therefore what you see is contributed by the red half of the spectrum, and if you look very carefully, you can see yellow, orange, and red. If you invert the prism or look through the bottom edge of the lens the colors are reversed, as in Plate 9.

Goethe describes a number of experiments of this kind and concludes that color is produced *because* there is a boundary between white and black.

The purity of white is bruised and discolored at the sharp edge. He identifies three primary sensations: yellow, blue, and "purple," which is his name for a color compounded of red and violet and which most observers actually consider even redder than the red of the spectrum. He arranges these on a wheel (Plate 10), putting between them orange, violet, and green, each imagined as a compound of the primary on each side. This color wheel has many uses for Goethe, for he associates different moods with it, and even different personalities. At the time he was writing the *Beiträge* he was also buying a fine house in Weimar and remodeling it. The social rooms of the house are on the second floor. He took out the hallways that originally connected them and opened doorways so that you walk in a straight line through one room after another. The doorways are generally open, presenting a vista of five or six rooms at a time. The woodwork is white, and the walls of each room are painted in a color from his wheel, strong and clear. As you look down the vista you see the colors of each room; and as I recall, they progress, one by one, around the wheel. Goethe's friends, returning from Italy, brought him plaster casts of classical sculptures, and these, gleaming white, add to the light in the main reception rooms. Goethe planned that each room should have its special mood, determined by color and décor. It is hard to situate oneself in the mind of a visitor in the year 1800, but even now, one feels the attempt at control.

4. Goethe's Theory of Color

Goethe continued his research into color, and as his writing lengthened his opinions congealed. His next work (1810) was the *Farbenlehre* (Theory of Color), more than nine hundred pages in two volumes.[2] It is divided into three parts: "Didactic," which describes his experiments and his theory of light and color; "Polemical," which tries, page by page, to demolish Book I of Newton's *Opticks*; and "Historical," presenting a long and very learned survey of ideas of vision and color from Pythagoras and Empedocles to the writers of his own time. Our main interest here is in the didactic part.

Goethe begins with the two primary, unanalyzable manifestations of color: light and darkness, or, to use the terms of color, white and black. They are inherent in the very nature of light; the other colors arise at boundaries. The boundaries may be sharp, as in the experiment with the black stripe, but usually they are indistinct, cloudy; and cloudiness is an essential ingredient of the theory, for it mixes light with darkness. A cloudy medium is conveniently produced by pouring a dash of milk into a glass of water. Hold the glass against a lighted surface: the color of the glass is reddish. Look through it at a dark background: the glass looks blue. Examples

---

[2] See the anonymous review, certainly by Thomas Young, *Quarterly Review* 10 (1814): 427–41, entitled "Zur Farbenlehre. On the Doctrine of Colours," which dismisses the whole thing in language of foaming contempt.

of this in Nature are common. Above us is slightly cloudy air against the blackness of space; this is why the sky is blue. Look toward the sun at sunset; it is red. Any transparent medium is cloudy to some extent and therefore able to generate color. "Transparency, regarded empirically, is already the first degree of cloudiness. The further degrees, up to the limit of an opaque white surface, are innumerable." Goethe considers that an optical prism is cloudy, and so the spectrum it casts on the wall is cloudy too, merging from color to color with boundaries ill-defined. And why does all this happen?

In his scientific studies Goethe seeks the unanalyzable origins of things. In biology it is the primeval plant, the *Urpflanze*, archetype of all later species—probably existing only in the supersensory domain, but possibly, just possibly, as a little bunch of leaves that might still live in some remote landscape. In light, the birth of color in cloudy media is the primitive phenomenon, the *Urphänomen*; like the *Urpflanze* it explains everything but is not itself explained. In the next section of this chapter we shall meet Dr. Thomas Young and find that he had already noted and explained the cloudy *Urphänomen*, but Goethe, even if he had known about this, would have dismissed it as tinkering with the unknowable.

There is a letter by John Keats written a few years later which devises an unexpected phrase to describe this point of view:

> At once it struck me, what quality went to form a Man of Achievement especially in Literature & which Shakespeare possessed so enormously—I mean *Negative Capability*, that is when man is capable of being in uncertainties, Mysteries, doubts, without any irritable reaching after fact & reason. . . . With a great Poet the sense of Beauty overcomes every other consideration, or rather obliterates all consideration.

Goethe's approach to Nature expresses the same ideal as the casts of late classical sculpture that surrounded him at home, images concerned not with the traits that define an individual but with universal forms that, once and for all, set the standards of beauty and define the aims of art. In the same way, his science sought formal principles which cannot be justified any more than one can justify the proportions of a perfect body but which, established at the Creation, shape and animate all Nature.

From Goethe's perspective, the faults of Newton are easy to see. The main one is to take the colors of the spectrum as primary and white as a mélange:

> From this point of view we see that a great mistake has been made in natural philosophy, for a secondary phenomenon has been put first and the *Urphänomen* second, a compound effect is taken as a simple one, a simple one as compound. . . . The student of Nature should allow the *Urphänomen* to rest in beauty and majesty. Let the philosopher admit it into his study,

and he will find that fundamental phenomena make a stronger foundation for new thought than isolated cases, general categories, opinions, and hypotheses.

One point at which Newton and Goethe collide is the usefulness of mathematics in explaining the world around us, and it is not hard to see why Goethe regarded Newton's work with confusion and contempt. In §6.1 I explained the traditional difference between a philosopher and a mathematician: philosophers explain what exists; mathematicians calculate what is going to happen. When they look at the heavens to forecast weather or the outcome of a projected marriage mathematicians often turn out to be wrong, while philosophers deal with eternal questions of cause and existence. When Galileo insisted on being named the Grand Duke's Mathematician and Philosopher he was blurring a boundary that the Scholastic world took for granted. Kepler went further, and Newton erased the boundary altogether. Publishing his calculations in the Royal Society's *Philosophical Transactions*, calling his great work the "Mathematical Principles of Natural Philosophy," writing hundreds of pages about gravity while finally having to admit that he does not know what it is: to anyone educated in the old school, these were the pretensions of a little man who gives himself airs.

But mathematics was essential to Newton's philosophy, for it defines the areas in which one can draw necessary conclusions. Take Snel's law for example, a simple mathematical formula. It does not apply to white light, for experiment shows that the different components of it are refracted in different degrees. It applies only to the limiting case of an absolutely pure color. Further, the law applies only to a ray, the limiting case of a narrow beam. Neither exists in Nature (though today a laser produces narrow beams of a color that is very nearly pure); both pertain to the way physicists think about Nature in terms of models. The real world is so complex, so full of interconnections, that nobody is looking for an exact theory of it; even if such a theory were possible it would be so complicated that nobody would feel enlightened.[3] In a model one simplifies: one introduces an infinitely narrow ray of a pure color, a perfectly straight ruler, a perfectly formed prism, a material that absorbs all the light that reaches it, a lens whose surfaces are parts of mathematical spheres. The model can then be analyzed as the world cannot, and the beauty of this kind of study is the way the properties of the model illuminate our experience of Nature.

Goethe did not understand mathematics and had no use for numbers or models. It was nothing to him that Newton had explained the colors of the rainbow and calculated where it appears in the sky. For Goethe, truth was

---

[3] The physicist Eugene Wigner, shown the result of an immense computer calculation, said, "It's nice to know the computer understands the situation, but I would like to understand it too."

unitary and total. It dealt directly with experience, and so his experiments and descriptions were really complementary to Newton's. It is as if Newton anticipated Goethe's attack when he carefully identified the color of light in terms of its refrangibility and generally avoided talking about "the phantasms of colours." Goethe and Newton work in different domains of thought and experience, each making assumptions, performing experiments, and drawing conclusions that do not necessarily follow.

Consider for example the discovery by Hooke that the colors of a thin film repeat periodically as the thickness of the film is increased. Goethe studies thin films and explains their colors in terms of increased cloudiness, but he never mentions the periodicity. Newton explores it with a brilliant experiment, but his interpretation in terms of "fits of difficult and easy reflection" has not held up. Goethe's studies help painters understand the special color of a shadow, whereas psychologists now consider that Newton's remarks on shadows[4] pay too little attention to impressions that originate in eye and brain. Goethe's mistake was to see no merit in work that tries to unravel the mysteries of Nature, a little at a time, without any clear idea of what it will find next. He paid heavily for this mistake, for the scientists of his time and ever since have tended to dismiss his discoveries because he explained them in terms that make no scientific sense. His friend Dr. Eckermann reports that

> his feeling for the theory of colors was like that of a mother who all the more loves an excellent child the less it is esteemed by others. "As for what I have done as a poet," he would repeatedly say to me, "I take no pride whatever in it. Excellent poets have lived at the same time as myself; poets more excellent have lived before me, and others will come after me. But that in my century I am the only person who knows the truth in the difficult science of colors—of that, I say, I am not a little proud, and here I have a consciousness of superiority to many.

What can we learn from Goethe's mistake? Perhaps something about what it is to be a poet. The wisest comment I have seen is by the German polymath Hermann von Helmholtz, whom we shall encounter again in the next chapter:

> Now in poetry, as in every other art, the essential thing is to make the material of the art, be it words, or music, or color, the direct vehicle of an idea. . . . The poet, feeling how the charm of his works is involved in an intellectual process of this type, seeks to apply it to other materials. Instead of trying to arrange the phenomena of Nature under definite conceptions, independent of intuition, he sits down to contemplate them as he would a work of art, complete in itself and certain to yield up its central idea, sooner or later, to a sufficiently susceptible student.

[4] For example, *Opticks*, p. 183.

Goethe's work on color has not entirely disappeared; one still finds a few references to it in modern books. But now it is time to cross the Channel and meet the man who began the serious demolition of Newton's rays and also, as we shall see in the next chapter, provided the first insights into the way the eye distinguishes one color from another.

5. Light Is a Wave

The theory that represents a beam of light as a bundle of rays, each consisting of a stream of particles, is known as the emission theory, and its partisans are called emissionists; but the terms were scarcely used in the eighteenth century, for what good is giving a special name to something that everyone already believes? Huygens, Hooke, Euler, and a few others had argued that light is some form of wave motion, but the gate on which they hammered was protected by two huge towers: one was the simple fact that sound waves turn a corner but light does not; the other was the authority of Isaac Newton. Of course Newton knew, following Grimaldi, that a little light strays into a shadow of a needle, but that can be explained if the needle exerts a force that pulls a light ray out of its straight line. Hooke had already described the colors formed in thin layers of mica as if they were some sort of wave phenomenon, but Newton explained them (§7.5) by giving his rays "fits of easy transmission and reflection." He knew very well that suggesting a fresh mechanism for each new effect is no way to make a theory. His queries were guesses thrown out at the end of a long scientific career to guide future research, but their unhappy result was that during almost a century following the publication of *Opticks* (1704), little was done to search more deeply into the nature of light.

In 1801 a London doctor named Thomas Young (1773–1829) went before the Royal Society with a lecture entitled "On the Theory of Light and Colors." It is a general account of the wave theory that treats Newton with infinite respect, quoting at length from *Opticks*, *Principia*, and the early papers to show (correctly) that their author was not as firmly anchored to the emission theory as generally thought. Young proposed to consider that light is a wave in the ether in exactly the same way that sound is a wave in air. This is not just an analogy but a virtual identity. The mechanics of ether being supposed exactly the same as the well-understood mechanics of air except that ether waves travel faster, everything that can be learned from experiments on sound can be applied almost without change to light.

Like other members of the Royal Society, Thomas Young was a wealthy dilettante. Unlike most of them he possessed fabulous intelligence and a capacity for hard work, but he was a dilettante nonetheless. There was little he could not do if he tried: as a boy he learned Latin, Greek, and a dozen other languages, mostly ancient. He taught himself a little calculus and read *Principia* and *Opticks*. Later he studied medicine at London, Edinburgh, Göttingen, and at Cambridge, where he was known somewhat derisively as "Phenomenon Young." Though he wrote poetry in Latin and Greek he never learned to express himself clearly in English. He must have been a very competent physician but he inspired little confidence in his patients, and his small practice allowed ample time for the studies for which he is famous.

The paper of 1801 was largely a tutorial exercise to let his audience know that the wave theory was still alive and could easily explain optical phenomena that required an emissionist to twist himself into strange shapes. The color formed in thin films is one, and Snel's law of refraction is another; a third is the fact that light of all colors travels at exactly the same speed (otherwise when a satellite suddenly emerged from behind Jupiter it would appear first in one color, then another). All sounds—high and low, loud and soft—also travel at the same speed, and Newton had already given a theory that shows why.

In Young's lecture is the proposition that is his main contribution to optics: "When two Undulations, from different Origins, coincide either perfectly or very nearly in Direction, their joint effect is a Combination of the Motions belonging to each." Later he named this the *principle of interference*. In optics it is interesting only if the undulations are of exactly the same frequency, and in practice this is achieved by having them both come from the same small source.

Consider for example the formation of colored fringes in a film of oil on a wet street. Light reaches the surface of the oil. Some is reflected there, some passes through and is reflected from the surface of the water beneath. The two reflected rays reach the eye together. If the light of the source is white, it is composed of many colors. (We will have to consider later just what "composed" means.) Figure 8.3 shows waves corresponding to one of the colors, say green. It shows two possible relations between the two reflected rays, and of course all the intermediate relations are also possible. In case A the two waves are said to be *in phase*, and in B, *out of phase*. If both waves enter our eye, what do we see? Young's proposition says: add the two waves together. If the crests of one coincide with the crests of the other (a), the resultant wave has a greater amplitude than either component. If the crests of one coincide with the troughs of the other (b), you are adding a positive number to a negative one so that if they are of equal sizes, as drawn here, the result is zero. The phase relation between the two waves depends on the thickness of the film, the wavelength of the light inside the film, and some considerations as to what happens at the surfaces that do not concern us here. If top and bottom of a certain film return the waves of green light in phase, the sensation of green will be reinforced while the waves corresponding to other colors in the white light will be more or less out of phase and will tend to cancel each other. Our eyes will

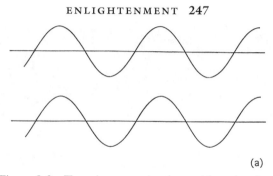

(a)

Figure 8.3a. Two sine waves in phase with each other.

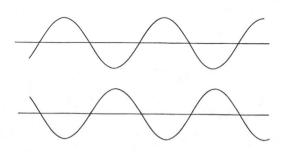

Figure 8.3b. The waves are out of phase. (b)

see mostly green. This is a rapid explanation of the colors that Robert Hooke saw in his mica flakes, that Newton observed in many experiments, and that Thomas Young was the first to explain correctly. But note the characteristic obscurity of his proposition as I quoted it. He does not mention that his words apply only to light of a single color, so that the two "undulations" must be of the same frequency, or what the word "combination" means. Everything he wrote, and by report everything he said in his lectures, was vague and offhand. If he did an experiment he rarely described it; if he calculated something he gave only the result, and it must have been partly because of his obscurity that he made so little headway against the emissionists.

Thomas Young's next lecture before the Royal Society, "Experiments and Calculations Relative to Physical Optics" (1803), continues his assault on the ray theory, only this time he has experiments that support what he is saying, including the simple but crucial one diagrammed in Figure 8.4a. Into a narrow beam of sunlight he introduces a narrow strip of cardboard, $1/30$ inch wide, and observes the shadow it casts on the wall of his room. He sees Grimaldi's colored fringes outside the area of geometrical shadow as well as what Newton had already seen, that some light is also bent into the shadow. For the emissionists it was puzzling, of course, that the bit of cardboard could exert a force that bends some of the rays one way and

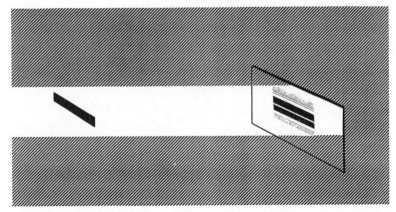

Figure 8.4a. A beam of light encountering a narrow obstacle is diffracted.

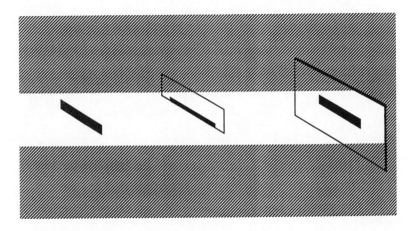

Figure 8.4b. If the upper part of the shadow is cut off, the diffraction pattern disappears.

some the other, but it was also a puzzle for a wave theorist: just what was happening? Figure 8.4b shows the next step: Young introduces a second piece of cardboard in front of the little strip, masking one edge of the shadow it casts. At once all the fringes disappear; therefore "these fringes were the joint effects of the portions of light passing on each side of the slip of card, and inflected, or rather diffracted, into the shadow." Light from both sides must cooperate to form a diffraction pattern. For emissionists the puzzle has become much more puzzling, but Young sets out to calculate the wavelengths of light of different colors required by his experiment and compare the result with the value derived from Newton's measurements on the colors of thin films. The agreement is good enough to show that the wave theory explains both. True to form, Young does not show his calculation, but a modern reconstruction shows he is assuming that each

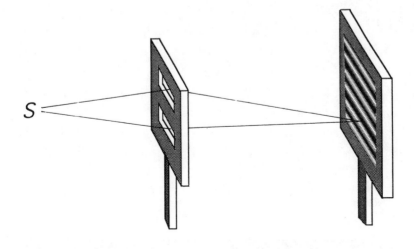

Figure 8.5. Young's experiment: formation of an interference pattern using two beams from the same source.

edge of the narrow strip of cardboard in (a) acts as a new source of waves and is adding these two waves to the wave that comes straight from the source in order to explain the effect produced where all three arrive together. The idea was perhaps suggested by Huygens, but the calculation is wrong, for Huygens's idea is that *every* point of the wave fronts AB and CD acts as the source of a new wave, and not just the points B and C. To do the calculation right required someone who knew calculus well enough to develop it in new directions, a task requiring much more than mathematics picked up from casual reading.

At the same time as he was talking to the Royal Society, Young delivered a long series of public lectures covering all branches of "Natural Philosophy and the Mechanic Arts," lectures so puzzling to most of his hearers that they were discontinued after two years; but as written up later they give a valuable survey of science at the turn of the century. Lecture 39 describes another experiment, so simple in concept that every modern student of optics knows it as Young's Experiment.

Figure 8.5 shows an arrangement in which light from a single source strikes a screen with two very narrow slits in it located a fraction of a millimeter apart. On the screen at the right is the pattern that is formed by the rays issuing from the two slits: an alternation of light and dark bands. To explain this, assume that each slit acts as a new source of light; Young's

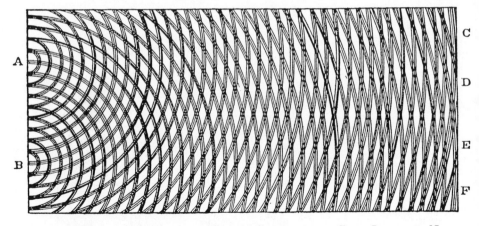

Figure 8.6. Young's explanation of the interference pattern. From *Lectures on Natural Philosophy* [1807].

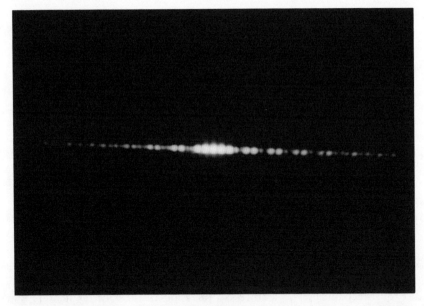

Figure 8.7. Two-slit interference pattern formed with laser light of a single wavelength.

principle of interference says "their joint effect is a combination of the motions belonging to each." Figure 8.6, reproduced from Young's book, shows the result. Suppose that the white bands surrounding each slit represent the crests of the waves coming from it and the dark bands represent troughs. If you sight along the page from the left you will see that the points marked (approximately) by C, D, E, and F mark lines along which the two sets of crests and troughs tend to cancel, while between them are regions in which they add. Thus by Young's principle the first set of points on the screen should be dark and the second set bright. Figure 8.7 shows the photographic record of an experiment using light of a single color. It is so easy to explain this result by means of waves and so difficult by means of particles that the experiment is usually taken as definitive. As we shall see in Chapter 10, it was only a century later, with the advent of the quantum theory, that the facts once more became hard to understand.

Why were there still emissionists after Young's simple demonstration? Mostly because there were so few experimental scientists around; I do not know of anyone who even repeated his experiment to make sure it had been done right.[5] Reverence for the memory of Newton was still stronger than hunger for new ideas, and very few opinions suffered any perceptible change, either in England or in France, until someone arrived who was far better equipped to observe optical phenomena and create a theory that explained them exactly.

Augustin Jean Fresnel (1788–1827) trained as an engineer and worked for most of his short life in the French Service of Bridges and Roadways. During vacations and some periods of leave that the Service allowed him, he worked with astonishing intensity and speed on optical research intended to destroy the emission theory and establish the wave theory in its place. He did not read English and knew almost nothing about Young's work and so did most of it over again, but more carefully; and unlike Young he wrote it up in crystalline form. Apparently the wave theory convinced him as soon as he heard of it. He got to work, and in 1819 one of his earliest papers won a prize that the Academy of Sciences had offered for a memoir on diffraction. Influential members of the Academy had hoped that someone would explain the phenomena on emissionist principles; what they got was a memoir of almost 140 pages that describes experiments of unprecedented accuracy (diffraction fringes are located to $1/100$ millimeter) and uses the theory of waves to give an exact and detailed explanation of what was observed. The essay's huge merit was that it explained the results of a large number of results involving interference and diffraction on the single assumption that light is a wave in the ether that obeys the principle of interference. The emissionists need rays with particles traveling along them, their "fits of easy transmission and reflection," the forces exerted on them by the obstacle that diffracts them, the absence of any force exerted by one particle on another, and many more such hypotheses; but even if one accepts them all, Fresnel points out, they still do not add up to a theory that accounts for the facts. In his paper Fresnel praises the natural order that produces the

[5] A question has been raised as to whether Young himself did it right (Kipnis 1991, Chapter 5). He describes the experimental arrangement and what he saw with it; the first description is so sketchy and the second so vivid that I think he probably tried the experiment in various ways and reported what he saw when he found one that worked.

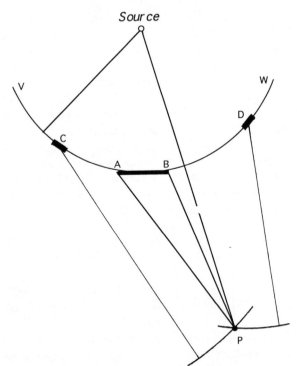

Figure 8.8. Comparing Fresnel's explanation of the diffraction pattern with Young's. AB is an obstacle in the light path. To find the intensity of light at point P, Young adds together waves coming from A, B, and directly from the source. Fresnel, following Huygens, takes every point of the arc VW except the stretch AB as a source of a wave contributing to the wave at P. (C and D are sample bits of the arc.)

maximum of effects with the minimum of causes, then continues:

> It is certainly hard to discover the reason for such an admirable economy. . . .
> But even if this general principle of the philosophy of science does not lead
> directly to a knowledge of the truth, it can still guide the efforts of the mind,
> turning it away from systems that relate the phenomena to too large a number
> of different causes, and causing it to prefer those which, depending on the
> smallest number of hypotheses, are the most fertile in their consequences.

The principle of interference was clear to Fresnel from the beginning, but he did not make Young's restriction that the rays have to "coincide either perfectly or very nearly in direction," and in the memoir he uses it for the first time to explain in mathematical form how Huygens's wavelets combine (Figure 7.13) to form a wave front. Figure 8.8 contrasts Fresnel's theory of diffraction with Young's. To explain how an opaque strip diffracts light Young superposed the three waves that follow the paths SP, AP, and BP. Fresnel, following the Huygens prescription, constructs the amplitude at P by dividing the wave fronts VA and BW into an infinite number of infinitely small segments, of which C and D represent two, and adding together the wavelets produced by all of them. The mathematics is delicate and is explained in Fresnel's memoir or in any modern text on physical optics. Once Fresnel's paper had been published (which took several years) and absorbed by the scientific community (which took even longer), the emission theory was nailed securely into its coffin, and Thomas Young's role as a pioneer was finally recognized.

In an aphorism that is often quoted, the German physicist Max Planck remarked, "A new scientific truth does not triumph by convincing its opponents and making them see the light, but rather because its opponents eventually die, and a new generation grows up that is familiar with it." In the 1850s old scientists still grumbled, but the emissionist theory finally crashed because it did not meet Newton's own standard: it did not lead to successful calculations.

Fresnel's analysis also laid to rest Newton's objection to the wave theory: if I can hear you around the corner of a building, why can't I see you? It turns out that as Grimaldi and his followers had shown long ago, light really does turn corners, and this can be understood if, following Huygens, one takes each point of a wave front as the source of a new circular wavelet. Each wavelet expands into the region of shadow, but all of them interfere with one another to produce almost (not quite) perfect cancellation, whereas in the illuminated region they reinforce one another. It turns out that the amount of light that penetrates a distance into the shadow depends on the wavelength of the wave: the shorter the wave, the less penetration. Light waves are a million times shorter than the sound waves of someone's voice. They do stray around the corner, but the light is so faint that unless you are as careful as Father Grimaldi you can't see it.

Augustin Fresnel worked while life lasted. Tuberculosis drew him into darkness in his fortieth year, but his name lives on. Every optical scientist uses his mathematics; almost every lighthouse uses the lens that he invented. Thomas Young lived longer and became a public figure, serving on boards and committees concerned with matters ranging from public health to the improvement of navigation, actuarial statistics, and the construction of ships. We shall see in Chapter 10 that he made a brilliant guess as to how eyes perceive color that turned out to be correct, but perhaps what gave him most satisfaction was the progress he made in deciphering the Egyptian hieroglyphs on the Rosetta Stone. Hermann von Helmholtz wrote, "He was one of the most acute men who ever lived, but had the misfortune to be too far in advance of his contemporaries. They looked on him with astonishment but could not follow his bold speculations, and thus a mass of his most important thoughts remained buried and forgotten in the

'Transactions of the Royal Society,' until a later generation by slow degrees arrived at the rediscovery of his discoveries, and came to appreciate the force of his arguments and the accuracy of his conclusions."

Even after Fresnel's prize memoir there remained the puzzle of polarized light. If you can find two pairs of polarizing sunglasses you can do experiments that at first are quite surprising. Put the pairs together and look through them both, then slowly rotate one. You will see a great change in the intensity of the light that comes through. You can think of this by supposing that the first lens prepares the light in some way—it is called the *polarizer*—while the second one is the *analyzer*. Obviously, polarized light has some directional quality in the plane perpendicular to the direction in which it is moving. Polaroid is a modern invention. Look through one of the lenses at sunlight reflected from a piece of glass (try it at different angles of reflection), and you will see signs of polarization. The reason polarizing sunglasses are used at all is that sunlight reflected from a road tends to be polarized, and they decrease the glare.

What is this new property of light? The analogy between light waves and sound waves breaks down here: you can't polarize ordinary sound. Newton suggested that rays of light have "sides," whatever that means; he never said how one side differs from another. In an *Encyclopedia Britannica* article in 1817, Young very tentatively proposed that light waves might have a slight sidewise motion in addition to their longitudinal motion and that this motion, like that of a stretched rope which one might shake up and down or right and left, might distinguish the direction of polarization. In 1821 Fresnel, probably independently, proposed that polarized light is different from sound in that its waves have no longitudinal motion at all; it is entirely transverse. But this simple hypothesis made it necessary to answer a fundamental question: because a ray of light is not at all like a rope, exactly what is it that is waving? Answers to this question will be discussed in Chapter 9, but first we will look at some advances in optical instruments.

6. Seeing Farther: Telescopes

With even the earliest telescopes, Galileo and his contemporaries made a series of stunning discoveries: Jupiter's satellites, sunspots, Saturn's rings, the mountains on the moon, the stars of the Milky Way. Nobody doubted that further marvels awaited the development of better instruments. Astronomers sought three improvements: a brighter image, a cure for spherical and chromatic aberrations, and more magnification. The first one has an obvious remedy: a bigger lens intercepts more light and brings it to the eye. But a bigger lens of a given curvature has more spherical and chromatic aberration than a small one. Newton had shown in Book I of *Opticks* that

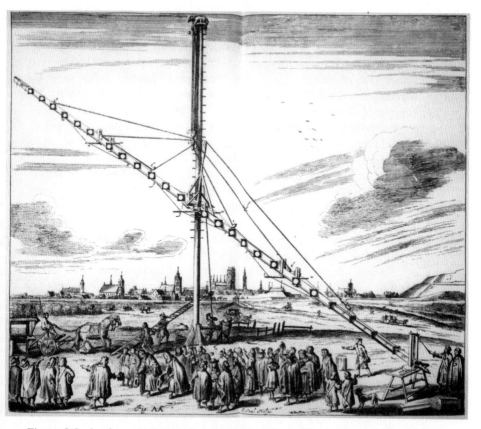

Figure 8.9. A telescope designed to minimize spherical and chromatic aberration. From Hevelius, *Machina Coelestis* [1673–79].

chromatic aberration is the more damaging, and so telescope makers concentrated on eliminating that. Figure 6.6 shows that the aberration is worse in light that goes through the outer part of a lens. Images can be made clearer by masking off this part; Galileo had done this even with the little telescopes he made in the early 1600s. But masking the edge decreases the brightness. One cure was to make the lens large but comparatively flat, but now the instrument had to be enormously long. Figure 8.9 shows one used by Hevelius in the seventeenth century, difficult to guide and impossible to use in even the slightest wind.

It had occurred to Newton that in a lens made of two different transparent materials, say glass and water, the formation of colors might be less, but in discussing this in *Opticks* he made a serious mistake and thought he had proved that such a thing could never work. In 1729, two years after Newton died, Chester More Hall, a gentleman barrister from Essex, remarked that light entering the eye passes through the eye through three humors, vitreous, aqueous, and crystalline, and argued that because there is no chromatic aberration, this must be the reason why the Great Design called for several media; therefore Newton must be wrong. The reasoning is incorrect—our eyes do have chromatic aberration, but we are used to it—but acting on it Hall combined two lenses of different kinds of glass. He found that images formed with them showed less color, and in the following decades a number of opticians, using cut-and-try methods, succeeded in making very useful lenses. When made in this way a lens is called *achromatic*. But to calculate the best design for such a lens requires a theory of lenses in combination and an accurate knowledge of how the indices of refraction of the two kinds of glass depend on color. Good measurements of refraction were made early in the nineteenth century, but it was another sixty years before mathematicians were able to say exactly how an achromatic lens should be made.

Newton's reflecting telescope was meant to solve the problem of color, but it tended to introduce a worse one, for the metal mirrors tarnished quickly and any attempt to clean them changed their shape. And if they were carefully polished back into a proper figure it was never the figure they started with; the magnifying power was changed, and new formulas had to be found for interpreting what was seen. There was another difficulty: a reflector shows clearly only the object it is aimed at, whereas every part of the sky that is seen when you look through a refractor (a telescope that uses lenses) is imaged almost equally sharply. And there was the question of cost: lenses were cheaper than mirrors.

As both kinds of telescope were improved, preferences shifted back and forth. In 1655 the Huygens brothers, using a twelve-foot refractor, discovered the first satellite of Saturn. In 1722 James Bradley, using a refractor like Hevelius's with 212 feet between the front lens and the eyepiece, measured the diameter of Venus; but soon, in search of still greater power, astronomers switched back to reflectors.

William Herschel (1738–1822) was born in Hanover into a German family of musicians. He had a good education for those days and absorbed it eagerly. At fourteen he was an oboist in a regimental band. At that time Hanover was a possession of the English crown, and in 1756 the English, involved in the Seven Years' War with France and fearing invasion across the Channel, brought in Hanoverian soldiers to help protect them. Herschel's unit was billeted in quarters near the south coast. He wrote later, "Here I applied myself to learn the English language and soon was enabled to read Locke on the Human Understanding." Later the unit returned to Germany where it was so thoroughly defeated by the French that it didn't need a band. William went back to England for the rest of his life. For many years he was director of music for the City of Bath, active as organist, conductor, and composer, but by the time he was thirty-five his thoughts had seized on astronomy. His brother Alexander had already joined him in the musical life at Bath, and now his sister Caroline (1750–1848) came to live with them, serving William with unselfish devotion. She managed his household; when his concerts needed a good soloist she sang; she copied musical scores, and in the course of time, as he needed help with his astronomical work, she became an astronomer. She read aloud to William while he polished his mirrors, she recorded observations at his dictation, she did the calculations that converted telescopic settings to give locations in the celestial sphere, and she observed what she was asked to observe. Only when he was away did she do any astronomical work of her own, but during those nights she found eight comets. For these the Royal Astronomical Society awarded her its gold medal and, shortly before her death at ninety-seven, she received another one from the hands of the King of Prussia.

From 1772 on, the three Herschels collaborated to make one reflecting telescope after another. The main task was to choose metallic alloys that were hard, shiny, and resisted corrosion and then grind and polish them to perfect figure. But also, as the mirrors became larger and the tubes longer, the machinery to raise them and steer them became heavier and more complex. The first successful instrument, seven feet long, proved much better than any at the Royal Observatory; the second, twenty feet, was finished in 1774 and with an occasional new mirror served William Herschel for the rest of his long life. A final huge effort produced in 1795 a forty-foot reflector (Figure 8.10). At best it gave beautiful images, but here the disadvantages of size became apparent. The mirror weighed a thousand pounds and the iron supports as much again, so that the strength of several men was required to aim it, and the mirror itself changed its shape as it sagged under gravity. In the end it was used only a few times and survived as a monument rather than as a tool for research. Many years later Oliver Wendell Holmes caught sight of it as he traveled from London to Windsor, "a mighty bewilderment of slanted masts, spars and ladders and ropes, from the midst of which a vast tube, looking as if might be a piece of ordnance . . . lifted its mighty muzzle defiantly towards the sky."

With the twenty-foot instrument Herschel discovered a planet beyond the orbit of Saturn. He named it after King George III and in return was named director of the Royal Observatory on a salary that enabled him to be a full-time astronomer; the planet's name, however, did not stick, and it is now called Uranus. He studied double stars and showed that some of them revolve around each other in orbits explained by Newtonian principles, suggesting that these principles are universal in their scope. His greatest project was to map the universe. Figure 8.11 is his chart of the Milky Way. The sun, by some long persistence of the Ptolemaic sense of things, lies near the center. But along with stars the telescope showed strange objects: spherical clusters of stars as well as other things that looked like puffs of smoke and were named nebulae, little clouds. Do the clusters belong to our own galaxy? (They do.) Are at least some of the nebulae "island

Figure 8.10. William Herschel's forty-foot telescope, made in 1795. From *Philosophical Transactions of the Royal Society* 90 (1800).

universes" like the Milky Way, isolated in space and seen from an enormous distance? They are; Herschel seems to have thought so but was never sure.

William Herschel was not able to establish the scale of his map. That was done by Friedrich Wilhelm Bessel (1784–1846), an astronomer and mathematician who had started out as a ship's clerk. He succeeded in measuring the parallax of a nearby star called 61 Cygni and showing that it was about $^1/_3$ arc second. This corresponds to a distance of about ten light years[6] (a

[6] Light travels at about 186,000 miles per second. A light year is the distance it travels in a year.

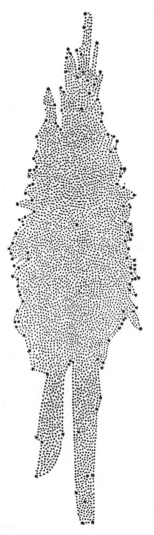

Figure 8.11. Herschel's map of the Milky Way. The sun is at the center.

newer value is six) and was the first step in the long and still unfinished task of measuring the universe.

After Herschel's labors astronomers returned to refractors, and by midcentury few reflectors were in operation. With the development of silver-coated glass mirrors the advantage swung slowly back toward reflectors, and there it remains. For gathering light from dim and distant objects huge mirrors are used. The one in the Palomar Observatory's Hale Telescope is fifteen feet across, but its field of view is narrow. To give a wider field, reflecting telescopes can be equipped with a sheet of slightly shaped

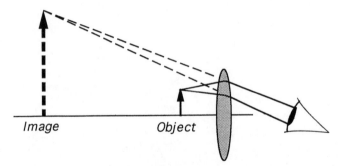

Image   *Object*

Figure 8.12. How a magnifying glass works. Two rays from the tip of the object arrow reach the eye, one of them having been bent more than the other. It appears to the eye as if they came from the image, which though farther away is much larger.

glass mounted at the front end. These correcting plates, as they are called, were invented in 1932 by Bernhard Schmidt of the Hamburg Observatory, and they allow a reflector to give a sharp image over its entire field.

After Herschel had drawn a primitive map of the Milky Way and Bessel had attached a scale to it, there remained the problem of the nebulae: are they island universes outside the Milky Way? How large are they, and how far off? These questions had to wait until 1924 when Edwin Hubble (1889–1953) used an eight-foot reflector on Mount Wilson to study the nearby Andromeda nebula. The telescope was so powerful that it resolved some of the nebula's individual stars, and by studying certain ones that vary in brightness Hubble was able to estimate its distance as 930,000 light years; later measurements doubled this figure. A century and a half after Herschel discovered our galaxy, Edwin Hubble had discovered the universe.

7. Seeing Smaller: Microscopes

The simplest microscope is the single lens we use for reading small type. Take a moment to study Figure 8.12, which shows how it works. Follow a pair of rays from a small object. Because the upper ray is bent more than the lower one, they arrive at the eye nearly parallel. (We instinctively hold the lens so that this will be so.) The brain constructs an image from these rays as if they came from a larger object a little farther away. The telescopes and microscopes mentioned here can all be explained by diagrams that are more complicated but in the same spirit.

Suppose we want to make a stronger magnifier. The trick is to use a fatter lens. The surface must be more curved. Compare a tennis ball and a marble. Both are spheres, but the marble's surface is more curved; to curve the surface of a lens we must make it smaller. Descartes illustrates a single-lens instrument in the *Dioptrique* [1637] that consists (Figure

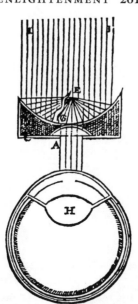

Figure 8.13. Descartes's microscope. A spherical mirror concentrates the sun's light on the specimen at E. The lens is at A; below is the eye. From *Dioptrics*.

8.13) of a little, highly curved lens A, with the edges masked to lessen the aberrations. But masking the edges reduces the amount of light that comes through, and so a spherical concave mirror is used to focus sunlight on the specimen, which is impaled on the spike E. And then of course, there is an eye. The trouble is, the lens is so small that one has to squint at it from close up. That is good enough for a casual glance, but not for making the kind of drawings that Hooke made for his *Micrographia*. He writes that he found the use of a single lens so offensive to the eye that he was obliged to add a second one, a magnifier called an eyepiece, through which he looked at the image formed by the first lens. Figure 8.14 shows such a microscope, essentially the Galilean model described in §5.6. The difficulty of this arrangement is that the eyepiece enlarges the errors and aberrations produced in the first lens, and when this is small and nearly spherical they are very severe. The early compound microscopes, as they are called, were more comfortable to use than single lenses but inferior in performance, and for the next two centuries the single lens was the chosen instrument for people who wanted to see things that are very small.

One student of *Micrographia* was Antony van Leeuwenhoek (1632–1723), a respected shopkeeper in Delft. He held civic offices, and we know that when the painter Vermeer died Leeuwenhoek was his executor. Leeuwenhoek knew neither English nor Latin but Hooke's pictures inspired him, and by the 1670s he was making superb lenses. Hooke's instru-

Figure 8.14. Hooke's microscope, showing arrangements for gathering light from a flame and concentrating it on the specimen. From *Micrographia* [1667], courtesy of the Chapin Library, Williams College.

ment shown in Figure 8.14 magnified about fifty times. His interest was in revealing to his audience the true appearance of a familiar object like a flea or the mold on a piece of cheese. Leeuwenhoek wanted to find something new. His tiny lenses were ground and polished, some from glass, a few from quartz crystals, and some magnified more than 250 times. The instruments were small and simple (Figure 8.15), hardly larger than a postage stamp, and he sold or gave away hundreds of them. Figure 8.16 shows Leeuwenhoek's drawing of a tiny crystal of silver precipitated out of a solution of silver in *aqua fortis* (nitric acid).

Leeuwenhoek kept the best lenses for his own use, and nobody knows how they were made. For fifty years his discoveries were reported to learned societies in simple Dutch. He discovered the protozoa that live in ponds and in our bodies. He discovered bacteria, species after species, and

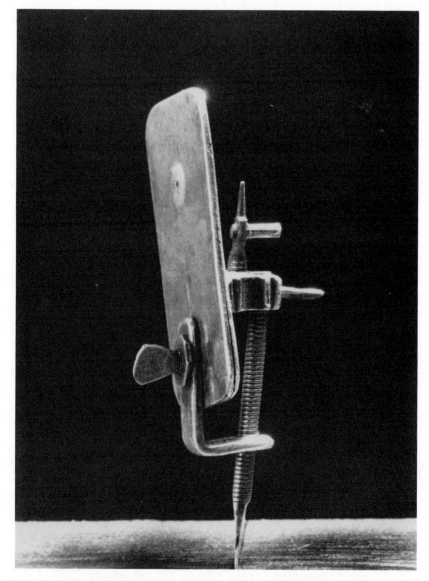

Figure 8.15. One of Leeuwenhoek's microscopes, courtesy of Museum Boerhaave, Leiden.

many kinds of little insects. He studied blood corpuscles and their circulation in capillaries, and compared the forms and activities of the spermatozoa of man and animals. He was an enormously productive scientist exploring an unknown world; to the Royal Society alone he wrote more than two hundred letters.

Ultimately the compound microscope caught up with the single lens,

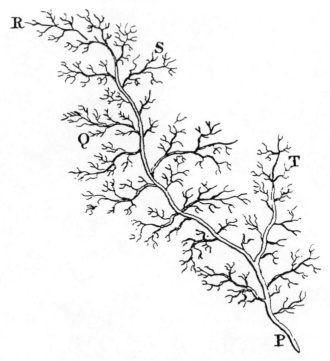

Figure 8.16. Leeuwenhoek's drawing of a microscopic crystal of silver. From *Philosophical Transactions of the Royal Society* 23 (1703).

and today it is the instrument of choice. But by the mid-1800s the theory of optical instruments had shown that there is a limit to the power of a conventional microscope. The broader the objective lens and the shorter the wavelength of the light used, the more clearly the microscope can see, but there are limitations on each. In §10.5 we shall see that some of the most powerful microscopes now use a beam of electrons instead of a beam of light, but also that unconventional instruments are being developed that escape the old limitations.

8. Learning from Color: Spectroscopy

Newton's analysis of white light showed that in the realm of color the eye is easily deceived. Presented with this light, it does not distinguish the colors it contains and reports that there is no color. We have noted that one cannot distinguish pure green from a light that is a suitable mixture of blue and yellow and has no green in it at all. It is very different with sound. A reasonably sharp ear knows quickly what instruments are playing in an orchestra, and this is more remarkable when we think that each

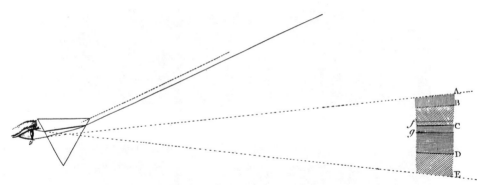

Figure 8.17. William Wollaston's observation of dark lines in the solar spectrum. From *Philosophical Transactions of the Royal Society* 20 (1802).

instrument produces not just one tone but also the overtones that create its characteristic timbre, so that perhaps a hundred tones are being heard and sorted out. Evidently an ear is a far more delicate analytical instrument than an eye, and if we are to learn anything about the colors present in a beam of light we must do as Newton did—pass it through a prism or some other instrument that separates them so we can look at them one at a time. But for scientific purposes it is not enough just to look at the colors; they must be identified in some measurable way, and once the wave theory had been accepted the obvious identifier was their wavelength.

A century after Newton's experiments Thomas Young deduced some figures for the wavelengths of light of different colors. He found that humans can see light from 39,180 waves per inch at the red end of the spectrum to 59,750 at the violet end. This is not a very wide range when compared with the range of audible sounds. In musical terms, the interval that we can see corresponds to about a sixth, say from C to A, whereas a healthy ear can hear nine or ten octaves. Why does the eye have such a limited range? We shall see in a moment that there is a reason.

At the same time as Dr. Young was calculating wavelengths, another doctor introduced an important improvement to Newton's way of looking at spectra. Newton had exposed his entire prism to sunlight. William Hyde Wollaston (1766–1828) had the idea [1802] of examining the spectrum of a beam of sunlight that had been narrowed by passing it through a slit $1/20$ inch wide and then looking at it while standing ten feet back from the prism. Since the slit cut out most of the light, he could look directly into the prism instead of casting the image onto a wall (Figure 8.17). The point of using a narrow beam is this: Consider two adjacent colors of the spectrum, say two particular shades of blue and green. If the sun's image is seen

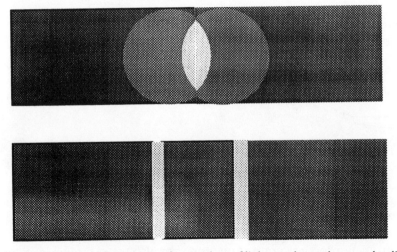

Figure 8.18. Top: circular images of two colors of light overlap and cannot be distinguished. Bottom: if the light has passed through a narrow slit the images can be distinguished.

through the whole prism (Figure 8.18) the two colors run together, whereas if the light arrives in a narrow beam they are separated and therefore more pure. Wollaston saw the sun's spectrum very clearly and found that the colors were separated by some dark gaps, indicated in Figure 8.17. When he looked at the lower part of a candle flame he saw several distinct bands of color and a brilliant narrow yellow line. Before him, a Scot named Thomas Melvil (1726–53) had recorded similar observations, seeing the same bright yellow line again and again in light from different substances. Melvil died young, and Wollaston became well known for his discoveries in optics, but he seems to have done no more with spectra. Optics is an orderly subject, in which one knows what is going on and can use geometrical reasoning to design and perfect an instrument, whereas someone who looks at the spectra of hot objects sees strange patterns with no signs of order. Different substances show different spectra, and even they depend on how the sample is treated. After Wollaston it was more than a decade before anyone else took up the study of spectra.

The spectra of the sun and various substances that are either burning or glowing in the heat of a flame have red light at one end and trail off in a dim violet at the other. It was natural to ask oneself whether these limits marked the ends of a spectrum or merely the limits of our eyes' ability to see it. As early as 1800 William Herschel took up this question. We detect sunlight in two ways: our eyes see its colors, and our skin feels its heat. To detect the heat, Herschel placed a small thermometer in the visible spectrum formed by a prism. It registered a rise in temperature. He moved the thermometer along the spectrum and past the end of the red light. For more than an inch along the dark screen it still registered heat. This was the first detection of infrared radiation. Beyond the violet end of the spectrum he found nothing, but the next year Johann Wilhelm Ritter (1776–1810) looked for that radiation by other means. He had noticed that a chemical compound called silver chloride darkens when it is exposed to light. It darkened when he exposed it beyond the violet end of the visible spectrum and showed that there is also ultraviolet light. Almost forty years later Louis Jacques Mandé Daguerre (1789–1851), William Henry Fox Talbot (1800–1877), and William Herschel's son John (1792–1871) simultaneously developed processes for fixing an image on a surface coated with silver chloride, and that was the beginning of photography.

Joseph Fraunhofer (1787–1826) was a poor boy with little education who after bouncing from one apprenticeship to another landed a job in an optical works that had been set up in the buildings of an eighth-century abbey at Benediktbeuern,[7] outside Munich. It had never recovered from the depredations of the Thirty Years' War and had finally, in ruinous condition, been secularized a few years earlier. It was a good place for a glassworks because it had ample supplies of wood for furnaces as well as water power for machines that shaped and polished lenses. The factory became famous because nowhere else was glass made so free from streaks and bubbles. Fraunhofer was soon running the optical department, and in his midtwenties he was in charge of the whole operation. In the design of instruments it is important to know the optical properties of the materials used, especially how their refractive indices depend on the wavelength of light. To study this property he realized that he needed light more nearly monochromatic than that produced by putting a prism into the path of the sun's rays, and so he reinvented the technique of the slit and in 1814 rediscovered Wollaston's dark lines, but his instruments were so much better that now he saw "*an infinite number of vertical lines of different thicknesses.* These lines are darker than the rest of the spectrum, and some of them appear entirely black." (Plate 11 was prepared by Fraunhofer; compare it with Wollaston's Figure 8.17, made only twelve years earlier.) Once he had measured the wavelengths of these dark lines, he used them to calibrate wavelengths in his studies of glass. He saw a dark line at the exact position of the previously observed yellow line, and marked the line on his chart of the spectrum.

Fraunhofer looked at Venus and saw nothing new, as was to be expected once one understood that he was seeing only reflected sunlight, but Sirius and other bright stars showed a variety of different spectra. For a few years he ran the optical works with one hand and wrote papers for learned jour-

---

[7] The name is familiar to lovers of late Latin poetry. When the abbey was secularized the contents of its library was sent to Munich, where one Baron von Aretin discovered an uncataloged book, an under-the-counter anthology of poems. These are mostly in Latin and concern love and religion and student life. Like birds in a cage, monks sang of an existence most of them had scarcely known. The songs were published under the title *Carmina Burana*.

nals with the other, but like Augustin Fresnel, he died of tuberculosis at thirty-nine. The dark lines in the sun's spectrum are named after him.

In the years after Fraunhofer's death clever means were devised to measure the continuations of the sun's spectrum into infrared and ultraviolet. Extensive dark bands were found on each end of the visible spectrum, especially in the infrared. We now know they occur because these radiations are absorbed in the Earth's atmosphere, largely by its water vapor. In fact, the narrow range of colors for which the atmosphere is perfectly transparent corresponds roughly with the range that our eyes can see.

After these discoveries many questions remained unanswered, some of them very puzzling: Why do burning or glowing substances sometimes produce spectra of bright lines and sometimes a continuous spectrum? Why does the pattern of bright lines vary from one substance to another? Why is the sun's spectrum a continuum from red to violet, crossed by hundreds of dark lines? It took a century after Fraunhofer's death in 1826 for satisfactory answers to be found, but long before they understood what was going on, physicists, chemists, and astronomers were using spectra to identify chemical substances. John Herschel studied the bright-line spectra of chemical substances heated in a flame and found he could identify even minute quantities by looking at their spectra through a prism. This would be a wonderful and easy way of identifying substances if it were really true that each substance makes its characteristic pattern of bright lines. At first it seemed not true, for the brilliant yellow that Melvil had noticed kept recurring. (It is the same yellow light that appears in the flame of a gas burner when soup boils over.) Only in the 1850s was it realized that sodium is the culprit. Sodium is one of the most common elements; it occurs as a contaminant in chemical samples unless they have been very carefully prepared, and the yellow line it produces is particularly strong. (Actually, there are two yellow lines; a good spectroscope shows a pair very close together.) Plate 12 shows the spectrum of neon gas. All the brightest lines are in the red-yellow region, accounting for its characteristic color.

Those are bright-line spectra. In 1836 David Brewster (we have met him before studying a crystalline lens from Nineveh) produced dark lines in the laboratory by passing light from a glowing hot surface through a flask of nitrous oxide ("laughing gas"). It was easy to guess that something similar happens when light from the glowing body of the sun passes through the cooler gases of its outer atmosphere. Often (but not always) the same line can be seen either bright or dark, according to how the observations are made.

During the decade after Brewster's observation several workers, especially Robert Bunsen (1811–99) and Gustav Kirchhoff (1824–87) studied the spectra of the chemical elements then known and concluded [1860] that the elements could be positively identified by their spectra. In the sun's atmosphere the authors could identify six elements definitely and several more tentatively. This was good news to those who had been irritated by the finality of some remarks made by the French philosopher August Comte, founder of the movement known as positivism. In 1835 Comte had mentioned as an example of the essential limitations of human knowledge that, though we may some day determine the shapes and distances of stars, we will never know anything about what they are made of.

Spectroscopy is an immense subject, and these brief remarks have shown only how it got started. In modern times it has led to the remarkable discovery of immense numbers of large and complex molecules floating around in space—molecules that until recently were thought to occur naturally only in living creatures. Where they came from and what they may have had to do with the evolution of life on Earth and its possible evolution elsewhere are questions that at the moment are unanswered.

Light enables us to know the universe, but what kind of a thing is it? It is of course a sensation in the eye, but we have seen that it has an existence outside ourselves. Newton thought it is a beam of particles, but what kind of particles? What are they made of? What size are they? What shape? How does a blue one differ from a red one? As this chapter ends, virtually everyone has conceded that Newton was wrong and light is a wave. But what kind of a wave? An ocean wave is not a thing; it is a property of the water, something that water does. Anyhow, if there is no water there is no wave. So if light is a wave, what is waving? By the end of the nineteenth century this was the physicists' most urgent question.

9

UNIFICATION

> To give a complete explanation of electric phenomena that
> reduces the laws of physics to the fundamental principles of
> mechanics, that problem has tempted many a searcher. But
> isn't the effort rather otiose, wouldn't it consume our forces
> with nothing to show at the end? If we knew there were a
> single solution, it would be a truth beyond price. But that is
> not how things are. No doubt, some mechanism to give a
> more or less exact imitation of electric phenomena could be
> invented, but if there is one, there is also an infinity of others.
> (Henri Poincaré, 1904)

1. A World Floating in Ether

FOR ARISTOTLE the existence of ether was a logical necessity. Of the four elements, Earth and Water naturally move downward; Fire and Air move upward and sidewise. When they arrive where they are going, they stop. But stars and planets move forever in circles; therefore they and the eternal, invisible machinery that guides them must be made of something else, and that substance is called ether. Further, there can be no empty space between the stars, for as we have seen, emptiness (*kenon*) is what does not exist; and obviously, what does not exist does not exist. Therefore the region above the terrestrial four elements must be full of something; this also is ether, and the heavenly bodies are glowing concentrations of it. Aristotelians could at least see the ethereal stars. Those who followed them began to think of stars as material objects and ether as a necessity more logical than physical. For Descartes, for instance, who *defines* matter as space and space as matter, empty space is a contradiction in terms, and space that seems empty must be filled with something not seen. And Isaac Newton, in the splendid paragraph that ends *Principia*, hints of an ether that explains almost everything he has not already explained:

And now we might add something concerning a certain most subtle spirit which pervades and lies hid in all gross bodies; by the force and action of which spirit the particles of bodies attract one another at near distances, and cohere, if contiguous; and electric bodies operate to greater distances, as well repelling as attracting the neighboring corpuscles; and light is emitted, reflected, refracted, inflected, and heats bodies; and all sensation is excited, and the members of animal bodies move at the command of the will, namely, by vibrations of this spirit, mutually propagated along the solid filaments of the nerves, from the outward organs of sense to the brain, and from the brain into the muscles. But these are things that cannot be explained in few words, nor are we furnished with that sufficiency of experiments which is required to an accurate determination and demonstration of the laws by which this electric and elastic spirit operates.

That marvelous fluid with so much to do has never been observed by anybody. One can say of course that the observed fact that light shines or my hand obeys my desire proves that it exists, but that kind of argument can prove anything at all. There is no evidence for ether beyond the facts that the hypothesis was invented to explain.

Isaac Newton, in arguing his philosophy, was faced by a delicate problem. The intellectual world of Giovanni Battista della Porta and his contemporaries was full of forces that act through empty space. If the wound cure quoted in §5.5 works at all, the ingredients must possess qualities that enable them to influence the world around them, even if we cannot see or understand how the influence is exerted. Qualities of this kind were called occult (meaning hidden), and one of the great achievements of the scientific renaissance of the seventeenth century was to begin to get rid of them. But what about gravity, or magnetism, or the electric force exerted by your comb on your hair, or other effects of the "most subtle spirit" that Newton has just described? He believed that these forces are just as real as the force you exert when you open a door, and yet they are not exerted by any visible or tangible means. Almost every European scientist of Newton's time regarded him as an excellent mathematician but accused him of dragging in occult qualities to explain what he did not understand. Newton did not know what causes gravity and said so, but he was careful to counter the criticism. In a letter to the theologian Richard Bentley he wrote

> that gravity should be innate, inherent and essential to Matter, so that one Body may act upon another at a Distance thro' a Vacuum, without the Mediation of any thing else, by and through which their Action and Force may be conveyed from one to another, is to me so great an Absurdity that I believe no Man who has in philosophical Matters a competent Faculty of thinking, can ever fall into it. Gravity must be caused by an Agent acting constantly according to certain Laws; but whether this Agent be material or immaterial, I have left to the Consideration of my Readers.

When Newton says "absurdity" I think he is saying that if one thing exerts a force on another it is *logically* necessary that there be some medium in between that transmits the force, just as if light is a wave it seems logically necessary that there be some medium that is moving. But the properties that we can ascribe to this medium are limited by the nature of the world as we know it. It must be transparent, that goes without saying. Further, the Earth circling the sun moves through space at thirty kilometers (about twenty miles) per second, and yet we feel no ether wind. In fact the Earth as a whole does not slow down as it forces itself through ether, century after century, at that immense speed. Newton knew this because his own dynamical theory of planetary motion agrees beautifully with observation if ether exerts no force; otherwise every planet would spiral inward toward the sun. Thus ether had to be a thin and imperceptible fluid, yet much more springy than air, since light travels almost a million times faster than sound.

Then came Fresnel's discovery that he could explain all phenomena of polarized light if he assumed that the wave of light is transverse and not, like sound, longitudinal. The calculations spoke for themselves, but in 1823 Thomas Young reacted with dismay: "The hypothesis of Mr. Fresnel is at least very ingenious, and may lead us to some satisfactory computations: but it is attended by one circumstance which is perfectly *appalling* in its consequences. . . . It might be inferred that the luminiferous ether, pervading all space, is not only highly elastic, but absolutely solid!!!" The point is that though transverse waves can travel over the surface of a liquid, in no way can they travel through the body of it. That requires the kind of rigidity that jelly has. How can the earth push its way through such a substance without being slowed down? Does ether belong to the world of manticores and unicorns? But other than the logical reasons why ether exists that I have mentioned, there is at least one good physical reason. Light carries energy. You can feel the energy of sunlight warming your skin. This energy took eight minutes to travel from the sun to you. Where was it during that time? The emission theory has a ready answer: it was carried along in the particles of light. But suppose you think light is a wave. When an ocean wave travels from Oahu to Molokai, the water does not travel, only a pattern on its surface. It carries energy, but the energy resides in the water, for a pattern is only a shape. Where does the energy of sunlight reside while it is in transit if not in some substance that extends from here to the sun?

Thus the great project of explaining the universe as a machine—the mechanization of the world picture, Eduard Dijksterhuis calls it [1961]—came up against a stone wall. In summary: if you assume there is an ether, and that it and everything else in the world are encompassed by the laws of Newtonian physics, the world becomes incomprehensible. But anyone in the early 1800s who suggested that this world contains entities that escape the vast Newtonian synthesis and are governed by entirely different mathematical principles would have been dismissed as an ignoramus. A hundred years later almost everybody understood that such entities exist, and now physicists are involved in a long, slow effort to find out whether there is a theory that once more unifies Nature under a single code of law.

Not knowing what else to do, the theorists gave up trying to explain why ether has no effect on something that moves through it. Perhaps it has an effect that a delicate experiment might some day reveal; perhaps the reason we do not experience an ether wind is that the Earth carries the neighboring ether along with it as it moves; perhaps there is some other reason. The theorists shrugged their shoulders and considered a simpler problem, to figure out the nature and properties of an ethereal substance through which the transverse waves of light would move so quickly. Easy to ask but not to answer, for the students of this problem, like theologians of the middle ages, had some choice as to what they assumed but no choice as to how their reasoning came out. They had to come out with Fresnel's mathematical theory of light or something essentially equivalent to it, since experiment after experiment had shown that Fresnel had it right. That turned out to be very difficult. Model after model was tried: crystals, particles that rolled around, a sponge of vortices like tiny smoke rings. Huge theories were constructed, monuments to mathematical labor, and abandoned. They had some value in teaching the workers new ways to calculate, and in §3 we shall see that in a sense one of them succeeded—not that it was right, but because it turned someone's thoughts in the right direction. This model had its conceptual roots in pre-Newtonian times.

2. Faraday Imagines a Field

William Gilbert (1540–1603), a native of Colchester, took his M.D. at Cambridge and established a successful practice in London. In Colchester he and some "philosophical" friends had performed scientific experiments; he went on with them in his free time, and in 1600, the same year that he was named personal physician to Queen Elizabeth, he published a book, *De magnete*, containing some of his conclusions. The main content of the book is the discovery that the whole earth acts as a huge magnet, but he has studied electric forces as well, and we must pause to examine his ideas as to how electric and magnetic forces are exerted.

Gilbert's language is Aristotelian: "In the production of any object in this world two causes or principles are at work, matter and form. Electrical impulses come mainly from matter and magnetic impulses from form; these two are very different. Magnetism is ennobled by many qualities of strength, while electricity is of lesser rank and strength; it is as if it were confined in some prison." All right, form is nobler than matter. A magnet is always a magnet,[1] exerting its force without diminution through paper or wood or glass or any metal except iron. An electric force moves a bit of paper, but it can be generated only by rubbing a sample of amber or wax or

---

[1] Nobody yet knew that magnetism can be destroyed by heating.

glass with a cloth, and when this is done the force, never strong, soon fades away. Rubbing, Gilbert suggests, liberates an emanation, or effluvium, that floats for a while in the space around the sample, but he is not sure whether the paper moves because the effluvium moves it directly or because it produces a little current of air. To explain how the effluvium is produced, Gilbert gives us some verbal play concerning the four elements and the four bodily humors that we need not follow.

The Earth as a whole, says Gilbert, has an effluvium, the atmosphere. The force it produces is gravity, but the Earth's magnetism is not an effluvium; its arises out of a property far more profound: the Earth has a soul. "The stars' condition would be miserable, the Earth's would be abject, if they were denied the high dignity of a soul that is granted to worms, ants, roaches, plants, and mushrooms."

It is the effluvium that concerns us here. Gilbert does not have much more to say about it, and a century later Newton, in Query 22 of his *Opticks*, is no wiser. He asks

> how an electrick Body can by Friction emit an Exhalation so rare and subtile, and yet so potent, as by its Emission to cause no sensible Diminution in the weight of the electrick Body, . . . and how the Effluvia of a Magnet can be so rare and subtile, as to pass through a Plate of Glass without any Resistance or Diminution of their force, and yet so potent as to turn a magnetick Needle beyond the Glass?

During the eighteenth century effluvia and ether began to fall out of fashion. Newton's universe changed little, and his unanswered questions remained unanswered, but mathematical techniques advanced enormously, and the mathematical physicists of the era found that they could calculate phenomena of gravitation and electricity and magnetism without making any hypotheses about ether or effluvia at all. Those things occurred only in arguments intended to explain the mathematics, but because they are unobservable one did not have to believe the explanations, and people began to forget about them. What was left was action at a distance, which, as we have seen, Newton had scornfully repudiated. The British physicists adhered to the Newtonian faith, but it was only after the 1880s that Continental scientists began once again to admit that there might be something, some *thing*, in the region around an electrified or magnetized body, and that it might be an entirely new kind of thing, foreign to Newtonian physics. Most had come to believe that with a few additions to take account of new phenomena, Newton's theories were all that would ever be needed to explain the physical world, and this conviction was so strong that even after new ideas took root they grew slowly in stony soil.

The person mainly responsible for the change of views was Michael Faraday (1791–1867), whose career shows vividly and sometimes sadly what happened when a man of genius was born entangled in the net of British class distinctions. He came out of a London slum, son of an impoverished black-smith. His formal education stopped with the elements of reading, writing, and arithmetic. Not being strong enough for his father's trade he served eight years as apprentice to a bookbinder and newspaper dealer. In the evenings he continued his education in meetings of a club of young working-class men struggling to improve their vocabularies, pronunciation, and general knowledge so that they might rise a little in society. He attended public lectures on scientific subjects, copied interesting passages from books that passed through the bindery, and performed experiments with the scanty means available to him. When he was twenty-one a friendly customer gave him tickets to scientific lectures to be given at the Royal Institution by the chemist and physicist Humphry Davy. He had already met Davy and been impressed by him; now he made careful notes, bound them neatly, and sent them to Davy, who offered him a job as a research assistant. The Royal Institution's building contained laboratories, a lecture hall, and private rooms upstairs in which Faraday, soon joined by his wife, lived until their old age.

As an experimenter Faraday was extraordinary, and in his way of thinking he was a natural philosopher of the old type. He had little to do with experts in the universities. His ideas were unlike anybody else's, and though they were expressed so that one could grasp the general idea, their essential physical content was always hard for other people to understand. Thus his experiments were always taken more seriously than his thoughts. He became the Grand Old Man of English science, but contemporaries tended to read his papers for the results he reported and skip what he thought they meant.

According to Faraday, matter is atomic (everybody agreed on that), and each atom has an "atmosphere of force" around it. This impressionistic phrase was used to denote an idea that had never been clearly defined and had no better name. It is now called a *field*, and in some situations one can almost see it. Figure 9.1a, copied from a plate in one of Faraday's papers, shows the pattern that iron filings form when they are spilled onto a piece of paper laid over a simple bar magnet. (This technique had been known since the seventeenth century.) The pattern is quite simple, and one can even calculate what it will look like if one makes the simplifying assumption that at or near each end of the magnet is a pointlike center of force. To mathematicians this familiar pattern was the graphic representation of a mathematical fact. To Faraday it mapped something that is physically present.

Faraday's *Experimental Researches in Electricity* [1839–55], the chronicle of his thoughts and discoveries, shows that for a long time he thought of ether as a gas of material particles, but this investment of hypothesis paid off poorly in explanation, and in 1852, his sixty-second year, he changed his mind: the "atmosphere of force" that iron filings serve to make visible is no kind of material substance. The field surrounding the magnet consists of *lines of force*, real but not in the way wires and strings are real. Perhaps they are some kind of strains in ether, but they do not behave like matter.

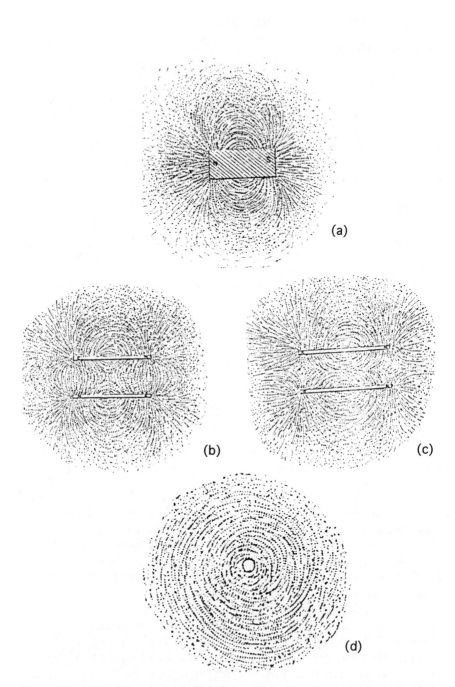

Figure 9.1. Michael Faraday's drawings of patterns of magnetic fields surrounding different sources: (a) a plain bar magnet; (b) two parallel magnets with poles opposed (north to south, south to north); (c) two bar magnets with poles parallel; and (d) a wire carrying an electric current, shown in cross section by the circle. From Faraday 1839, vol. 3.

What are their properties? Put two bar magnets together head-to-tail (Figure 9.1b), so that they attract each other.[2] Lines of force reach from one to the other, pulling them together. The lines, whatever they are, must be like rubber bands, in tension. Then why do they bell out into space—why don't they collapse close to the magnet, as rubber bands would do? It must be that they repel each other laterally. Now set the magnets head-to-head (Figure 9.1c). You can see how the repulsion pushes them apart. There are also electric fields, set up by an electrically charged object such as a comb with which one has just combed one's hair on a dry day, and electric lines of force have similar properties.

James Clerk Maxwell, whom we shall encounter soon, writes in 1873,

> Faraday, in his mind's eye, saw lines of force traversing all space where the mathematicians saw centres of force attracting at a distance: Faraday saw a medium where they saw nothing but distance: Faraday sought the seat of the phenomena in real actions going on in the medium, they were satisfied that they found in it a power of action at a distance impressed on electric fluids.

From a formal point of view, though, Faraday's explanation by means of fields is just another way of saying what mathematicians say with formulas, for as I have just indicated, the forms of Faraday's lines of force can be calculated if anyone needs to. "As I proceeded with the study of Faraday," Maxwell writes, "I perceived that his method of conceiving the phenomena was also a mathematical one, though not exhibited in the conventional form of mathematical symbols."

Faraday was not inclined to be dogmatic, and he never insisted on his view of force; it is only when we think about it that we see how radical it was. In the Newtonian theory that every student, then and now, is persuaded to accept as the starting point of physics, the first definition learned is that a force is a push or pull. Push and pull are verbs that here stand in as nouns: a force is an action. Faraday proposes that it is more like a thing than an action. If I push a pencil with my finger, Newton's force acts where finger touches pencil; Faraday's force moves in space, for if I carry a magnet into another room the lines of force go with it. Newtonian physics does not make sense if a force is considered as a thing.

Is there are any connection between electric, magnetic, and gravitational forces, or do they fall out of Nature's closet as three separate surprises? This is a more important question than it might seem at first, and since Faraday's time, answering it has been the goal of an immense program of physical research. The general term for the program is *unification*.

I have mentioned noble but ultimately futile efforts to explain light as a Newtonian motion of an ethereal fluid. In 1820 a Dane named Hans

---

[2] I am sure you know that a magnet like this has a north end and a south end, named because if you suspend it so as to make a compass, the north end points north. If you bring two such magnets close together you will find that unlike poles attract each other, and like poles repel.

Christian Oersted opened a new path when he discovered that an electric current moving along a wire produces a magnetic field. Figure 9.1d shows the magnetic lines of force wrapping around the wire. It was not clear then, but it is clear now, that the current is just a cloud of electric charge moving along the wire, and so electricity, if it moves, makes magnetism. Can magnetism make an electric current flow? In 1831, after many attempts, Faraday found that a moving magnet, or indeed any arrangement that makes a magnetic field change, can produce a current in a loop of wire;[3] and further experiments proved that the reason the current flows is that the changing magnetic field produces an electric field. On the experimental level, the first steps toward a unification of electricity and magnetism had been taken, but nobody as yet had anything to say about what those two are, or any ideas as to how they are related.

The connections between electricity and magnetism did not escape the attention of practical men. When an electric current produces a magnetic field, that field can cause a piece of iron to move; therefore an electric current is able to produce a mechanical force, and ingenuity soon gave birth to the electric motor. Because a magnet moving past a wire, *or* a wire moving past a magnet, produces a current in the wire, ingenuity produced a generator.

These discoveries suggested to Faraday that electric and magnetic fields are more than just forces. They have also a dynamism, and it is when we study what happens when they change that we begin see their connection. It was not long before the term "electromagnetic field" was used to connect them in language as well. Faraday was convinced that there are other connections:

> I have long held an opinion, almost amounting to conviction, in common I believe with many other lovers of natural knowledge, that the various forms under which the forces of matter are made manifest have one common origin. . . . This strong persuasion extended to the powers of light, and led, on a former occasion, to many exertions, having for their object the discovery of the direct relation between light and electricity, and their mutual action in bodies subject jointly to their power; but the results were negative. . . . These ineffectual exertions, and many others which were never published, could not remove my strong persuasion derived from philosophical considerations; and therefore, I recently resumed the enquiry by experimenting in a most strict and searching manner, and have at last succeeded in *magnetizing and electrifying a ray of light, and in illuminating a magnetic line of force.*

The melodramatic words he has stressed actually mean that if polarized light[4] is sent through a piece of glass[5] placed in a strong magnetic field

---

[3] This was discovered independently by Joseph Henry (1797–1878), a professor at Albany Academy in New York State [1832].

[4] Polarized light was mentioned in the previous chapter and will be further explained in §10.4.

[5] Later he finds the same effect, in varying degrees, when the glass is replaced by various other transparent solids and liquids.

directed parallel to the beam, the beam's polarization is rotated, and that for this purpose the magnetic field produced by a current is as effective as that produced by an iron magnet. It is not clear, and Faraday does not try to explain, what is the connection between light and magnetism, except that transparent matter must be present to mediate it.

Readers of scientific history often wonder what moves someone to undertake a difficult experiment, perhaps to try again and again, fishing where no fish are known to be. But the skilled angler does not drop the fly at random on the surface of a stream. Mind penetrates the troubled water, it sees the fish below and knows the habits of its kind. Faraday sometimes compares himself with a fisherman, and when he speaks of "philosophical considerations" I think he is referring to that skill.

Having found that magnetic lines of force influence the polarization of light passing through glass, Faraday speculated on possible relations between light and magnetism. In a note headed "Thoughts on Ray-Vibrations" he suggests, very tentatively, that light is a transverse wave in lines of force analogous to the waves that travel down a stretched string if someone shakes the end of it. Faraday has nothing to say about the lines themselves, whether they are electric or magnetic or in some sense both, but if he is right—if light is an aspect of the electromagnetic field—then ether no longer has a necessary role. We need no longer puzzle over the questions it raises, and it can be surrendered to history.

At this point, is light a substance or an accident? Once more, the old question encourages us to think carefully. Faraday's waves are an accidental quality of lines of force, but consider what happens when I strike a match. Do waving lines of force then spurt outward at the speed of light? If lines of force are things, the answer seems to be substance. It is not really fair to push Faraday's tentative notion so hard, but there is a third possibility: that the lines represent some new mode of existence that cannot be called substance.

In his old age Faraday experienced intervals of depression and his mental powers declined. Toward the end he wrote in his notebook: "15,809 Let the imagination go, guiding it by judgement and principle, but holding it in and directing it by experiment." The right time to say where you are going is when you have finished the journey.

3. Maxwell: Electricity Plus Magnetism Gives Light

Faraday's ideas, radical and partly untestable, did not penetrate quickly, but people were learning more about light all the time. Optical instruments were invented which extended the range of microscopes, telescopes, and spectroscopes. Polarized light was studied with increasing refinement, and in 1850 Léon Foucault (he of the pendulum) drove a final nail into emission theory's coffin by measuring the speed of light in water. Descartes and

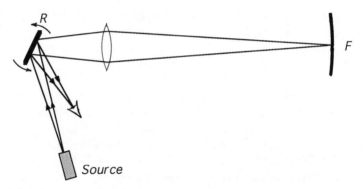

Figure 9.2. Léon Foucault's apparatus (1852) for measuring the speed of light. The mirror R rotates very quickly. When it is in a certain position light from the source flashes down to the mirror F and back again to the eye. By this time the mirror has turned through a small angle, and measuring it gives the speed of light.

Newton and their followers needed it to go faster; Huygens and Fermat and the other wave theorists needed it to go slower. Foucault [1850] managed to make a direct measurement of the time it took a flash of light to travel about ten feet down a tube of water and back again. Figure 9.2 shows the principle of the apparatus, though the actual design was more complicated. The measurement did not need to be very exact, because the expected difference in speeds was quite large. It showed that light travels more slowly through water than through air.

By the mid-1800s, mathematical physicists in England, Germany, and France had found equations that accounted for various phenomena of electricity and magnetism, but equations are only marks on a piece of paper; what do they mean in terms of our experience of Nature? One can answer that mathematics has nothing to do with experience, but the questing mind tries again with an example: if you show me the mathematical theory of fluid flow and I understand the mathematics, it describes something I have seen many times in different circumstances, so that as I read it I can imagine a fluid in motion. The equations of planetary motion call up the image of a little ball circling a big one. What am I supposed to imagine when I read the equations that describe an electromagnetic field? It was desperately important, especially to British physicists of the time, to see in their mathematics an image of something familiar, something that material substances do, something mechanical. Many tried; I choose the one who did it best.

James Clerk Maxwell (1831–79) was born into a family of landed gentry in the Scottish Lowlands. The estate covered some six thousand acres, including a number of tenant farms, and Maxwell and his wife spent several months there every year. He graduated from Cambridge in 1854 and after a brief period of teaching in Aberdeen was appointed professor at King's College,

Figure 9.3. Oliver Lodge's version of Maxwell's ether-theory explanation (see text) of the production of a magnetic field by an electric current.

London, where he did most of the work to be described here. At thirty-five he retired to Scotland, where he lived for a few years, working out a theory of gases, writing a long and wandering treatise on electricity and magnetism, and, assisted by his wife, doing experiments. In 1871 he was invited to Cambridge to organize what is now known as the Cavendish Laboratory for research in physics. There he served as professor for the short time remaining to him, dying in Cambridge in his forty-eighth year. He is buried in Scotland.

At midcentury, physicists faced three great questions: what are electricity and magnetism, what is the relation between them, and what is ether? William Thomson, another Scot, noticed that the equations for heat conduction are the same as those that describe a magnetic field. If one replaced a bar magnet with a large block of matter containing a source of heat at one point and an equal source of cold (if I may use the expression) at another, the flow of heat between them would be exactly along the lines made visible by iron filings in Figure 9.1. Does this mean that magnetism is a kind of heat? Not at all; only that there is an analogy. Maxwell noticed an analogy with the motion of a gas through a substance that resists it, something like tightly compacted steel wool. But the magnetic field is not a gas moving through steel wool, either. In 1861 Maxwell produced a mental model to show how an electric current produces a magnetic field. It represents the field by little whirling eddies in the ether that, once started, continue to revolve without friction. Figure 9.3, from a book by Oliver Lodge [1889], shows a later version based on a hint from Maxwell. It reduces ether, electric current, and field to a crude little mechanism. The eddies that form magnetism are shaded in the drawing and marked "+." The thing like a saw represents the electric current, which when it moves sets the eddies next to it whirling. If it were not for the wheels marked "−," the second row of eddies would turn in the opposite direction from the first. The wheels marked "−" are what engineers call idlers. Since there is an idler between every pair of eddies, all the eddies turn in the same direction. If the saw is moved toward the right, the eddies below it will turn clockwise, and those above it counterclockwise. The magnetic field is imagined as threading through the axes of these eddies, and thus, in three dimensions, it will wrap around the current just as it does in Figure 9.1d.

Maxwell's model explains, or embodies, Oersted's experiment. In addition, it suggests why light passing through glass in a magnetic field gets its polarization rotated: it is carried around by the turning eddies. As Maxwell worked he kept this model in his mind. It was a way, admittedly over-mechanized, of making sense of phenomena and relationships that otherwise made no sense to the imagination. Electric charge, current, electric and magnetic fields—he did not know what any of these things were. He had no better way to imagine them than this; it was fruitful of insights, and it worked for him.

In France they approached these questions differently. Inspired by Fresnel, who preached that the ideal theory starts from clear and simple hypotheses and attains its goal by delicate mathematical analysis, the French looked with astonishment on a contraption with saws and idler wheels. A little later the historian and physical chemist Pierre Duhem wrote a book called *Aim and Structure of Physical Theory* [1914] which praises the "straitness of mind which makes the Frenchman eager for clarity and method, and it is this love of clarity, order, and method which leads him, in every domain, to throw out or raze to the ground everything bequeathed to him by the past, in order to construct the present on a perfectly coordinated plane." British scientists are cut from different cloth:

> In the treatises on physics published in England, there is always one element which greatly astonishes the French student: that element, which nearly invariably accompanies the exposition of a theory, is the model. . . . Here is a book [Lodge's book] intended to expound the modern theories of electricity and to expound a new theory. In it there are nothing but strings which move around pulleys, which roll around drums, which go through pearl beads, which carry weights; and tubes which pump water while others swell and contract; toothed wheels which are geared to one another and engage hooks. We thought we were entering the tranquil and neatly ordered abode of reason, but we find ourselves in a factory.

It may be that purity of method had something to do with the comparative sterility of French theoretical physics in the seventy-five years before Duhem wrote those words.

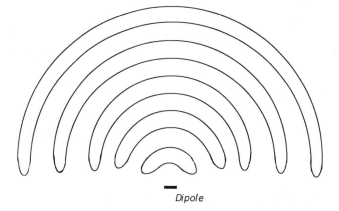

Dipole

Figure 9.4. Lines of electric field produced by a dipole, a short stretch of wire with a current rapidly running back and forth along it. The loops move outward at the speed of light. (The lower half of the diagram is omitted.)

Maxwell's essay, "A Dynamical Theory of the Electromagnetic Field" (1865), is the one that is remembered. He writes that "it has to do with the space in the neighborhood of electric or magnetic bodies, and it may be called a *dynamical* theory, because it assumes that in that space there is matter in motion, by which the observed electromagnetic phenomena are produced." The theory swims in ether. He mentions the undeniable fact of light waves. "We therefore have reason to believe, from the phenomena of light and heat, that there is an æthereal medium filling space and permeating bodies, capable of being set in motion and of transmitting that motion from one part to another, and of communicating that motion to gross matter so as to heat it and affect it in various ways." He does not try to explain how Earth moves through this medium without friction.

Maxwell then writes down the formulas, most of them already known, that connect electric field with electric charges, and magnetic fields with electric currents, and also those that connect electric and magnetic fields with each other. These are the formulas that express the mathematical unification of electricity and magnetism. Not a word about the properties of ether in all this; he is concerned only with the fields and their sources. He works out the theories of several well-studied effects to show the formulas are correct, he expresses the mathematical relations in the most general and abstract form, and then he is ready for the last section. It is called "Electromagnetic Theory of Light," and it shows that electric and magnetic fields do not necessarily remain attached to the charges and currents that are their sources, for a changing electric field can by itself give rise to a magnetic field, and vice versa. If energy is delivered to the field by moving electric charges, some of it detaches itself from the charges and journeys off into space, carried by waves in the field. Figure 9.4 is the upper half of a picture of the first few closed loops of electric field leaving an oscillating source at the center. The source is a very short dipole antenna in which a current is

Electric

Magnetic

Figure 9.5. Electric and magnetic fields in a wave of plane-polarized light moving toward the right.

made to flow back and forth. Magnetic fields wrap around the axis, but drawing them would confuse the picture. No energy is radiated along the axis of the dipole; this explains why the antennas of radio and television stations are vertical towers: nobody wants the power to go straight up or down. Maxwell's equations show that the electric and magnetic fields lie across the wave's direction of travel, so that Faraday's picture of waving lines was wrong. Figure 9.5 shows the configuration of fields in a small region of the radiation traveling along the previous figure's vertical axis.

Though in Maxwell's theory a moving magnet can start an electromagnetic wave, a moving electric charge is the usual source. The sun or a match flame or a piece of hot metal emits light, but none of these things is electrified as a whole. It must be that there are electric charges inside the atoms that compose them. An atom that emits light must contain an electric charge that oscillates back and forth or whirls in some tiny orbit. You may remember Galileo's speculations in §5.7 concerning light that bursts forth from inside an atom, or the "musical atoms of Leonhard Euler" mentioned in 8.2 that emitted light waves by vibrating. We are getting closer to seeing how this happens.

Electromagnetic waves move through space as ocean waves move across the water, and their velocity can be calculated from the results of electrical measurements. From the numbers available to him Maxwell calculates the speed of light as 310,740 kilometers—about 193,000 miles—per second. He then quotes a recent measurement by Foucault showing that light travels 298,000 kilometers per second (the modern value is about 299,800). "The agreement of the results seems to shew that light and magnetism are affections of the same substance, and that light is an electromagnetic disturbance propagated through the field according to electromagnetic laws." The "substance" is of course ether, which without entering the equations has been in the background all the time. Since Newton's growling insistence that pure action at a distance was an absurdity, a pall of ether had hung over the British Isles, and now there was more reason than ever to insist that it is real. The sun emits light, it travels, it arrives. In what form did it travel? One can interpret Maxwell's theory as saying that it was in the field; but a field, in the minds of most physicists, was a mathematical abstraction, a symbolic representation of something that happens in the ether. Energy is something mechanical; sunlight concentrated by a mirror can ignite paper or boil water. Maxwell's theory seemed to require the "substance."

Nevertheless, Maxwell found the formulas that describe a light wave (they are still regarded as correct) without making any assumptions about ether. Valuable information comes out of these equations, but they refer only to measurable fields and visible light and the spectrum's prolongations toward longer and shorter wavelengths. They say nothing about ether; they answer none of the questions that had perplexed scientists for so long—what is its nature, how can it support transverse waves and still allow the planets to circle the sun, what is its relation with the fields that inhabit it and whose effects we can observe? From the perspective of those who sought the truth about these profound matters Maxwell's theory was a disappointment. Today it is regarded as the single great advance made in physics between Newton and Einstein. Three areas of knowledge—electricity, magnetism, and light—which at the beginning of the century appeared entirely unrelated, were by the end of it unified into a single body of knowledge. To make this statement clear, suppose an experiment shows that we have been mistaken all along about some property of light—suppose for example that the waves turn out to be partly longitudinal. Then everything we think we understand about both electricity and magnetism would have to be revised.

Instead of a theory based on intuitive notions of how an ethereal fluid might behave, Maxwell gives us one based on mathematics. That involves much more than a change in the language of description. The field in Maxwell's theory, instead of being explained as a thing or a property of a thing, is, at any point in space and instant of time, characterized only by a number. Now ask: is it possible to have a place where there is no field at all? I think Faraday would answer that if no lines of force are present, there is no field. But to say that a number is zero is not the same as saying that it does not exist. A bank balance can be zero without ceasing to exist. One can reasonably say that whether or not it is a property of ether, the field is everywhere.

Substance or accident? The old question gets a new answer. The field does not go anywhere; waves that travel through it are changes in its configuration. Accident.

There was never any doubt in Maxwell's mind as to the reality of ether. The index of his two-volume *Treatise on Electricity and Magnetism* [1873], written toward the end of his life, contains no entry for it, but that is a scientific book, and the statements in it are supposed to be verifiable by experiment. He has nothing to say about ether in such a context, but in his last year he wrote a beautiful article, "Ether," for the Encyclopedia Britannica (9th ed.) that tells the history of the concept and explains the roles that the ethers of light and electromagnetism must play. The electromagnetic theory of light is presented as the theory which seems to show not that these ethers exist—that is assumed—but that they are the same. He speculates on the nature of ether and is inclined to favor the model caricatured in Figure 9.3, in which the field consists of "molecular vortices" that revolve without friction and perhaps drift around like the molecules of a gas. (There is no mention of the idlers between them.) The vortices are pure speculation, but on the last page of the article we read what is certain:

> Whatever difficulties we may have in forming a consistent idea of the constitution of the æther, there can be no doubt that the interplanetary and interstellar spaces are not empty, but are occupied by a material substance or body, which is certainly the largest, and probably the most uniform body of which we have any knowledge.

Maxwell's theory explains so many different facts of electricity and magnetism that there is no serious doubt that it is largely correct. It predicts waves and tells how fast they will travel, and because the speed came out close to that of light Maxwell deduced that light is electromagnetic. It is a natural deduction, and today we still think it is right, but he gives no experimental evidence to support it. The theory itself suggests experiments, however, and the next section will tell how some of them were done.

4. Hertz Finds the Electric Waves

During the era of Faraday and Maxwell there was not much physical research going on in the United States, but in 1842 Joseph Henry, now at Princeton, studied electric sparks produced by the discharge of an electricity-storing device called a Leiden jar. He found that the jar did not simply empty out its electric charge; instead, the charge surged back and forth, reversing its direction many thousands of times in a second, during the brief moment before it died out. (We now know that this is true also of lightning discharges.) Henry used the magnetic field produced by the oscillating current to magnetize a needle, and he reached his conclusion by noting that after the discharge the needle was sometimes left magnetized in one direction, sometimes in the other. Continuing his study, he noticed something even more remarkable:

> A single spark from the prime conductor of a machine, of about an inch long, thrown on to the end of a circuit of wire in an upper room, produced an induction sufficiently powerful to magnetize needles in a parallel circuit of iron placed in the cellar beneath, at a perpendicular distance of 30 feet, with two floors and ceilings, each 14 inches thick, intervening.

Though he didn't know it, Joseph Henry [1886] had produced and detected electromagnetic waves. Maxwell was eleven years old, and Henry, by publishing his work in this country, could be sure that it was unlikely to be appreciated, or even read, on the other side of the Atlantic. It was many years before his work was noticed.

Maxwell's "dynamical theory" appeared in 1865, but even today it is hard to read, and at the time its content was so surprising, not to say disappointing, that twenty years passed before it attracted much attention outside a small group of British experts. Nevertheless, the experiments that firmly established the claims made in that paper were made in Germany.

Heinrich Rudolf Hertz (1857–94) was born in Hamburg; his father was a lawyer who later became a senator of the Free City. At the University of Berlin the shy and modest young man came to the attention of Helmholtz, who encouraged him to enter research. He turned out to be good at theory and experiment, and at the unusual age of twenty-eight he was appointed professor of physics in the university at Karlsruhe. He read Maxwell's papers and tried to separate their essential content from the surrounding accumulation of arguments inherited from Faraday and his predecessors. He would have liked to regard the theory as purely mathematical; "Maxwell's theory," he once said, "is Maxwell's system of equations," but it is not that simple, for a physicist always has in mind a tactile or visual image of what equations are about. In his mind ether was still there: "Take away from the world electricity, and light disappears," he wrote later, "remove from the world the luminiferous ether, and electric and magnetic actions can no longer traverse space." For a while longer, very few physicists thought they could live and ply their trade without the help of that substance.

At Karlsruhe, Hertz did some experiments that clarified the physical content of the theory and helped to establish it in people's minds. Maxwell says that an oscillating electric current will produce electromagnetic radiation, and also that the more rapid the oscillation the more radiation it produces. We have just seen that Joseph Henry, and he was followed by many others, had shown how to generate electric high-frequency oscillations in a circuit containing a spark discharge. By pushing the design of the circuit, Hertz was able to get frequencies of almost a billion cycles per second (a billion hertz, as one now says). As Maxwell had predicted, the oscillations produced an electromagnetic wave. By being clever Hertz was able to find both the frequency and the wavelength of the wave and thus deduce the speed with which it traveled. According to Maxwell's theory those waves bear the same relation to light that low notes on a piano bears to high notes (though in musical terms his electric oscillations were about twenty octaves lower than those of light). All sounds travel at the same speed, and as accurately as Hertz could measure, his waves traveled at the speed of light.

I have already mentioned how William Herschel and Johann Ritter extended the range of light wavelengths beyond the visible into infrared and ultraviolet; now came the prospect of producing and detecting radiations over a vast range of frequencies and wavelengths. Using electromagnetic radiation to transmit a message was an obvious idea, and ingenuity soon produced the radio. The first person to do it on a commercial scale was an Italian, Guglielmo Marconi. In 1896 he succeeded in sending a message to someone a few miles away, and five years later he sent one across the Atlantic. People talked of establishing a radio link with Mars and learning Martian.

Heinrich Hertz's experiments verifying Maxwell's theory produced one great surprise. As I have mentioned, the electric waves he studied originated in the oscillations of a loop of electric wire containing a spark gap, and he detected the waves in an identical, secondary loop that he could move around the laboratory. When a spark jumped across a gap in the primary loop a smaller spark could be seen in the secondary one if the room was dark. The primary spark was very bright. To make the little spark more visible he put up a screen between it and the primary, but when he did this he noticed that the little spark was weaker. Trying various arrangements, he found that the influence, whatever it was, traveled in straight lines from one spark gap to the other as if it were some sort of radiation. To see what kind of radiation it might be he tried stopping it, using dozens of different materials. It was stopped by all metals, by paraffin, rubber, shellac, glass, wood, paper, ivory, horn, animal hides, feathers, and a number of liquids; but a few other materials let it through: crystallized sugar, rock-salt, and especially quartz. Hertz tried interposing a quartz prism between the spark gaps. Now he could see the visible spectrum of the spark, but he found that it was the ultraviolet part of the spectrum that produced the effect he had seen. There was nothing in Maxwell's theory that suggested or explained any of this, and other workers soon discovered a further fact that made it even more mysterious: ultraviolet light is effective in strengthening a spark only when it falls on the negatively charged side of the gap. This phenomenon, soon named the photoelectric effect, remained mysterious for almost twenty years, until Einstein made a radical suggestion that explained it, but only, of course, if one accepted the suggestion. In §10.1 we will see what happened.

5. The Vast Spectrum of Light

I have mentioned that our eyes respond only to the relatively narrow range of wavelengths, from about four hundred to seven hundred nanometers (billionths of a meter), for which the atmosphere is most transparent. An ultraviolet or infrared eye would see only a few miles, and so we and the animals concentrate our powers where seeing is best. In the direction of longer wavelengths air is somewhat opaque out to a wavelength of about ten centimeters—about four inches—beyond which it is once more transparent. This is the range in which Hertz did his experiments and in which the world's radio and television traffic flows. On the ultraviolet side of the spectrum the opacity of the atmosphere, due largely to the ozone it contains, protects us from the carcinogenic effects of the sun's radiation, and at shorter wavelengths still are X rays.

X rays were discovered in 1895 by Wilhelm Conrad Röntgen (1845–1923), a professor in Würzburg. Some experimental hints made him suspect that high-voltage electrical discharges in an evacuated glass tube might be producing some kind of disturbance in the ether outside the wall of the tube. To detect it he prepared some paper coated with barium platino-cyanide, which glows when exposed to penetrating radiations, and to exclude extraneous light he enclosed the tube in a cardboard box. When he first applied a voltage to the tube, the screen was on a bench some distance away, and he was surprised to see that it was glowing. He moved the screen farther away, and it still glowed. He found that the radiation penetrated a thick book, a piece of wood, even thin sheets of metal, and that it exposed a photographic plate wrapped in black paper. He held something in front of the coated paper and noticed the shadows of the bones in his fingers.

These discoveries were made in November 1895. Röntgen quickly performed one experiment after another to find out what the rays would do and what their nature is. At the end of December he published a paper summarizing his work, and the world's newspapers took notice. Within a month X rays were being used in medical diagnosis. What are these rays? In his first communication Röntgen tentatively suggests that they may be longitudinal waves in the ether, but later study showed that they are less exotic: electromagnetic radiation, like light, except that their wavelength is several hundred times shorter—in musical terms, about nine octaves further up the scale.

The first applications of X rays were almost all in medicine—broken bones and things like that—but as more powerful and better-controlled sources became available they turned out to be useful in many branches of science and technology. Some of these developments will be mentioned in §10.5.

In Montreal at the same time, Ernest Rutherford and his students studied the rays that come out of naturally radioactive substances like uranium and discovered that they are of three kinds, which they named alpha, beta, and gamma. Alpha and beta rays turned out to be familiar things in an unfamiliar context: alphas are helium atoms without their electrons (though Rutherford did not yet know that atoms have electrons), and betas are electrons. These two radiations are electrically charged, but in 1900 Paul Villard, ex-

The Electromagnetic Spectrum

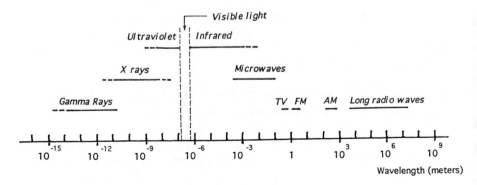

Figure 9.6. The electromagnetic spectrum.

perimenting in Paris, found that gammas are not. He suggested, correctly, that they are of the same nature as X rays but of even shorter wavelength. Gamma rays come from radioactive minerals and are now produced in other ways, but the most energetic ones are cosmic rays from outer space. It is not certain how they originate, but most likely they are born in the explosion of a supernova.

All these radiations, discovered in different ways, are unified in the concept of an electromagnetic field. Figure 9.6 shows the different bands of the spectrum spread out along the scale of wavelengths. The range is so vast that the scale has been compressed: at each marker the wavelength is ten times what is was at the preceding one. You will see that long radio waves are something like 10^{22} (that is, 1 with twenty-two zeros after it) times as long as the shortest known gammas, and yet, as accurately as can be measured, all these radiations move at the same speed through empty or nearly empty space.

The sun, with a surface temperature of about six thousand degrees centigrade, floods the Earth with its radiation. What color is it? Everyone knows the answer, white. Everybody should also know that that is not the end of the story: the sun gives out a range of colors, with a maximum in the green part of the spectrum, but also some infrared and ultraviolet. What then is white?

To approach this question let us consider another physical phenomenon that is more familiar: what is noise? Start by considering the sound of a piano playing a single tone. If it is the A to which many orchestras tune, the frequency is 440 hertz. That is the fundamental frequency that the piano string produces, but the string also makes overtones, at twice the fundamental frequency, three times, and so on. These are what give the instrument its char-

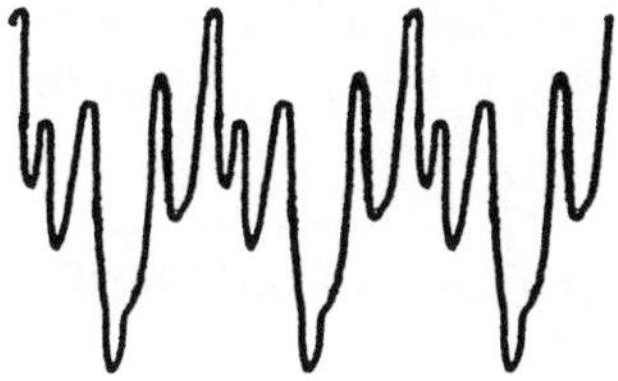

Figure 9.7. Pressure wave produced by a violin playing an A.

acteristic sound, and the way they combine tells an expert whether the source is a Steinway or a cheapie. White is optical noise. If we understand the nature of acoustical noise we will be able to understand what white is.

The ear recognizes a musical instrument by sorting out the various frequencies present in its sound. If you give me a bag containing a mixture of different kinds of beans I can sort them out, too. This is *not* what the ear does. Think of an eardrum, the little membrane that vibrates in and out as the air pressure on it fluctuates. What enters the inner ear is a single train of fluctuations, but they do not have the simple form of the wave in Figure 8.1. The motion of an eardrum hearing a violin play that A is graphed in Figure 9.7. The inner ear contains a device that does for sound what Newton's prism does for light: it separates out the tones contained in that motion and tells the brain how much there is of each; then that great mystery, consciousness, recognizes the tone, the instrument, the composition, and perhaps the player. Suppose there are several instruments sounding. If the players hold a single note or chord, the pressure wave will be more complicated but the pattern will still repeat, as in the figure, and the ear will sort it out. What happens when a truck goes past? Now the eardrum moves in and out in a random way. There is no pitch, and the motion does not repeat itself. There are so many tones that the ear can no longer sort them out, and the resulting sensation is called noise.

When Isaac Newton thought of a beam of white light he saw it as a mixture of particles corresponding to all different colors. For him the prism sorted the particles like beans. If light is a wave, there is nothing to sort: what reaches any point on the retina is a single wavering thread of electromagnetic field. White is not a mixture of colors in the sense that Newton meant. What, then, is it? I would say it is a color, as much as any other color, only the field of white light entering the eye, instead of varying in a way that is periodic and analyzable, varies at random. But it is more complicated than this. There is not just one field any more, for fluctuations reaching different parts of the retina will ordinarily not be related in any way. If one considers the matter carefully there are refinements—randomness is not easy to define—but I think it is not an oversimplification to say that—roughly, at least—what noise is to the ear, white is to the eye.

After these excursions to the ends of the spectrum and back we must return to the question of the nature of light. It is very nice to be able to say that it is a varying electromagnetic field in a certain range of frequencies, but what is that field? And what, if anything, has it to do with ether?

6. Trying to Find the Ether

> ... the great problem of the nature and properties of the ether which fills space, of its structure, of its rest or motion, of its finite or infinite extent. More and more we feel that this is the all-important problem, and that its solution will not only reveal to us the nature of what used to be called imponderables, but also the nature of matter itself and of its most essential properties—weight and inertia. . . . These are the ultimate problems of physical science, the icy summits of its loftiest range.

With these words, spoken before a scientific meeting in 1889, Heinrich Hertz referred to the most uncomfortable part of physical theory. There on the mountaintop sat a gross and ugly contradiction: that theoretically ether must be at the same time a tenuous fluid and an elastic solid, but for an experimenter like himself that was not all. The first challenge was its unobservability, for until you can observe a thing you can only guess at its nature.

A number of experiments were proposed in these decades in which the ether ought to make known its existence and at least some of its properties. I will describe the one that was the most straightforward in principle, that had the greatest sensitivity, and that in the end proved the most convincing.

Albert Abraham Michelson was born in Strzelno, Poland, in 1852. A few years later his family, fearing the ethnic violence that raged through the Polish villages, decided to try the United States, and hearing of the gold in California and the fortunes being made, they went there. After a harrowing journey the family settled as merchants in a California mining town and later in another in Nevada. Michelson's youth was passed in the atmosphere of an old-fashioned Western: cowboys and claim jumpers, bandits and sheriffs, bullets and bottles flying through the air. The education available in these settings was only so-so, and in 1869 he traveled to Washington to personally ask President Grant for an appointment to the Naval Academy. After graduation and two years at sea he was appointed instructor in physics at the Academy, where he started on the optical experiments that would make him famous. His first project was to measure the speed of light. Simplifying Foucault's method, he got an answer that is only 0.06 percent away from the one currently accepted. The technical achievement gave him a reputation, but he wanted to make scientific discoveries, and he got himself assigned for two years' study and work in Europe. In Berlin and Potsdam he made his first attempt to detect the Earth's motion through the ether of space.

The idea of Michelson's experiment is that if light is a wave in an ether that fills the universe, and if the Earth (and perhaps the whole solar system) is moving through it, we should be able to detect an ether wind. Michelson proposed to do it optically. The experiment depends on a simple idea. Assume that as Maxwell's theory and experience tell us, light moves through ether at a rate that is always the same. Imagine a lake, with a motor boat in it that moves at some fixed speed. The boat travels half a kilometer down the lake, turns around by some instantaneous magical process, and returns to its starting point. This takes a certain time, T. Now bring the boat to a river half a kilometer wide, and send it across and back in a straight line. This takes longer than T because the boat has to point its nose a little upstream to keep from being carried down. Finally, send the boat half a kilometer downstream and back. This takes longer still, essentially because whereas the boat moves faster on the way down and slower coming back, the journey back takes longer than the journey down, and so the average speed is less.

If the river is ether and the boat is light and the speed of the Earth in its orbit is thirty kilometers per second, the difference between the across-and-back time and the down-and-back time is very small; for a table-sized experiment it is about 10^{-16} second, but Michelson's idea was to make the light time itself. The waves of visible light fluctuate about 5×10^{14} times a second,[6] and so if two identical beams start out at the same moment, one going across the ether stream and the other parallel to it, the first one will arrive back about 5×10^{-2}, or $1/20$ of a wavelength before the second. Michelson's great invention, which he called an interferometer, was designed to detect this difference by looking at the interference pattern formed when the two returning waves were superposed.

Figure 9.8 is a highly simplified diagram to show the idea of the experiment. A beam of light from the source shines on a lightly silvered glass plate a, where about half the light is reflected to a mirror b and back down through a to the eye. The rest of the light goes through a to the mirror c, and back to a where again part of it reflects into the eye. What I have illustrated as the eye is actually an arrangement of lenses that superpose the two beams so that they interfere and form an alternation of light and dark fringes as in Young's experiment in which the beams come from two different slits.

The observation consisted in looking at the fringes, rotating the whole apparatus through 90 degrees, so that what used to be the faster beam is now the slower one, and seeing if the fringes have changed position. I won't go into details—they can be found in Michelson's paper [1881] or in Swenson [1972]—but I should give an idea of the care required if the experiment is to give results. It failed in Berlin because in spite of every

[6] 10^{-16} is a decimal point followed by fifteen zeros and a 1; 5×10^{14} is 5 followed by fourteen zeros.

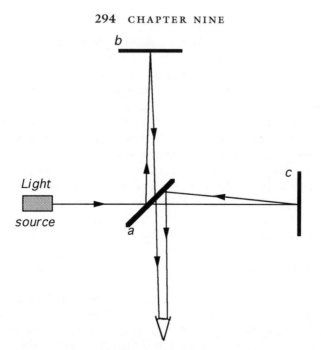

Figure 9.8. Michelson's interferometer, explained in the text.

precaution the vibration of traffic shook the instrument. In the village of Potsdam, where he tried again, things were quieter in the middle of the night, but even so, when someone stamped on the pavement three hundred meters away from the building the interference fringes wobbled and disappeared. Michelson saw no trace of motion through the ether and resolved to build on his experience and make a better instrument and try again.

Michelson resigned his commission and joined the faculty of the newly established Case School of Applied Science, now Case Western Reserve University, in Cleveland. Here he was joined by Edward Morley, a distinguished chemist, and together they did a better job. The principle was exactly the same, but the apparatus was awesome. Michelson designed the optics. Because the size of the effect depends on the length of the optical path, he used mirrors to make the beams shuttle back and forth four times between the plate a and the eye. To minimize effects of tremors Morley used a basement laboratory. He mounted the whole apparatus on a two-ton sandstone slab, five feet square and a foot thick, which he set on a ring-shaped wooden support that floated in a ring-shaped vessel containing mercury. Friction was so small that the apparatus, once started, rotated slowly for hours without anyone touching it. The experimenters walked round and round, looking into the eyepiece and taking readings as it drifted along. Once the apparatus was ready Michelson and Morley took readings for five days during July of 1887. If there were any ether drift to be seen they would have seen it easily, but they saw nothing. They were dismayed and astonished. The great experiment had failed.

It is a rare thing when the Lord bends down and speaks to his children, but on this occasion he did. Clothing his Word in the language of Nature, he told those two men that they had blown the ether away; but they didn't hear him.

The result of the experiment was spread around by letter before the published paper appeared, and it produced consternation among the ether's friends. How could there be nothing? But a way can be found to get around the result of almost any experiment. Think: if you drive in a car with the windows shut you don't notice any wind. Suppose the Earth as it moves carries with it a certain body of ether. Then in a basement laboratory there would be no ether wind. Perhaps the ether in the basement was carried along by the surrounding earth. In 1897 Michelson repeated the experiment at some distance above ground. Nothing. Later Morley tried it in Cleveland Heights, three hundred feet up. Nothing.

There were other ways to explain the failure. Two physicists, George Francis FitzGerald in Dublin and Hendrik Antoon Lorentz in Leiden independently proposed that the effect of the ether wind would have been observed as expected had it not been masked by another effect. FitzGerald suggested the idea in a single paragraph in the American weekly *Science* [1889]; later and probably independently Lorentz worked out a detailed theory. The Michelson-Morley experiment involved a race between two beams of light which finished in a dead heat though one had been expected to move more slowly than the other. The suggestion was that one did indeed go more slowly than the other, but it went a shorter distance and so arrived at the same time. This was because the effect of the ether wind blowing through the molecules of the interferometer, of the laboratory, of the whole Earth, is to move the molecules a little closer together along the line of motion. Things contract as they move through the ether but the effect can never be observed, for any measuring instrument by which one might hope to observe it has contracted in the same degree itself.

For several years the ether's friends found themselves faced with this faintly plausible hypothesis as the only way out of the swamp into which Michelson and Morley had thrown them. Of course, there was always the possibility that the experiment had been wrong. In 1922 Dayton C. Miller repeated the experiment with a very large interferometer hauled to the top of Mount Palomar in California. He found a small but positive result. Soon afterward, Auguste Piccard repeated the measurement at fifteen thousand feet above the French-Swiss border in the gondola of an immense, slowly rotating balloon, with results that were ambiguous but perhaps positive. Only believers still believed. The Michelson-Morley experiment was performed at a time when the men who had worked hardest on ether theory were getting old and had nothing but failure to report; and for younger people, who had not been baptized in ether, the Cleveland experiment simply settled the matter.

There was only one physicist of distinction who kept the faith, and that was Michelson. He was thirty-four when he established that ether cannot be found; he made delicate optical measurements for forty-four more years and to the end of his days did not believe there could be a wave without some material substance to do the waving.

7. Einstein Draws Conclusions

As I have said, people in England had faith in ether, but at mid-nineteenth century it was otherwise on the Continent, especially in Germany, where theories based on a purely mathematical description of the interactions that omitted all mention of ether had been very successful. There, in Maxwell's time, ether was generally regarded as a quaint superstition. Ludwig Boltzmann (1844–1906), an Austrian physicist whom Einstein greatly admired, wrote of his own upbringing, "We all as one might say absorbed with our mothers' milk the idea of electric and magnetic fluids acting directly at a distance." Because Maxwell's explanations required an ether even if his equations didn't, few Germans took his theory seriously until Hertz's experiments in the 1880s verified it, but then his belief that the phenomena implied an ether encouraged some of them to change their minds.

The fundamental doctrine of ether conceives it as a fluid that fills the universe and is, roughly or exactly, at rest with respect to the average motion of the stars. That means one can meaningfully say that a certain object is at rest and some other one is in motion. The aim of Michelson and Morley was to detect the earth's motion. They failed, and we have seen how FitzGerald and Lorentz faced the question "What if the world is such that the motion cannot be detected?"

Albert Einstein (1879–1955) was born in Ulm, a South German farming and industrial center, into the family of a perennially unsuccessful electrical contractor. As a child he was lucky enough to have an educated friend who fed him grown-up books; at twelve he found Euclid's *Geometry*, "that little holy book," and was hooked. High school in Munich was intolerable and he went off to Italy, where his family had moved, to study for admission to the Technical High School (really a scientific and technical university) in Zurich. He was accepted on the second try. At seventeen he gave up his German citizenship and after several years as a stateless person became a Swiss. He graduated in good standing but had become known as a rather perfunctory student (Figure 9.9), more interested in his own ideas than those set before him, and no one offered him the kind of teaching

Figure 9.9. Albert Einstein in Zurich, c. 1905 (photographer unknown). From Lotte Jacobi Archives, University of New Hampshire; used by permission.

position that he might have expected. For two years he made a living tutoring and teaching part time, until in 1902 he got quite a good job as an examiner in the Swiss Federal Patent Office in Bern. The job was varied, interesting, and not tiring, and in evenings and intervals of work his studies went forward. The next year he married Mileva Marič, a schoolmate who had failed to get a degree. They had two sons, plus a daughter born before the wedding, but they separated after a few years. For Einstein there followed professorships in Prague, Zurich, and Berlin while Mileva stayed in Switzerland. In 1932 he came to the United States as a professor at the Institute for Advanced Study in Princeton and remained there until his death.

That perfunctory sketch of Albert Einstein leaves out everything of importance: the ideas that came to him, the work he did. Of his more than three hundred scientific papers only three will be discussed here. The paper to be discussed first [1905b] opened the special theory of relativity and encouraged physicists to picture the world in a new way. Another [1911] introduced ideas that led to the general theory of relativity. These two involve the observed properties of light, without regard to what kind of thing it is. That question occupies the third paper [1905a], which will be discussed in §10.1.

In Bern, Einstein and two of his friends united to found a three-member learned society they called the Olympia Academy, which met regularly to discuss questions of literature and science and philosophy. They studied Hume, they read Plato and Dickens, and a writer to whom they paid special attention was Henri Poincaré (1854–1912), arguably the greatest mathematician of his time as well as a sensitive and learned critic of science. One of the Olympians, Maurice Solovine, mentions Poincaré's book *Science and Hypothesis*, published about 1902,[7] as one that "engrossed us and held us spellbound for weeks." Opening it one can see why, for Poincaré is finding internal cracks in the beautiful intellectual structure of physics that had been smoothed and polished and brought nearly to perfection during the nineteenth century.

Experiments detect no motion through the ether. Poincaré suggests (Chapter 7) that though it makes sense to say that some object A is in motion with respect to another one, B, it may be meaningless to say that A is moving.[8] He calls this proposition the principle of relativity. He questions whether ether exists (Chapter 10), but a survey of the traditional arguments leave him firmly convinced that it does. One of his reasons is that the mathematical structure of Maxwell's equations seems to show that they are anchored in space (Chapter 13), and to what could they be anchored but an ether? Lorentz had understood this point, and we have seen how he supposed that a body traveling through ether is shortened in the direction of its motion. He labored mightily, but his conclusions, based on hypothesis after hypothesis, did not convince Poincaré.

---

[7] It is hard to know when the books by Poincaré mentioned here were published, since their early editions (Flammarion) bear no dates.

[8] In a less extreme form this idea had been around for a long time. Copernicus had to persuade his readers that just because they could not detect the Earth's daily rotation or its annual revolution they must not conclude that these motions don't occur; to show what he meant he quoted Vergil's *Aeneid*: "Forth from the harbor we sail; the city and harbor move backward." Before Copernicus, Alhazen and Witelo had made the same point.

The same peculiarity of Maxwell's theory had occurred also to Einstein when he was about sixteen. What would someone see, he asked himself, who ran alongside a light wave at the speed of light? It would look like a field with crests and troughs standing still in space. But according to Maxwell's theory such a thing cannot exist. It turns out that the theory can be correct only for one class of observers who, if there were an ether, would be the ones standing still in it. For any others moving with respect to them, it predicts things that do not happen. That is what I meant by saying that the theory seems anchored in space.

By 1905 Einstein had been thinking about this and related paradoxes for a decade, and then he found a way around them. He may have been helped by the appearance in 1904 of another book by Poincaré, called *The Value of Science*. Chapter 2 of this book reprints an earlier essay [1898] called "The Measure of Time," in which the author's radical style of questioning is directed at a still more fundamental part of the mental apparatus we have constructed in order to understand the world—the concept of time. First he points out that "we have no direct intuition of the equality of two intervals of time." If we want a rough notion we can refer to a clock, but no clock is infinitely accurate, and equality means equality, not something rough. Clocks indicate time but do not define it. Then Poincaré raises a more disturbing question: when we say that two events A and B are simultaneous, or that A happens before B, exactly what do we mean? If they are causally connected, and it makes sense to say that A causes B, then it also makes sense to say that A happened first, but causality is usually established by priority, and as Hume had already made clear, there is no sure way to avoid a vicious circle. On the other hand, if no causal connection exists, there is no way at all to say in exact language that A and B are or are not simultaneous. Poincaré concludes: "The simultaneity of two events, or the order of their succession, or the equality of two durations, are defined so that the statements of natural laws may be as simple as possible. In other words, all these rules, all these definitions, are only the fruits of an unconscious opportunism." I mention this remarkable essay because it foreshadowed Einstein's most original idea. The essay is cited in *Science and Hypothesis*, which we know the Academy read and studied, and he could have looked it up. Perhaps, though, he did not, for as I have said, he generally worked things out for himself.

"When I saw that time is the culprit," Einstein said many years later, "the whole thing took only five weeks." "The whole thing" was the special theory of relativity, and it is described in a paper [1905b], containing no reference to anyone else's work, called in English "On the Electrodynamics of Moving Bodies." I will shorten the thirty-page paper into less than one. It starts with two assumptions. The first is the principle of relativity, and the second is that, as you will remember from the discussion of Maxwell's theory, the speed of light is a universal number that can be measured or calculated once for all, independent of the motion or any other property of its source. But the principle of relativity denies that it makes any sense to talk of the motion of the source. One can talk only of the relative motion of source and observer, and so the speed of light is also independent of the observer's motion: it is the same for all observers. If you try to visualize this situation by moving your hands and imagining something that travels from one to the other, you will find it impossible. It is certainly not what happens if two people throw and catch a football while they are running in different directions on a field. It cannot happen. The paradox is so obvious and so simple to state that to insist on it requires a radical rethinking of the act of measurement, and this caused Einstein to analyze what it means to make measurements of space or time, and what is the meaning of simultaneity.

The analysis showed that certain statements that people had been (and still are) accustomed to make—"This pencil is five inches long," "The time is exactly a quarter past three," "This coin has a mass of six grams"—cannot be given unambiguous meanings. They acquire meaning only when the relative position and motion of pencil or clock or coin and the person making the statement are specified, and the meaning varies from one such person to another. In a world of space and time defined according to these new meanings, Maxwell's theory makes perfect sense and is no longer anchored to any particular observer, but Newton's dynamics, the backbone of physics for three hundred years, no longer fits. Following the ideal of unification Einstein corrected it into a new dynamics, and when this was done he found that the mass and dimensions of an object are not absolute but depend on its state of motion, that as Poincaré had already surmised in Chapter 9 of *The Value of Science*, absolutely nothing can move faster than light; and that the terms "energy" and "mass" are only different names for the same thing.[9] But all that lies outside the property lines of this book and must be read about elsewhere.

Now a very short comment on the general theory of relativity. In 1911, while Einstein briefly held a professorship in Prague, a line of argument of the following kind occurred to him, though he must not be blamed for the way I explain it. Suppose I am sitting at my desk and suddenly my body feels much heavier. Two possible explanations occur to me. First, it may be the sensation we feel when an elevator starts up. In that case the elevator causes us to accelerate upward by exerting a force on our feet. I conclude that some huge machine may be lifting my house into the air, faster and faster. The other possibility is just as extravagant: that an object of gigantic mass has been added to the Earth so as to increase its gravitational pull. Before people in white coats come to take me away, I have time to ask

[9] That is, the famous formula $E = mc^2$, relating a quantity of energy to the corresponding quantity of mass, is analogous to the formula that says to multiply a length expressed in feet by twelve to find the length in inches.

myself whether any scientific measurement could tell which of these possibilities is correct. To answer that question Einstein proposed what he later called the principle of equivalence: no, the two are absolutely indistinguishable, by any means at all, to anyone confined in the room.

In the special theory Einstein invented a doctrine of space and time tailored to Maxwell's electromagnetic theory, and then he altered Newton's theory of the dynamical behavior of matter until it fitted the new doctrine. Now the situation was reversed. It had been known since the seventeenth century that Newtonian physics automatically obeys the principle of equivalence, but what about the electromagnetic field? If physics is to be unified, that must obey it too, and so the motion of light, just like the motion of a ball, must be affected by gravity. Einstein calculated how much a ray of light would be bent passing near the sun. The result was a very small angle, perhaps too small, he says, to be measurable with existing techniques.

He continued with his calculations, trying to find a way to fit his new perception into his earlier theory of relativity. The result was the general theory [1916]. The principle of equivalence is still part of it, but the predicted bending of light comes out twice as large as in the first crude calculation. The only chance to observe it optically is when the sun is eclipsed, for only then can one see stars close to it. An eclipse specially favorable for the observation took place May 29, 1919, in the South Atlantic Ocean, and two British expeditions observed it, one in Brazil and the other on an island off the coast of Spanish Guinea. The reduction of data took several months, and in November the leaders of both expeditions reported results with large uncertainties but in general agreement with Einstein, and not with the result of a naive application of Newton's laws. Both the London and New York *Times* explained in awed tones that the Newtonian era had ended and a new one begun. Wherever Einstein traveled, people thronged auditoriums to hear him lecture on a subject and in a language of which they understood not a word.

Two other predictions of the general theory are tested by observations of light. The first is that a clock in an intense gravitational field such as is found at the surface of a very dense star registers time more slowly than a clock on Earth. An atom is a little clock, and so one would expect a given line in the spectrum of light from such a star to be displaced in the direction of lower frequencies, that is, longer wavelengths. It is a small effect, but by now it has often been observed. The second is that not only is a flash of light passing near a massive object such as the sun deflected from a straight path, it is also momentarily slowed down. This effect also is extremely small, amounting to 250 millionths of a second in a flash that leaves the Earth and reflects back from a planet on the other side of the sun, passing close to the sun both times. It is impossible to observe this with light, but luckily the sun produces no radio waves, and a radar signal, the same as light but with longer wavelength, has made the long journey and returned a little late, as predicted.

The general theory banishes the concept of gravitational force and shows that its effects can be accurately reproduced by arguments that invoke curved spacetime. More recently, people have discovered that it predicts and explains entirely new phenomena such as black holes, which have only the roughest counterparts in Newtonian theory. From the 1970s, increasingly persuasive evidence of the existence of black holes has accumulated, and in §10.5 we shall see that there is now almost ironclad evidence of them. But these are pages from another story . . .

Let us return to some of the questions raised in this chapter and see how they were answered by the two papers I have mentioned. What is light? Let us suppose it is a wave in the electromagnetic field, described by Maxwell's equations. What is that field? Einstein (following Maxwell) could have told how the field is produced and how it is detected; but as to what it is, mathematics does not answer a question like that. What carries the energy of a light wave during the eight minutes it takes to travel from sun to Earth? The field, as Faraday had intuited it, is something with physical reality, that pushes and pulls electrified objects and can carry energy. Relativity answers: Faraday is right. But what about ether?

Einstein, in his paper of 1905, had no need to assume that ether exists, but he never claimed that it does not, and in 1920 he gave a lecture called "Ether and Relativity" at the University of Leiden which starts with a critique of mechanical conceptions of ether along the lines that have been developed here but then points out that the idea of a space that is absolutely empty and thus devoid of qualities has its own difficulties. Suppose I spin a top. Spinning it does not appreciably change its shape or any other of its internal properties, but one observable quality changes very much: now it will stand without being supported. We have to say that there is a real difference between a top that is spinning and one that is not, that the difference lies not in the top but in its relations with space, and that for any such relation to exist space must therefore have properties. The observable qualities of a top located here and one located in another place or moving uniformly in the straight line between them are all exactly the same, but with rotation it is different. By this I mean: imagine yourself in a closed room, full of apparatus with which you can experiment if so inclined. Put the room far away from gravitating matter so that there is no up and down, and do not let anyone exert forces from outside that would change its state of motion. The principle of relativity then states that there is no way, by making observations inside the room, to tell where it is or how it is moving. But if the room revolves you will get dizzy and spill your coffee and know at once.

Is it necessary to talk of ether in this context? Of course not; ether is just a word. But the general theory, which enlarges the special theory by including gravity, gives empty space a very specific set of properties that are determined partly by the presence of matter and partly by the form of the universe as a whole, and these properties are expressed in mathematical terms. Einstein proposes that it is within the tradition of physics that after a suitable warning we are allowed to give the name ether to the ensemble of the properties of space, so that

> we may suppose there are extended physical objects to which the idea of motion cannot be applied. They may not be thought of as consisting of particles which allow themselves to be separately tracked through space. . . . The hypothesis of ether in itself is not in conflict with the special theory of relativity, but we must be on our guard against ascribing a state of motion to it.

Recall Faraday's idea of a vast domain of physics in which Newtonian principles do not apply.

In the next chapter we will continue with our examination of the role of light in the conceptual revolutions of the twentieth century but first, I cannot end this book without some account of what has been learned since the time of Descartes and Newton concerning the eye and its connection with the mystery called consciousness.

8. Eyes and Brains

Something sits on the grass outside my window. At a glance I see its shape and color, how far away it is, that it is a sparrow. I know this without having to think about it, yet clearly there has been a process of sensing and recognition. By 1800 it was understood that the first step in the process was the formation of an inverted image of the scene on the retina of each eye. What came after that was a mystery, and though we know more today, the core of the mystery remains.

Let us start at the retina; what happens there? Much happens there, but we can begin with the registration of color. (I say registration rather than, for example, perception, because the word perception is probably better used for something that happens in the brain.) In §8.5 I mentioned a lecture, "On the Theory of Light and Colours," that Thomas Young delivered before the Royal Society in 1801. Most of it is concerned with showing that light is a wave, but there is a paragraph on color vision that illustrates the agility of his reasoning. He is sure that light is a wave and that we detect it when it sets the minute structures of our retinas into vibration.

> Now as it is almost impossible to conceive each sensitive point of the retina to contain an infinite number of particles, each capable of vibrating in perfect unison with every possible undulation, it becomes necessary to suppose the number limited, for instance, to the three principal colours, red, yellow, and blue, of which the undulations are related nearly as the numbers 8, 7, and 6, and that each of the particles is capable of being put in motion more or less

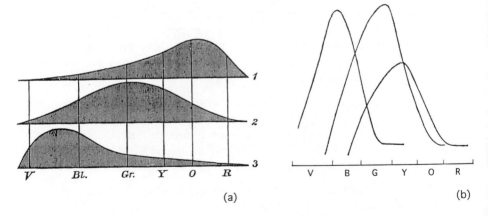

Figure 9.10. (a) Hermann Helmholtz's conjectured sensitivity curves for the three types of cone cells in the eye (from Abney 1895), compared with (b) a modern measurement. Redrawn from Brown and Wald 1964.

forcibly by undulations differing less or more from a perfect unison; for instance, the undulations of green light being nearly in the ratio of 6½, will affect equally the particles in unison with yellow and blue, and produce the same effect as a light composed of those two species.

There are a few more words; then he goes on to talk about something else, but it turns out that except for minor details he was exactly right. There are indeed three kinds of receptors (two would not suffice), and this is the basic mechanism by which color is registered.

Scientists of Young's time got this proposal mixed up with his advocacy of light waves, and it was generally ignored until it was rediscovered fifty years later by Hermann Helmholtz, whose curiosity took him into many branches of science and whose talents were largely responsible for the emergence of Germany as the leading scientific power in Europe. He measured the speed of nerve impulses (about thirty meters, or ninety feet, per second), he discovered how the ear analyzes tones, he invented the ophthalmoscope that optometrists use every day, and his *Handbook of Physiological Optics* (1856–67) collected all that was known on the subject and added much that was new. Figure 9.10 compares his guess as to the sensitivity of the three kinds of receptors to different wavelengths of light with the results of modern measurements, which show that the sensitivities are centered roughly on violet, blue-green, and yellow-green.

Suppose you look at light of pure spectral blue. The violet and blue-green receptors will respond about equally. Now replace the blue light with a suitably balanced mixture of violet and green that excites the two kinds of receptors in the same way. As Young promised, you will see the same blue as before.

One might think that red and yellow, which differ in wavelength by only about 20 percent, ought to look pretty much alike. This is about the same difference as between B and D on a piano, and most people, if they heard only one of these, wouldn't be able to say which it is. That red and yellow look so different to our eyes is because they affect (mostly) two different systems of nerves. This will not surprise people who, following Democritus and Galileo and Newton (see the quotation that begins Chapter 7), distinguish primary from secondary qualities. If two primary qualities are somewhat similar this does not guarantee that the secondary ones are similar at all. There is, after all, not much objective difference between a caress and a tickle. We see here the origin of the painters' discovery that three primary colors are enough to reproduce any color of the spectrum, even when there is no agreement as to what the primaries are.[10] As Maxwell wrote in 1872, "So far as I know, Thomas Young was the first who, starting from the well-known fact that there are three primary colors, sought for the explanation of this fact, not in the nature of light but in the constitution of man."

The receptor cells in the retina that respond to light were first seen by the German anatomist Gottfried Trevianus in the 1830s, but they are exceedingly small, only a few thousandths of a millimeter across, and he failed to notice a strange thing: the retina is inside-out (Plate 13), with the receptors at the back, next to the retinal wall, while the nerves that carry information to and from them lie between them and the source of light. It turns out, though, that there is a reason: receptor cells need continual servicing and renewal, and this is done by cells in the retinal wall.

There are two kinds of receptors, called rods and cones. Rods have little to do with the perception of color, while cones are the receptors that Young invented. By the 1960s microscopy had developed to the point where several research groups, among them that of Paul Brown and George Wald at Harvard, were able to study the color and chemistry of the dye inside an individual cone and show that there are just three kinds. Brown's and Wald's measurements of their color sensitivities are shown in Figure 9.10b. Rods are about a hundred times as numerous as cones; they are also much more sensitive, and so they take over the task of vision when there is not much light. Walk in a moonlit garden. You see the forms of flowers quite clearly but not their colors; in the dark all cats are gray. This is not to say rods are indifferent to color. Like cones they have their favorite, and it is blue, but as we know from the experience of moonlight, what we perceive is tones of gray. Another retinal property is easy to see: there is green grass outside the window beside my chair, but if I look straight ahead

[10] The mixture of paints follows different principles from the mixture of colored lights. As an extreme example to illustrate the difference, imagine an idealized red paint that absorbs all colors except red, which it reflects, while a green one absorbs all but green. Mix them and they would absorb everything and look black. Real paints are not so precisely targeted. Mix real red and green paints and you get something dark and muddy, but not black.

so as to see it only out of the corner of my eye, it has no color. This is because there are very few cones at the periphery of the retina. The center of the retina is called the fovea; here cones are close together, there are almost no rods, and vision is sharpest. If you need to see something clearly in the evening, though, look a little to one side of it.

So much for the reception of light at the retina. What happens next? The receptors generate graded electric signals—the brighter the light, the higher the voltage—and these pass through an intermediate level of cells that edit and combine them to begin the process of interpretation. Though there are a million fibers in the optic nerve that conveys signals from eye to brain, this is only about one fiber for a hundred receptors; therefore much editing and collating has been done before a signal moves toward the brain. For this purpose there are several types of cells in the intermediate level, with connections crosswise (Plate 13) as well as up and down. If I look at a dandelion it seems that the signals from some of the nearby cones are compared so as to generate not several messages but one, "This little area is yellow."

As well as judging color, the crosswise connections incorporate data from the rods, and they also generate signals that describe shape and motion. (Remember, I called it intuition amounting to genius when Galen identified the retina as part of the brain.) The output from the bipolar cells (Plate 13) goes to the ganglion cells, which code it into the usual kind of signal that passes along nerve fibers for transmission to the brain, no longer a graded electrical voltage but a succession of short pulses. There are as many as a thousand of these per second if the light is bright. The working of this system illustrates the principle that its German discoverer, Johannes Müller (1801–58), named the law of specific nerve energies: that a nerve impulse reports nothing about the external world but only the condition of the sense organ to which it is attached. It is the brain that interprets the impulse in terms of sound or color or taste, according to the organ from which it comes and the region of the brain at which it arrives.

Thinking this way one might imagine that an image is constructed point by point in the brain, but this is only partly true, for the eyes are sending several different kinds of information down the optic nerve. If a butterfly moves across my field of view I do not deduce its motion from the changing colors at each point of my retina. There are specialized receptors in my eyes that register its motion as motion, by seeing how its outline moves.[11]

There are other specializations. In 1959 David Hubel and Torsten Wiesel at Johns Hopkins made the astonishing discovery that there are cones in

[11] The brain, however, must interpret these signals. There is a case of a brain-damaged woman who has no perception of motion at all. Crossing streets is difficult and dangerous for her. "When I'm looking at the car first it seems far away. But then, when I want to cross the road, the car is suddenly very near." (Zihl, von Cramon, and Mai 1983; quoted in Sacks 1995.)

a retina which look not at points but at the orientations of short boundary lines; for this purpose there are nerve cells that fire when they see / but not when they see \, and so on around the circle. It seems that we construct the outline of a shape by detecting and putting together these little segments. There are cells that fire only when what they are looking at is part of a human face. There are other cells that detect change by comparing a signal now with what it was a moment ago. All the signals travel down the optic nerve and reach a part of the inner brain called the thalamus, which acts as a switchboard, sending them for interpretation to appropriate neural networks located in different parts of the brain. The signals that actually create the image of something I am looking at go to the visual cortex, a layer of cells about two millimeters thick on the back surface of the brain, which creates a map of what is being seen. It is a map in the literal sense that signals from two points close together in the field of the eye arrive at two little areas close together in the visual cortex. But in each of these areas different kinds of information are being registered: the orientation of a contour passing near it, whether the contour stops there or goes on, whether it is moving and if so in what direction, what the color is on each side of the contour. And furthermore, the signals from both eyes are delivered to this tiny area, processed separately, and then combined so as to produce the sense of depth and distance. What is created in the visual cortex is therefore not a map in any ordinary sense. (Most of this knowledge comes from studying people whose brains have been damaged in one region or another and seeing what abilities they have lost; the rest comes from putting tiny needles into the nerves of anesthetized laboratory animals and detecting the electric signals.) Another area of the brain, located near the eye, receives the coded pulses that summarize this visual information and says "butterfly"; another locates the butterfly in space.

Let us stop to consider for a moment the judgment of color. Isaac Newton and Thomas Young associated color with wavelength according to a definite rule: for example, light with wavelength in the neighborhood of 450 nanometers (millionths of a millimeter) is perceived as blue. But suppose I am sitting outside in the last rays of sunset reading a book with a blue cover. The reddish sunlight contains very little of the color blue; therefore when I look at the cover not much blue light comes to my eyes, and one would expect the cover to look nearly black. But still it looks blue. A color center in my brain is constructing the sensation of blue out of data received from the whole surrounding visual field as well as from the book. It perceives that there is not much blue in the sun's light and it compensates.

Here is another example. Ordinary color film marked *Daylight* is made so as to reproduce an outdoor scene in the colors we are used to. Use the film indoors under incandescent light and the result is terrible: the colors are mostly shades of red and orange. What goes wrong with film when you take it indoors? The answer is that it is merely reproducing what would be seen by an eye if it brought its color sense in from outdoors. But the eye does not do this: it observes that light from an electric bulb is different from the light of the sun, and the visual mechanism adjusts its entire color response so that the colors look as nearly as possible as they would out-doors. Goethe knew from a study of painting that the perceived color of a thing is not just a question of wavelength. It depends very much on the color of its surroundings, and this knowledge was perhaps at the core of his passionate rejection of the whole Newtonian theory.

The content of the last few paragraphs can be summarized in a single sentence: do not think that vision has any similarity to the action of a television camera which sends its pictures along a cable as electronic signals; it is much more complicated than that.

To watch one aspect of the task of giving meaning to a perceived image I will ask you to find the hidden man in a picture. Very likely you will see it at once, but if you do not, try to observe the process of your search, the formation and rejection of hypotheses, that goes on in your mind as you look. The figure is Figure 9.11. How long did it take you to find the man? The same process goes on every time you look at anything. As Alhazen has already said very clearly, a distribution of light and dark, color and motion, acquires a meaning only in the light of knowledge and experience. Obviously this is also true of animals. It seems that human beings do something more, but what is it? Here we have arrived at the question of human consciousness, concerning which much has been written but little is known.

Figure 9.11. Find the hidden face.

We have wandered a short distance up the steep path marked "Vision" and must now return for a final try at the one marked "Light." Perhaps some-where at the end of this path is an inscription that tells what light is, but I have never been there. Instead, consider cheese. Give instructions for making it, but you still haven't said what it is. We know how to make light by rapidly oscillating an electric charge, but that does not help us to understand its essential nature. Perhaps it is a fundamental substance of some kind. What kind? Here the scientist is on firmer ground, for that question is best answered by trying to say what it does, not what it is. Now we are talking about experiments and the conclusions drawn from them. But the conclusions do not come easily, for the experimental evidence gathered during the last century has been strange, it has often seemed contradictory, and it has puzzled everyone who has considered it carefully. Here the upward path is very steep and should be labeled "Proceed with Caution."

10

WHAT IS LIGHT?

The quantum is the crack in the armor
that covers the secret of existence.
(John Wheeler)

1. Light Seems to Be a Particle

PUT YOUR HAND near a warm surface and feel heat coming out. If the surface gets hotter it glows red; hotter still and it turns orange or yellow. The sun's gaseous surface has a surface temperature about six thousand degrees Centigrade (eleven thousand Fahrenheit), and emits radiation in the form of optical noise that our eyes report as white. Suppose you want to study the radiation from some solid material at a given temperature. The best way is to prepare a sample of it that is hollow with a little hole leading to the outside, so that by looking through the hole with your eye or some instrument you can observe the radiation from the hot material uncontaminated by any other radiation. One might at first suppose that different materials at the same temperature would emit different amounts of radiation, or radiation of a different color, but experiment shows that this is not so. Not only is the total amount of radiation the same, but its color, its distribution across the spectrum, is exactly the same: cast iron or burned toast, it doesn't matter. There is no way to identify a material by looking at its radiation in this way. The law that connects temperature with amount and color of the radiation is universal; and, that being the case, because there are not many universal physical laws, it is something to study carefully and, if possible, to understand. But first, what is the law?

Max Planck (1858–47), a student of Helmholtz, had colleagues in Berlin who fed him new experimental data almost daily, and in 1900 an ingenious guess based on deep knowledge of heat and temperature gave him the formula that correctly describes the intensity of radiation emitted at any wavelength from any material at any temperature. Having guessed the formula Planck then tried to explain why it is true.

Explaining consists in this: state some clear physical assumptions from which the formula can be derived unambiguously by a straightforward mathematical calculation. It is easy to see now why this turned out to be a difficult task: Planck had opened the door to a new kind of physics that neither he nor anyone else knew anything about; they did not even know it existed. In a few weeks, working backward from the result to the assumptions, he had an explanation. Because at this point he knew more about matter than radiation, he concentrated on the material substance of the box from which the radiation comes, and because it does not matter what the substance is, he imagined an artificial one: the walls of his imaginary box were lined with millions of tiny imaginary oscillators. Each one was something like a little ball carrying an electric charge and attached to a spring; they were all tuned to different frequencies and continually exchanged energy with the radiation around them. If he understood the behavior of the oscillators at a given temperature—how many of them have what energy—he knew how to calculate the spectrum of the radiation in the box. But to understand what the oscillators were doing, he found it necessary to take a radical step. Standard methods of calculation gave the wrong answer, so he borrowed a novel method that Ludwig Boltzmann had used in calculating the properties of a gas. The crucial point was this: A gas molecule can have any value of energy, from zero to a big number, but for computational convenience Boltzmann divided the energy scale into narrow strips (a common trick in calculus) and pretended for a moment that all the molecules whose energies lie in a given strip have exactly the same energy. Later in the calculation Boltzmann allowed the width of the strips to be arbitrarily small, which made everything all right. Planck made a similar calculation, except that he left the widths finite. He wrote the width as $h\nu$, where ν (the Greek letter nu) is the frequency of some particular oscillator and h is a universal number that came to be called Planck's constant. The calculation then led to the correct formula. Much of it was based on well-established physics and not hard to follow. Planck had derived the formula, but what did the derivation mean? His writings do not help. In a letter written to a colleague years later he says that adopting Boltzmann's trick was for him "a purely formal assumption and I really did not give it much thought except that, no matter what the cost, I must bring about a positive result." Formerly there had been one puzzle, how to explain Planck's radiation formula. Now there was another, the explanation.

The theory of heat radiation was not at that time a very important topic of physics, and Planck's discussion, though it came out with numbers that agreed beautifully with experiment, involved a calculation that made no sense to anybody. It was discussed among German professors but otherwise passed almost unnoticed. One of the few others who did notice was Albert Einstein, Technical Expert, Third Class, in the Bern patent office, who five years later submitted a short paper, "Concerning a Heuristic Point of View of the Creation and Transformation of Light" [1905a], to the journal *Annalen der Physik*.[1] In it he mentions Planck but does not follow

[1] This paper was submitted in March. Another very important paper giving evidence for the still-disputed reality of atoms and molecules was submitted in May, and the relativity paper was submitted in June. Altogether, including potboiler reviews of the literature, Einstein published twenty-six items during the year 1905.

his calculation, and in fact chooses to describe the spectrum of thermal radiation by a formula that is less accurate but a little simpler than Planck's. It suits his purpose because he shows that in this approximation several formulas that describe radiation at a particular frequency are the same as those calculated in an entirely different way for an ordinary gas, provided that one interprets Planck's quantity $h\nu$ as the energy of a gas molecule. Here Planck's constant enters physics through the door of analogy, not as part of an unexplained computational trick but as determining the energy of an actual thing, for if one adopts Einstein's "heuristic point of view," he has shown that the radiation inside Planck's imaginary hollowed-out sample of matter behaves in some ways like a gas of particles. The total energy in any sample of radiation at any particular frequency is $h\nu$ (one particle) or twice that, three times that, and so on, but it never has any fractional value. In modern usage the energy is said to be *quantized*, and Einstein referred to the parcels of energy as *quanta*.[2]

One of the obligations of any scientist with a new idea is to set it on the big black block and suggest an experiment that will chop its head off. Planck's hypothesis concerned some oscillators that were purely imaginary. Einstein's quanta are a hypothesis about the nature of light, and he suggests a way to test it. Remember Heinrich Hertz's discovery (§9.4) that ultraviolet light facilitates the production of an electric spark from a metal surface, a phenomenon later named the photoelectric effect. Others had found that ultraviolet light causes a metal surface to acquire a small positive electric charge, and with the discovery of electrons (which carry negative charge) in the 1880s it was guessed that ultraviolet light was somehow removing electrons from the surface. If Einstein's quanta are real things, this becomes easy to understand: the quanta, like bullets, were knocking electrons off one by one. The higher the frequency of the light (that is, the shorter its wavelength), the higher the energy of the quantum. One would expect that a single electron might catch all the energy of one quantum, but none could catch more; therefore, Einstein suggests, we have a relatively straightforward experimental test of his theory: change the frequency of the light and see how the energy of the fastest electrons changes. Easy to talk about; very hard to do. Much later [1914] Robert Millikan in Chicago succeeded (to his dismay) in verifying the hypothesis that he had hoped to disprove, and later refinements of the experiment gave a good numerical value for Planck's constant. But by that time there were other reasons for accepting Einstein's idea, and Millikan's experiment was hardly news.

Einstein's quanta also give a simple explanation of the penetrating power of X rays. I have mentioned that their wavelength is in a range a few thousand times smaller than that of light. Therefore the energy of each quantum is a few thousand times greater. This is why they pass through our

---

[2] The word "quantum," which in late Latin means a small portion of something, was in common use in Germany at the time.

bodies, and why we need to be protected from the damage they cause to living cells.

But what was going on? Thomas Young's experiment, the mother of all experiments in this domain, had shown a hundred years earlier that light is a wave. It was something that everybody knew, a thousand experiments had verified it. How could a wave be a particle? Einstein didn't know. In the 1905 paper he urges readers to keep an open mind, but like Newton before him, he resonates to the idea that if the fundamental nature of matter is atomic, light may also be atomic, for Nature is conformable to herself. When we pick up a piece of matter we perceive not the atoms but only the gross qualities contributed by a huge number of them, and ordinarily when we experiment with light we see only the average effect of a huge number of vibrations so quick we cannot follow them. Perhaps light waves are only some average property of large numbers of quanta. In an address before scientists and doctors in 1909 Einstein said,

> No one can deny now that there is a mass of evidence to show that some fundamental properties of light are much better understood by Newton's emission theory than by the wave theory, and therefore I think that the next phase of the development of theoretical physics will produce a theory of light that in some way unites them.

In the same year he writes to his friend Jakob Laub, "This question of quanta is so extraordinarily important and difficult that everyone should be concerned with it." But everybody wasn't. In those days each country had its scientific interests and its point of view, and its workers tended not to read much in other languages. The kind of physics I have been describing was a German-language affair, and it was not until 1911, when an international conference supported by the industrialist Ernest Solvay brought together the leading physicists of Europe, that people outside the German orbit woke up to the existence of an immense challenge to their intelligence, imagination, and experimental skill. Soon they had a new discovery to chew on.

2. Niels Bohr Holds an Atom in His Hand

One of the participants in the Solvay Conference was Ernest Rutherford (1871–1937), leader of an active experimental group at the University of Manchester. I have already mentioned his fundamental discoveries in radioactivity, and a few months before the conference he had invented the nuclear atom. In his model, electrons, very light in weight and carrying a negative electric charge, orbit around a relatively massive positively charged nucleus in a manner that recalls the structure of the solar system. This invention is arguably the single most important event of the twentieth century, but at the Solvay meeting Rutherford seems to have said nothing

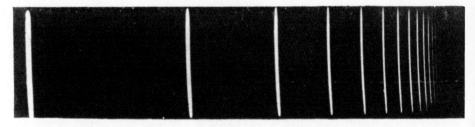

Figure 10.1. Part of the spectrum of atomic hydrogen.

about it. Perhaps this was because most of those present were makers of theories. Rutherford was an experimenter. His atomic model was strongly suggested by experiments he had done, but from a theoretical point of view it made no sense. According to Maxwell's theory an electric charge moving in a circle emits light, and therefore after the first creation the whole universe would have glowed for a moment as every electron in every atom spiraled inward toward whatever fate awaited it as it hit the nucleus. If the model contained any truth there had to be some physical principle that keeps this from happening, that keeps the world in a state of stability. An idea was needed, and at this point the quantum entered through yet another door.

Niels Bohr (1885–1962) came to Manchester from his native Denmark to do experimental research on atomic physics with Rutherford, but he soon decided that he could make his best contribution by trying to understand what Rutherford was saying. Attempts to account for his atom's paradoxical stability filled the air of Manchester; Bohr sat among the physicists and gathered scraps of insight from which, in 1912, he put together a theory of the hydrogen atom. This was the natural atom to select, for in Rutherford's scheme it consists of a nucleus orbited by a single electron and is the simplest atom possible. The electron could be imagined as sailing around the nucleus as the moon sails around the Earth. Any heavier atom would have more electrons which would get in each other's way and have a much more complicated motion. The spectrum emitted by a gas of hydrogen atoms has a simple pattern (Figure 10.1), and this is what Bohr proceeded to explain. Planck had introduced his number h as part of a computational scheme with no clear physical basis. Einstein had revived the Newtonian idea that light is made of particles and had proposed that their energy is given by a simple formula involving h. Bohr now proposed that h plays a third role in physics: it is something mechanical, and it enters the description of the electron's motion as it circles the nucleus. It enters in such a way that the electron's energy, its speed in its orbit, the size of the orbit, and the orbit's ratio of length to width are quantized: they can have a certain list of values but nothing in between.

So much for the electron's supposed motion in its orbit. Atoms and electrons are too small to see; how can we have any idea that what Bohr says is true? Consider, he says, what happens if an electron orbiting the nucleus with one of its allowed energies drops into another orbit with a lower energy. The energy it loses must be got rid of in some way, and Bohr proposes that it is emitted as a single quantum. The energy of this quantum is related to its frequency of vibration by Einstein's formula, and so its frequency and wavelength can be calculated. The wavelength can also be measured in a spectroscope. There is only a certain list of energy states in which an electron can start, and an electron in any one of them has only a certain number of lower states into which it can drop; thus one can list every transition that can possibly occur. It may be a long list, but if each transition produces light of a definite frequency and wavelength, the spectrum of hydrogen will consist of a series of well-defined lines, with darkness (or relative darkness) in between. Without introducing any new numbers ad hoc, Bohr calculated the wavelengths of the lines shown in Figure 10.1, as well as many others not shown.

For a gas or vapor to radiate at all, the atoms must be supplied with energy in some way. In the laboratory this is usually done by passing an electric current through it at high voltage, but the same can be done by making it very hot. Remove the energy supply and the atoms begin to radiate their energy, dropping from states of higher energy to states of lower energy, then lower still until the process ends with each atom in its lowest possible state, called the ground state, from which it can radiate no more. The air you are breathing consists of uncountable billions of molecules, of which all but the most negligible number are in this state. Bohr's theory also provides a figure for the diameter of a hydrogen atom, about a ten-millionth of a millimeter, which fits the experimental evidence. Thus the quantum introduces an element of order and stability into the material world: all atoms of a given substance, in their ground state, are exactly alike.

What I have written seems at first sight not to apply to our favorite light source, the sun, even though it is made mostly of hydrogen, for we have seen in Plate 10 that its spectrum is composed of dark lines against a continuous spectrum, the opposite of Figure 10.1. To understand the continuous spectrum think of the ringing of a bell. You supply energy, strike the bell or shake it, it emits certain tones for a while and then is once more quiet. This is an analog of the way a single atom emits light. Now fill a bag with bells and shake it. All this does is make a noise, and the reason is obvious: with the bells all touching one another none gets a chance to vibrate freely. This is what happens in the luminous layer of the sun called the chromosphere, where the gas is not so dense that the atoms actually touch one another all the time but it is dense enough so that they often collide. They produce not pure colors but the optical equivalent of noise, and that, as we have seen in §9.5, is the sun's white light.

Then what causes the dark Fraunhofer lines that cross the sun's continuous spectrum? This is not hard to imagine. Suppose there is an isolated atom somewhere between us and the sun's bright disc. The sun bathes it in quanta of every frequency. Inevitably, one of these quanta will have exactly the right energy and frequency to drive the atom from its ground state to some state of higher energy. In doing so the quantum is subtracted from the beam, leaving a gap in the continuous spectrum. The atom may after a while radiate an identical quantum, but that departs in any direction at all and is very unlikely to be aimed into our spectroscope. Place not one but many atoms high in the sun's atmosphere, above the chromosphere, and make them of many different kinds, and you have the Fraunhofer spectrum. This explains the discovery by Robert Bunsen and Gustav Kirchhoff [1860] mentioned in §8.8, that the dark lines in the spectra of the sun and other stars coincide exactly with some of the bright lines produced by hot gases in the laboratory, and this is how we find out what stars are made of.

Bohr went on with a long campaign of brilliant intuition and guesswork to explain, at least roughly, why the chemical elements have the properties they do, but good calculations were difficult—not just because the atoms and compounds are very complex systems but also because there was something unsound at the core of the work, as finally became clear when Max Born and Werner Heisenberg succeeded [1923] in calculating one of the energy levels of the next most complicated element, helium, the one with two electrons. Their result, computed with greatest care, was grossly in error. Something was wrong with Bohr's theory. By good luck it had yielded the hydrogen spectrum correctly and encouraged people to think that the truth lay somewhere down that road, but in the end it could not be patched up, and, for a moment, people stood around waiting for something to happen.

Experiment delivered another devastating blow to received ideas when Arthur Compton of Washington University in St. Louis [1923] studied what happens when X rays hit a material target. Maxwell's theory told exactly what ought to happen: the oscillating electric field of incoming X radiation should set the electrons of the target into motion and make them radiate the energy back again. The size of the effect can be calculated, but the result was contradicted by experiment. Compton found that the only way to understand what happened was to think of a beam of X rays as a swarm of quanta à la Einstein, some of which hit electrons in the target. Exactly as in Einstein's theory of the photoelectric effect, it is like one billiard ball hitting another; the electron speeds out of the target in one direction, and the quantum, with slightly diminished energy, leaves in another. The calculation was simple, and it agreed very well with the measurements. Here was another situation in which Maxwell's theory is wrong and a simple calculus of particles takes its place. From now on people began to think of light quanta as particles like any others, and to agree with the usage that spoke of electrons and protons they were soon being called photons.

Experiments like Compton's pounded their lesson into skeptical heads: light acts sometimes like a wave, sometimes like a particle. But how can it be both? We have seen in §9.3 that Maxwell's field fills all space, whereas a particle fills at most a very small part of it. And besides, it isn't *either* wave *or* particle, for the energy of the particle is exactly determined by the frequency of the wave. Somehow it seemed to be *both* wave *and* particle. It is not easy to explain one's way out of this apparent contradiction, which arises from Nature's reluctance to adapt itself to our linguistic habits and our modes of thought. It is we who must adapt, but the experience of the last hundred years shows that such adaptations aren't easy.

3. A New Kind of Understanding

The subject of this book is light, and so I shall continue talking about it, but I must not give the impression that the puzzle of waves and particles pertains only to light. In 1923 a previously obscure French physicist named Louis de Broglie (1892–1987) published a series of notes suggesting that the wave-particle "duality," as it seemed then, was characteristic of electrons also. He showed that if electrons are waves as well as particles the Bohr theory becomes easier to understand, and suggested that one could do experiments with electrons that are analogous to those that prove the wave nature of light. At first Einstein was the only established physicist who was interested by these ideas, writing to a friend that the young man had "lifted a corner of the great veil," but as the results of the experiments came in showing that not only electrons but later other material particles and even whole atoms have wavelike properties just as light has, the rest of world started to pay attention, and in about three years a new science called quantum mechanics had been developed by Werner Heisenberg (1901–76) and a few others. Its first principle, already obvious from the facts I have mentioned, is that the wave-particle "duality" is not a special property of light but pervades the physical world. Newton believed that the uniformity of Nature is most apparent at the atomic level of size and structure, but he would hardly have imagined that, on a still smaller scale, there is more similarity than difference between a ray of light and the smallest particles that make up a brick.

Though "duality" is a word that was often used as people first encountered the experimental facts, it was an unfortunate term, for it implies two kinds of nature. Nature is one, and it is up to us to understand it that way, if we can. As Heisenberg wrote [1971],

> Those who have truly understood quantum theory would never even dream of calling it dualistic. They look upon it as a unified description of atomic phe-

nomena, even if it has to wear different faces when it confronts experiments that are different when described in everyday language. Quantum theory thus provides us with a striking illustration of the fact that we can fully understand a connection even if we can only speak of it in images and parables.

At that point in history scientists who wanted to understand the situation and talk about it in a rational way were forced into new ways of thinking. Let us focus for a moment on the word "understand." In one sense it is impossible to understand how something can be a wave and a particle: we ask imagination for a mental picture, and it produces nothing. But we can also understand the simpler aspects of Nature mathematically. A mathematical theory can predict what will happen in a given situation but does not tell us what words to use when we describe it. The words "wave" and "particle" refer to mental images that help us understand the calculations and the experiments, but we should not take the words any more literally than we do when someone refers to Wellington as the Iron Duke. Our factual knowledge of the world derives from events. When we perform an experiment we produce an event. An experiment shows what *happens* under certain circumstances; it does not show what anything *is*. That is a question for philosophers. It is one thing to say that light is wavelike because it can be made to produce interference phenomena, but quite different to say that it produces interference phenomena because it is a wave.

Heisenberg was using "understand" in a new sense. If there is a theory that allows someone skilled in mathematics to calculate correctly the result of an experiment that has not yet been done, if the calculation starts from principles that have been tested in different ways and found satisfactory, and if on close acquaintance the whole thing seems reasonable, then in Heisenberg's sense it has been understood. As to what seems reasonable, that is as much a matter of personal judgment as the reading of a poem. From the beginning of the theory, in the 1920s, there have been many opinions on this point—as there should be, for it is both important and difficult, and the ensuing debates have taught us much. One thing should be understood, though. Considered as a way of calculating numbers to be compared with experiment, quantum mechanics seems flawless, as far as it goes. In almost seventy-five years there has not been a single instance when a correct application of it has been found to give an incorrect answer, and we must therefore try seriously to understand what it says.

Let us reconsider two of the classic experiments on light, those of Young and Compton. Thomas Young designed his experiment (§8.5, Figure 8.5), the one in which a beam of light passes through two slits and produces light and dark fringes on a screen, to show that the two beams issuing from the slits interfere with each other. He asked light, "Do you behave like a wave?" and light answered, "Yes." A century later, Arthur Compton asked it, "Do you behave like a particle?" and again light answered, "Yes." Sup-

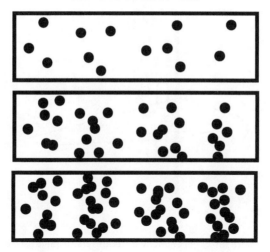

Figure 10.2. Accumulation of photons in the formation of a two-slit diffraction pattern.

pose someone invents an experiment that asks both questions at once. What will light answer then? But even though that question has been around for a long time, nobody has succeeded in devising such an experiment. It seems the world is designed so that we can ask one question or the other but not both at once. Niels Bohr described the situation with the word *complementarity*. Suppose I visit an art gallery and look at a sculptured figure. Standing in front of it I cannot see the back; for that I have to walk around to where I no longer see the front. I cannot see both at once. Bohr calls two such views complementary: I piece together my understanding of the figure from observations that cannot be made at the same time from the same place. Wave and particle are in the same sense complementary descriptions of light.

Think about Young's experiment from the particle point of view. We have a beam of photons, some of which pass through one slit, some through the other. Figure 10.2 shows three stages of the gradual formation of a pattern as more and more photons arrive. It seems there is an influence that prevents them from arriving at some regions of the screen and bunches them together in others. Perhaps our first reaction is that of Grimaldi or Newton: look for the force that is doing it. Because the separation of the fringes on the screen depends on the distance between the slits, the force must involve both beams of photons; it must be a force exerted by the particles of one beam on those of the other that depends on how far apart they are. There is an easy way to check this hypothesis: put a photographic film at the screen, do the experiment in such very dim light that there are almost never two photons in the apparatus at the same time, expose the whole thing for a month, and develop the film. In this situation one particle cannot influence another, and yet it is found that the result of the exper-

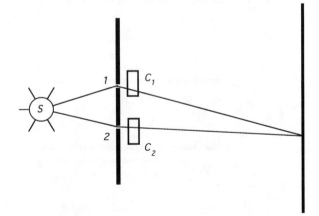

Figure 10.3. The two-slit experiment, with counters installed so as to find out through which a photon passes.

iment is unchanged, still the same light and dark fringes. The explanation by forces doesn't work.

How could one prove that a photon goes through only one of the two slits? Put some kind of photon counter into each beam, as in Figure 10.3, and let the light be so faint that we can study what happens to the photons one at a time. There are three possibilities: counter C_1 registers and C_2 does not, C_2 registers and C_1 does not, or both register. It is never found that both register. Suppose C_1 detects a photon and C_2 does not. Then if nothing goes through slit 2, how can it make any difference if slit 2 is there? And yet the particle through slit 1 tends to land in regions on the screen that are determined by the position of slit 2. A paradox? Let us study the situation more closely.

Analysis of how photon counters actually work shows that if a counter is installed in one of the beams, and if it works properly, it changes the phase of the wave in that beam so much that no interference pattern appears at the screen. We were trying to understand the interference pattern which presents light as a wave, but now, with the counter in place, there is no pattern to understand. Take away the counter and the pattern reappears, but now we cannot know what happened to the photon. Counters count particles. If you install it you find the particle, but there is no wave. If you take it away there is a wave but no sign of a particle. You may choose whether you want to find wave or particle phenomena; in either case you will find what you are looking for. Nature is behaving simply and predictably, but we have trouble finding a way to talk about what it is doing. To clarify discussion Bohr proposed to redefine the word phenomenon: you should not just say what you saw, for example, a wave or particle phenomenon; you must describe the experiment in which you saw it. The whole thing is the phenomenon, and it is only in this sense that you should talk of wave and particle phenomena.

A glance at Figure 10.2 shows that photons seem to be arriving at random, subject to a constraint that makes them statistically more likely to land on some parts of the screen than others. Well, economics is a statistical subject full of constraints. I am unlikely to buy two houses in the same year, but there is nothing random in my decision to buy a house. Every economic transaction is for some reason. The molecules of a gas move at great speeds in different directions, seemingly at random, but each molecule has a reason for being where it is, moving as it is. What is the reason that sends one photon to one part of Figure 10.2 and another to another? Quantum mechanics answers clearly: there is no reason. This does not mean that we are too lazy or too dumb to figure out where a photon will go; it means that it is impossible. In practice, of course, we would have trouble if we tried to choose a large gas molecule and follow it around and analyze its collisions and near misses so as to predict where it will go next, but as one imagines the forces and motions the prediction is possible, reasonably accurately, *in principle*, which really means in imagination. Quantum mechanics says that at the smallest scale imagination deceives us. For a large molecule rough calculations could in principle be made; for something as small as an electron or a photon, it is impossible. The theory gives energies and mechanical properties in the atomic domain with astounding accuracy, but in other situations its results are only statistical. If you measure the behavior of a very large number of particles you find that the statistics have been correctly predicted. If you are interested in a single occurrence (as for example when one bets on a horse) there is a calculus of individual cases, but its results are probabilistic.

In the early days of quantum mechanics this edict was hard to accept. For many physicists, among them Einstein and Planck, physics was almost by definition the science of an objective reality governed by exact causal principles. The pioneers have all departed, and their successors, not finding any weak point in the theory, have come to accept it, though sometimes with reservations. In one of his last papers the American physicist Richard Feynman voiced some private doubts,

> Turning to quantum mechanics, we know immediately that here we get only the ability, apparently, to predict probabilities. Might I say immediately, so that you know where I really intend to go, that we have always had (secret, secret, close the doors!) we have always had a great deal of difficulty understanding the world view that quantum mechanics represents. At least I do, because I'm an old enough man that I haven't got to the point where this stuff is obvious to me. . . . You know how it always is, every new idea, it takes a generation or two until it becomes obvious that there's no real problem. I

cannot define the real problem, therefore I suspect there's no real problem, but I'm not sure there's no real problem.

The lesson thus far is that we can talk reasonably about what is going on if we do not demand exact descriptions of what happens on the microscopic scale, if we are careful in our use of words, if we remember to describe the world in terms of its phenomena. The first task of the quantum in physics is to make theory agree with experiment, which it does, to an astonishing degree. The second is to help us conceive that other world— not the world built out of our perceptions, but the quantum world that is imagined as underlying those perceptions and responsible for them. The truth, Democritus told us, is hidden in the depths, and we should be careful before we assume we know anything about those depths. As the British biologist J.B.S. Haldane said, the world is not only queerer than we suppose, it is queerer than we *can* suppose. We must be ready for that, and must look next at some questions that are not cleared up just by watching our language.

Let me summarize the situation physicists must learn to live with. Their science concerns itself with, among other topics, entities loosely called particles, waves, and forces. A particle is a kind of thing, a wave is a kind of motion, a force is a kind of push, the last two are more like verbs than nouns. It has turned out, though nobody expected or wanted it that way, that the three are tightly linked at the level of theory. If you have one you also have the other two. Light shows all three natures in a specially clear way. Different experiments can show the behavior of either waves or particles, and the electromagnetic field is what builds atoms into molecules and holds them all together. It is also the medium through which we sense the world. The field conveys sight directly, but also touch, taste, smell, and hearing are all conveyed indirectly, through contacts mediated through electromagnetism. The question is often asked, "Is there perhaps some other field, a whole universe of activity, that does not interact with our kind of matter and exists unknown to us?" The phrasing of the question excludes a scientific answer. There would have to be at least some slight linkage to the world we perceive, some definite, hard-edged phenomenon that cannot be explained with the kind of physics we know. At present there is no candidate that would convince a skeptic.

4. Spooky Actions at a Distance

In 1947 Einstein wrote to his old friend Max Born that he could not seriously believe in quantum mechanics "because the theory cannot be reconciled with the idea that physics should represent a reality in time and space, free from spooky actions at a distance." He was thinking of a strange kind

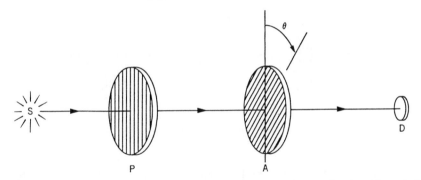

Figure 10.4. Photons from the source S encounter two sheets of Polaroid on their way to the detector D.

of phenomenon predicted by quantum mechanics, whose existence he was perhaps the first to notice. The simplest illustrations of the effect involve polarized light, so let us start with a crash course on the subject. If your toy box contains two pieces of Polaroid or a couple of pairs of polarizing sunglasses you can quickly verify what I say.

Figure 10.4 shows a source of light, S, the sun or an electric bulb, emitting a light ray toward the right. It encounters a piece of Polaroid,[3] P, where about half the photons are stopped and half go through. According to quantum mechanics this is a purely random matter; each photon has a 50 percent chance of going through. The axis of P is oriented up and down, and the beam that emerges from it is, as one says, polarized vertically, meaning that the electric and magnetic fields are oriented as in Figure 9.5. If now the ray encounters a second piece of Polaroid, A, the probability that a photon goes through A depends on the angle θ: if θ is zero and the two axes are parallel, the photon will certainly go through A. If the axes are perpendicular it will certainly not, and for positions in between some photons will go through.[4] The detector D is drawn to show what I mean by "goes through": it registers a photon's arrival with a click or a flash or a record on a piece of paper.

Now consider the arrangement shown in Figure 10.5. Here the source is an unusual kind of atom—such as calcium-40, which, when one of its electrons jumps from a higher energy state to a lower one, sometimes gives off two photons instead of the usual one, and whatever their direction of polarization may be, it is the same for both. A little quantum-mechanical calculation[5] shows that with the axes of Polaroids A and B parallel as shown, what happens is very simple: if the right-hand photon goes through A, the other photon will certainly go through B; either they both go

---

[3] I will assume an idealized Polaroid; the real thing is not very different.

[4] To be exact, the probability is $\cos^2\theta$, and the reason that a photon from an unpolarized beam has a 50 percent probability of going through Polaroid P is that the average of $\cos^2\theta$ over all values of θ is ½.

[5] E.g., Park 1992, p. 350.

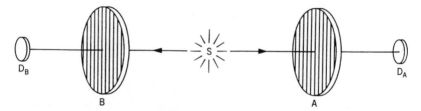

Figure 10.5. Pairs of photons encounter two Polaroids on their way to different counters.

through, or neither does. Simple to say, but strange to think about. Since the beam reaching A is unpolarized, each of its photons has a 50 percent chance of going through and being counted, and similarly, one would suppose, at B. The experiment can be arranged so that B is many yards away; I suppose miles would be possible. One might then think that whether or not the photon reaching B goes through it would be independent of what had happened to the other photon at A—how could it possibly know what had happened—so that the chance that both would go through and the two counters would register simultaneously would be only 25 percent. The experiments are difficult and have been carried out at several centers, notably at the Institut d'Optique in Paris by Alain Aspect and collaborators [1981 and later reports], and the figure 50 percent is correct.

Consider for a moment what this implies. Situate A in the laboratory, and put B far away. Choose an orientation for A; its axis need not be vertical. Suppose that D_A counts. Then D_B is certain to count if B's axis is parallel to A's; that is, you can change the counting rate at B by turning Polaroid A one way or the other. The photon with which A interacts is not the one that goes to B, and it is hard to see how doing anything at A could affect what happens when a photon arrives at B. To add pepper to the soup, Aspect and his collaborators [1982] changed the orientation of A in the brief moment *after the two photons had left* S, and the result was the same.

The technical word for what has been seen in this experiment is *nonlocality*, defined as the situation that exists when an action at one place influences an action at another place but there is no apparent link of cause and effect between them. There is an ancient belief that if I sit in my house sticking pins into a doll I can cause my enemy in a distant town to double up in pain. That also is nonlocality.

Human ingenuity has thought of one way in which the modern examples of nonlocality could be explained causally. Suppose that quantum mechanics is wrong and there is no question of probability in the whole matter at all. Suppose a photon moving toward A possesses some definite property, a key that either will or will not open the door when it reaches A, and that B carries an identical key. If D_A counts, then D_B will also count and nonlocality disappears. This so far undiscovered key would be some property of a photon expressible by a number or a symbol; such properties are called *hidden variables*. If a hidden variable were to be uncovered, quantum mechanics would be in for a very hard time, and that is the reason for the fame of a remarkable paper written many years ago [1964] by John Stewart Bell (1928–90), an Irish physicist who for most of his career was involved in the design of particle accelerators. The thing about hidden variables is that because they are hidden from us, there is no way to know how they might work, but Bell proved from the laws of statistics that no matter how they work, they cannot explain the results of experiments like Aspect's. The situation is left this way:

> If quantum mechanics is true, there are no hidden variables, and Nature contains nonlocal actions.
> All experiments thus far support quantum mechanics.
> Therefore, as far as is known, Nature contains nonlocal interactions.

Some scientific questions are interesting and some uninteresting, where the words are defined as follows. After an uninteresting question has been settled, it is settled. After an interesting question has been settled, it keeps popping up again. The question of nonlocality, in the opinion of many physicists, is interesting.

5. Instruments and Techniques

The discoveries of twentieth-century physicists have forced us to look with wondering and increasingly skeptical eyes at the material world. They have forced us to think carefully about the relation between the image of this world that we carry in our heads and whatever it is that is actually "out there." They have also provided deep new insight into the world's structure and how it works. Technical skill and scientific knowledge go hand in hand. Most of the discoveries of the last hundred years have depended on the development of new experimental tools, and these tools have generally been made possible by new developments in science. The next few pages will tell what has happened to telescopes, microscopes, and X rays and then say something about a new tool, the laser.

Telescopes

In the decades following the Herschels the trend was toward the biggest mirrors that money could buy, but soon the telescope builders found themselves faced with two inherent limitations. The first is weight. Glass bends, and so as a telescope turns to look at different parts of the sky its mirror subtly changes shape. The second limitation is posed by the atmosphere itself, which is not an optically uniform substance. As wind blows or heat currents rise, its density and temperature vary from point to point and moment to moment, with the result that the incoming rays are refracted unevenly. Think of the quivering outlines of a house looked at across a sunbaked field. The image shakes because the air above the field is unevenly heated and moves unevenly. Its index of refraction varies from point to point and from instant to instant, so that light does not travel through it in a straight line. If one looks at stars on a cool, calm night the disturbance of light rays is much less, but a telescope makes it seem larger, so that a distant star seems to jump around as we look at it. The result is that the image formed in a big reflector is actually not much sharper than the one an amateur sees with an instrument in the back yard. The main advantage of the big instrument is that it gathers much more light, and can see things that are much fainter and farther away. As we look outward in space we look backward in time, and it is those distant objects that have the most to tell us about the origin of the universe. Strategies have been developed to deal with the limitations of weight and atmospheric refraction, and here are some of them.

The Hubble Space Telescope, launched in 1990, deals with both limitations, since it is in a weightless environment above the atmosphere. On the other hand, it cost a billion dollars. After it was aloft astronomers discovered a design error of $1/10{,}000$ inch ($1/25$ the diameter of a human hair) in the shape of the mirror. This produced a spherical aberration that required expensive refitting, but by now the HST has provided several important new discoveries. In 1995, for example, astronomers using it found at the center of a nearby galaxy a whirlpool of matter that circles at a rate up to a million miles per second, thus establishing beyond reasonable doubt that at the galaxy's core is something hugely massive and yet very small which is probably a black hole. Because of the HST's initial expense and because the service man calls only once in three years it will probably be a long time before another large orbiting telescope is built.

Another way of coping with the effects of weight was developed in the Lawrence Berkeley Laboratory in California and embodied in the Keck telescope, which at present is the largest in the world. It is part of the observatory at the top of Mauna Kea on the Big Island of Hawaii where the altitude, almost fourteen thousand feet, places it above much of the atmospheric turbulence. Its mirror is almost thirty-three feet across, four times the size of Edwin Hubble's mirror on Mount Wilson, and it is formed of thirty-six hexagonal segments independently supported and realigned every two seconds by a computer-controlled mechanism. It was expensive, but it works very well and cost about a twentieth as much as the orbiting telescope.

A more radical approach to the problem of the atmosphere is to use a flexible mirror supported on hundreds of little pistons that subtly distort its shape in response to fluctuations in the incoming light. Because the fluctuations are so quick, the pistons must be able to readjust the surface at least one hundred times per second, and plans exist to increase this figure to two thousand. If the telescope is looking at a bright star or something close to one, it can sample the star's light at each position and instruct the pistons how to move so as to produce a clear image. At the end of 1995 the eighty-year-old Mount Wilson telescope was refitted in this way to produce images almost as sharp as those of the HST. Its handlers claim that if there were a nickel two hundred miles out in space (and close to a bright star) they could read the date on it. But bright enough stars are rare, and distant objects do not produce enough light for this kind of instant diagnosis, and workers at several laboratories are developing arrangements that use a laser beam shining out through the telescope to produce an artificial bright star in front of the telescope. This is a very active field of research and development, and anything more I might say would be antiquated by the time you read this book.

The telescopes discussed up to this point all make use of visible light, but as we have seen, that is only a small fraction of the whole spectrum of radiation. X rays and gamma rays also carry their messages to us from outer space, and on the side of long wavelengths there are also radio waves. They are not light, but some of them originated as light.

The light that God ordered up in the beginning faded at the end of the first day, but the old writers make it clear that the darkness of night was not the same as the darkness that had prevailed when there was no such thing as light. Modern astrophysicists, trying to reason backward from the universe as we know it to determine how it started, are fairly generally agreed that the universe began with a great flash of light and other particles now known irreverently as the Big Bang. All the kinds of particles except a few are unstable; they very quickly disintegrate and leave behind only the stable forms of matter and energy. Photons are one of those forms, and in the first moments there must have been a lot of them. Where are they now? Why isn't the universe bathed in light? The answer is that it is indeed bathed in light, whose nature and amount are determined by the universe's expansion. The expansion is something we are fairly sure of. In the first place, we observe that some of the nearer galaxies, and all the distant ones, are moving away from ours. In the second place, there is this light. Once it was intense; now it is dim. Once it had a spectrum (of the kind described by Planck's formula) corresponding to a very high temperature; now the temperature is low, only 2.7 degrees above absolute zero. The change, in quantity and quality, reflects the cosmic expansion. In quantity because a given amount of radiation is spread through a much larger volume; in quality because it follows from Maxwell's equations that as the universe expands the length of every light wave expands in the same degree, so that what were once the short wavelengths of ultraviolet light have lengthened until they are measured in millimeters. This cosmic background radiation, known as CBR, was first detected in 1965 by Arno Penzias and Robert Wilson at the Bell Laboratories as a continuous hiss arriving from every direction at a large antenna. Though this ancient light is faint it is nearly everywhere, and in the room where you sit each cubic centimeter of space contains several hundred of its photons.

There are other radio signals, originating in clouds of hydrogen gas, and big telescopes are built to receive them. They are our principal source of information about the central parts of our galaxy because there is so much gas and dust there that almost no visible light gets through. A generation ago all we knew about the heavens came from observations in visible light. What we have learned since then from studying other forms of radiation is more than the sum total of everything that was known before.

Few people look through the big optical telescopes any more. For a century astronomers preferred to take photographs and study them at leisure. Now the film in the camera is replaced by a CCD, standing for charge-coupled device, which is far more sensitive than film and displays the image as a digitized electric signal flashed to the television screen of an astronomer who may be controlling the telescope from a thousand miles away. Does this mean that astronomers no longer react to the magic of a night sky? Remember that all of them started small, and that something they saw through whatever telescope they could buy or borrow or make encouraged them to spend a lifetime among the stars.

Microscopes

There has been no further development of optical microscopes on the scale of that of telescopes. The atmosphere presents no problems, but there is another factor. In improving microscopes the main impulse is toward higher magnification, and at a certain point one encounters a limit, for as light enters the instrument it is diffracted à la Grimaldi at the rim of the lens and so no longer moves in a straight line. The result in practice is that an optical microscope cannot see clearly any detail smaller than about two hundred nanometers,[6] roughly half the wavelength of visible light. Atoms, which one might like to look at, are a thousand times smaller than that. No improvement of lenses can get around this limitation, and so people are starting to develop lensless instruments. I mention a recent one developed by Kumar Wickramasinghe and collaborators [1995] at the IBM research laboratories. The optics is very recondite, but the idea is simple: a delicately suspended probe with a very sharp tip about five nanometers in diameter scans slowly back and forth beneath the sample, one or two nanometers away from it. Above it, a laser scans it also, and an interferometer analyzes the light reflected from the probe and the nearby surface of the sample. From this the height of the surface can be calculated, and as the scanning proceeds the computer slowly produces a picture of the ups and downs of the surface. These pictures now show details at the scale of one nanometer, about ten times the diameter of a typical atom.

---

[6] Nanometer is defined in the "Technical Terms" section at the back of this book.

Another way to reach higher magnification is to use shorter waves. I have told in §3 how in 1923 Louis de Broglie suggested that the wave-particle "duality" is not just a property of light but is found in all particles. As with light, the higher the particle's energy the shorter its wavelength. The wavelength of an electron driven by one hundred volts (about the voltage of household appliances) is of atomic size, and by using higher voltages it can be made smaller. A microscope using electrons instead of light was an obvious idea. Electrons do not go through glass, and so lenses cannot be used, but electrons, unlike photons, carry electric charge and can be steered by electric or magnetic fields. The first electron microscopes of both kinds were built in Germany in 1932, and their power and versatility have steadily improved. By now they produce excellent images half a million times the size of the original, showing details a few nanometers in size; but this is a book about light, so we can stop there.

X Rays

Crystals have been collected and admired for centuries, and by 1900 they had been found to have remarkable optical properties. For example, the way a crystal refracts a beam of light depends on the beam's direction with respect to the crystal's symmetry axes. In the early days some investigators thought that crystals are a special form of matter in which the atoms have an orderly arrangement like the bricks in a wall; others thought it was more likely that the atoms are situated at random but are all somehow oriented in the same direction. In 1912 Max von Laue in Munich suggested that if the atoms are arranged like bricks, and a crystal is put into a beam of X rays a diffraction pattern should show up analogous to the pattern formed by the pair of slits in Thomas Young's experiment, although since the structure being investigated has regularities in three dimensions instead of only one, the pattern would be more complicated and require more art to interpret it. Experiments were successful, showing that X rays really are waves and at the same time that crystals are orderly and repetitive arrangements of atoms.

As the years went on people learned how to investigate structures more complicated than those of ordinary crystals, but to do this the method requires three great technical improvements: sources must be much stronger, they must produce only a single wavelength, and interpreting the data requires immense calculations. The X rays used today are trillions of times stronger than in the early experiments, and their wavelength can be carefully controlled. Computers measure the patterns and chew their way through huge calculations.

The technique of X ray diffraction is used to study the arrangement of atoms in intricate chemical compounds, and its most spectacular and useful applications have been in determining the structures of proteins and other large biological molecules. An early success was the famous double-helix structure of DNA found in 1953, but a double helix is not a particularly complicated form. By the 1990s it was possible to map biological molecules consisting of thousands of interconnected atoms, work that is necessary before anything can be done to cure diseases that arise from malformations of protein structure or of the genetic material that causes it.

In medicine X rays are still used to study broken bones and treat cancers, and a new development has been the kind of imaging known as a CT (computerized tomography) scan in which results from a large number of exposures from different directions, no single one of which shows anything much, are combined by a computer to give a clear picture of an internal organ.

Finally there is a much newer scientific field, X ray astronomy. Because X rays do not travel far in the atmosphere one has to go above it to detect whatever comes in from the stars. The first attempts used relatively primitive detectors mounted in rockets that soared briefly above the Western desert. Later came specially equipped satellites, which revealed that space is filled with a diffuse X ray glow that comes from everywhere and that there are also more specific X ray signals, some of which we can interpret. They are emitted steadily by the sun, and there are more distant sources that from time to time emit violent bursts a few seconds long. These come from the central cores of some distant quasars and galaxies, from pulsars in our own galaxy, and from black holes.

Black holes are supposed to be objects much more massive than the sun which gravity has contracted to very small size. Their gravitational fields are strong enough to keep light from escaping from them (hence the name) and also strong enough to pull matter toward them and swallow it up. If the black hole is formed from the gravitational collapse of a single star, and if, like many stars, it has a companion that orbits closely around it, it scoops up some of that. Also, as mentioned above, there is evidence that the cores of some galaxies contain huge black holes. These have masses up to a billion suns and devour stars and clouds of gas if they come too close. In either case the captured material usually does not drop right in; more commonly, before the final capture, it enters a gravitational whirlpool and traffic problems develop. Huge masses collide and are raised to temperatures so high that X rays come out of them, fluctuating randomly over wide ranges, and from observing these we try to guess what is going on. If the intensity of X rays from a distant source changes greatly in the course of an hour, this implies that the size of the emitting region is light hours at most—smaller than the solar system—and strengthens the belief that black holes actually exist.

Lasers

There is no doubt that the main event of twentieth-century optics was the invention and development of lasers. A laser is a device for producing a beam of light all of whose photons have essentially the same wavelength and move in essentially the same direction, so that the pencil of light that issues from it is almost monochromatic and shows very little broadening as

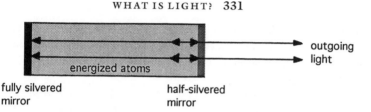

outgoing
light

energized atoms

fully silvered
mirror

half-silvered
mirror

Figure 10.6. Schematic diagram of a laser. As photons reflect back and forth between the mirrors, laser action produces more photons. The outgoing light is what escapes through the half-silvered mirror.

it travels through space. Lasers were impossible in 1900 because they can be conceived and understood only through the quantum theory of radiation, but it was twenty-five years after the invention of that theory before the first laser was made. The possibility of it was discussed in the 1950s, and the first successful one was demonstrated by Theodore Maiman of the Hughes Research Laboratories in California in 1960.

The operation of a laser depends on a property of radiating atoms that was apparently first mentioned by Einstein in 1917; it is called stimulated emission. Suppose an atom has been energized somehow into one of its states of higher energy. Left alone, it will quickly reduce its energy by radiating a photon. (At the most fundamental level of explanation, this is for the same reason that a stone released from one's hand falls toward the ground. The stone too is reducing its energy.) But there is another thing that can happen to this energized atom. If during the brief moment after it has been energized and before it spontaneously releases a photon it is hit by a photon of exactly the same wavelength as the photon it is about to emit, it is stimulated to emit its photon at once. This photon will be a clone of the photon that hit it, and the two will move off with the same wavelength in the same direction. Where there was once one photon there are now two, exactly alike. Suppose now there are many atoms. Presently the two photons will hit two more atoms; then there will be four photons exactly alike and as the process goes on an avalanche develops. This is laser light.

Figure 10.6 is a schematic diagram of a laser. The "lasing" atoms[7] are distributed between two mirrors. One of them is an ordinary mirror that reflects almost all the light that reaches it; the other is thinly silvered so that it reflects some light and lets the rest through. When the atoms have somehow been energized (by a flash of light, by an electric impulse, or some other means) photons appear as the atoms start to get rid of their energy. As these photons hit lasing atoms more photons will be produced, and by chance a few of them will move in the direction of one of the mirrors. As these photons and their progeny reflect back and forth between the mirrors (a few of them always escaping through the thinly silvered one) their number enormously increases, and the laser will begin to produce its light. In

[7] The word "laser" is an acronym for "light amplification by stimulated emission of radiation," and the verb "to lase" is an obvious and convenient back-formation from the noun.

the earliest lasers there was no way to maintain the supply of energized atoms, and so the laser produced only one brief, very bright flash at a time, but soon it became possible to energize the atoms continually so as to produce a steady beam.

Lasers have many uses and have been developed in appropriate shapes and sizes, from those used for cutting and welding hard metals to the tiny laser inside the device that reads from a compact disc without ever touching it, purely by light. Pulsed lasers are used by surgeons to cauterize tissue, especially inside the eye, or by astronomers to measure the exact distance to a point on the moon. Continuous lasers read prices from bar graphs in the supermarket and light up the optical fibers that are carrying an increasing amount of traffic on the Information Superhighway (§6.4).

6. Then What Is Light?

From meditations on the nature of reality to the ubiquitous laser; now perhaps it is time for a farewell look at an old question. Substance or accident? But why rattle the bones of a philosophy that has been dead for five hundred years? I do it because the old question gives a certain unity to the story that has been told here; because, as you may have noticed, the question seems to be interesting. A wave is an accidental quality, a wiggle in some medium, whereas a particle is more like a substance. Now it appears that wave and particle are only metaphors used to talk about an underlying unity. What that unity *is*, physics doesn't say; but the mathematical discipline known as quantum field theory accurately describes what it *does*. As the name implies, this is a theory of fields, and it correctly accounts for the experiments that show wave or particle phenomena. The electromagnetic field, for example, fills all space. It is present even in a perfectly dark room. When we think of the apparatus of Figure 10.5 as immersed in this field, it is perhaps less surprising that what happens in one region of it affects what happens in another, even though there is no way to represent the process in terms of traveling particles. Imagination is so wedded to things like particles that it tries in vain to depict the situation in terms of them and cries, "Paradox!" when it cannot. The paradox is not in Nature but in human mental behavior. Nature has revealed to us the precious secret that light is an attribute of a field—an accident, as people used to say. When we say it is impossible for a polarization measurement made on photons at A to affect anything that happens at B, far away, we are both speaking and thinking in the language of particles, of substance. To actually understand what is going on we have to use quantum mechanics. That theory has been crafted as a science of qualities, not of things. The trouble is, we live mentally in a world of things.

Everyone knows what things are. They have size and shape; they move from one place to another. I don't want to restrict the word, so I won't list any more properties that might prompt a hunt for exceptions; you know what I mean. In this sense it is hard to claim that a beam of light or electrons[8] or other elementary particles is composed of things, for then Young's experiment performed in dim light (§3) would require us to say that the particles of the beam go through both slits at once. Atoms are made of elementary particles. One can do Young's experiment with a beam of atoms; on the other hand it is also possible, using a beam of electrons, to photograph an atom. Should we say that an atom is a thing? One can argue it either way. The fathers of chemistry saw atoms as things and explained much, but in the mind of a modern chemist they are often smeared around and not in definite places. Baseballs are made of atoms and nothing else. Is a baseball a thing? Let's not ask stupid questions.

The physics of Newton and Maxwell is called classical physics. Except for changes brought by relativity theory, the ideas of classical physics are pretty much as they were at the death of Maxwell. John Bell begins one of his essays with the words "Theoretical physicists live in a classical world, looking out into a quantum-mechanical world." A world of things, in which something either happens or it doesn't, and a quantum world that is described in a very different language. How do the two worlds touch—what is the window between them? How do accidents become substances as you enter our world from the quantum world—how does probability become certainty? We think of quantum mechanics as the most fundamental insight we have into the workings of Nature; a reasonable program would be to start from quantum mechanics and derive the description of our classical world. In almost seventy-five years some of the most brilliant physicists of the century have worked on this question. It has not been settled, though there are by now several ideas as to how it might be. The following remarks are intended to clarify the question, not answer it.

The numbers called real are the familiar ones, as in 12 miles, 6.28832 volts, $1.69; but mathematicians use a more general kind called complex numbers, of which the reals are a special case. Complex numbers involve the square root of −1. Because this cannot be expressed in terms of reals this number got called imaginary and denoted by i, though as mathematicians use it it is no less real than the reals. Real numbers suffice for building a house, for buying it or insuring it. Classical physics is imagined, experienced, and calculated in terms of real numbers. A laboratory is a room in that classical world from which one looks out at the quantum world, and every instrument in it yields its results in terms of real numbers.

Quantum theory involves complex numbers in an essential way. We probably know only a little of how the universe is designed, but we know

---

[8] "Light" is singular, "electrons" is plural. I'm sorry that there is not a singular word denoting a beam of electrons as one speaks of a beam of light, but when electrons were first discovered in the 1880s they were imagined as particles. Their wave nature only emerged forty years later, and by then the usage was fixed.

that the design is simpler and easier for us to grasp in terms of complex numbers than real ones. The equations of quantum mechanics are written that way, and they are definite about what they say—they do not involve probability or statistics. These enter only when one seeks to translate the understanding gained from complex-number calculations into the language of our perceived world in which all numbers are real. Why does this happen? Earlier in this chapter I quoted Heisenberg: "We can fully understand a connection though we can only speak of it in images and parables." I have tried to show that words like "wave" and "particle" represent images. Now comes a parable. The friends of Socrates decided to surprise him with a basket of fruit. Peaches and cherries were in the market, and they knew he liked both. Which did he like better? They couldn't ask him, so they asked his slave to remember which fruit he requested every morning. After a week the slave reported: peaches five times, cherries twice. Apparently Socrates preferred peaches. His friends gave him a basket of them and he was very grateful, but when they considered what they had done, they realized that seven mornings was a small sample, that they would have had to wait much longer than a week to be sure, and that in fact absolute certainty could never be reached that way. If only they could have read Socrates's preference directly; but they couldn't ask him, and besides, the answer was hidden in the complexities of his inner nature. The best they could do was to replace an unanswerable question by another question that had a statistical answer. Again and again our questions, asked in the only way we know how to ask them, force the world to give us a statistical answer. It is easy to think of other situations. Trial by jury is one; the formation of an economic policy is another.

All right, what can we say about light? Many of the word's former meanings have evaporated. It is no longer understood as truth, or the earthly counterpart of an absolute Good, but the aura of these old meanings still clings to it. And we have seen that its electromagnetic nature still allows us to call it a universal causal agent. It is a sensory experience, it is a medium of perception, and physicists call it a quantized electromagnetic field. But what is that? We understand the mathematical theory very well, but how to translate its rarefied jargon into the rich and specific language of the classical world in which we live our lives? A quantized field has a mathematical description in terms of waves or in terms of particles. It is a matter of convenience, for the two descriptions are in every way equivalent. If we want to talk about it with someone not interested in calculation, or to understand it ourselves in terms of the words and mental images that we have learned from a life spent in the classical world, what should we say? Perhaps only quote Niels Bohr: "It is wrong to think that the task of physics is to find out how nature *is*. Physics concerns what we can say about nature." But of course we learn more all the time, and we learn how to say it better. I hope that when we can begin to say it clearly the story will be simple enough to tell in a book like this.

Figure 10.7. Light and shade on a column of the Temple of the Wingless Victory in Athens. Pencil drawing from D'Espouy, *Fragments d'architecture antique.*

7. Light and Shade

Every element of classical architecture was involved with the play of light and shade. Look for example at Figure 10.7, a loving nineteenth-century drawing of the base and capital of a column of the little temple of the Wingless Victory on the Acropolis of Athens. It is worth examining each part to see how it takes the light. There is a marvelous contrast between sharp shadows and graded ones. Each vertical fluting catches it differently.

The scotia moldings around the base are mostly concave; see how they turn from sun to shade. In the egg-and-dart molding of the capital, soft modulations of light and shadow alternate with the sharp accent of the darts.

After light, the eye rejoices in shadow, and the mind gives it meaning. Inside a Greek temple the light was always subdued, and in some Egyptian temples the Holy of Holies was kept very dark. Religion and literary convention make much of the light that dispels darkness, but that is not really what it does, for the power of light is finite. It only pushes darkness further away, and we follow it where it goes. That is something strange about humanity. Only a few of us want to pitch our tents in a plain bathed in the sun's vertical rays. Most of us move naturally toward shadow. It is where art and science build their houses, churches are nearby and many come to live close to them; others go on to seek the mystery and challenge they find in darkness. There will always be enough darkness for them. Truth lies in the abyss, said Democritus, and the philosopher Karl Popper [1962] was more specific: "The more we learn about the world, the deeper our learning, the more conscious, specific, and articulate will be our knowledge of what we do not know, our knowledge of our ignorance. For this, indeed, is the main source of our ignorance—the fact that our knowledge can only be finite, while our ignorance must necessarily be infinite." I have tried not to write this book as a chronicle of light driving out darkness, of good methodology supplanting an old one that was bad, of enlightened effort bringing answers to stubborn old questions. To some extent that has happened, but mostly the questions themselves have changed. The ancients saw themselves standing on the motionless Earth surrounded by the things and activities of nature, each one considered only to the extent that it affected the person standing there. If they asked what light is, they expected an answer that explained vision and understanding. If Aristotle said that a thing moves only as long as a force acts on it, he was generalizing from a hundred little human experiences each day. The mental perspective that sees us as the thinking part of nature's larger sphere of existence is commonly credited to Descartes, though we have seen signs of it in earlier writers. The modern theory of electromagnetic radiation makes no reference to human perceptions, and in fact I have suggested earlier in this chapter that it is when one tries to introduce observers into the quantum picture that things become hard to understand. To put it simply, the Cartesian program makes physics too easy. Light becomes a thing, the eye's lens is a thing, the retina and optic nerve are things, the brain . . . The edge of light moves, but shadows remain.

Attentive readers of this book may have noticed how seldom I have used the words "science" and "scientific" in it. It is because their meanings have shifted so much. In the old days it was thought possible that Democritus was wrong. When Aristotle spoke of the "science that we are seeking" he must have assumed it was there if we could find it, a science that consisted of what is known for certain. At its core would be unquestioned principles and conclusions that follow from them by unquestionable reasoning; at its periphery would be observations and conjectures waiting to be brought into the circle of the known. Sensory experience deals only with qualities, and only with the individual case. It can never establish a general statement. Universal truths originate elsewhere. In practice, of course, ancient science combined received opinion with the evidence of the senses just as we do today, except that our received opinion has been somewhat refined by long experience, and the evidence of our senses, aided by instruments, is more trusted, both to confirm and to refute. And today, a long experience of disappointed hopes has tended to make us so skeptical about claims of certainty that almost nobody talks about it any more.

What has science really produced? First, a collection of things: transistors, magnetic resonance imaging (MRI), the 747 jet plane, steps along the road that will be left behind and forgotten. Then, a collection of procedures, recipes for the best way to reach certain goals. Many of the best examples would be in public health and industrial chemistry, but we can expect both goals and procedures to change as time goes on. Then there is a collection of propositions like Maxwell's equations and the equations of quantum mechanics, which make people confident in varying degrees that they understand some aspect of Nature, but nobody thinks these will survive except as pieces of some deeper knowledge. Primarily, and perhaps most universally, science has become a procedure of guess and test, of trial and error, of honest action that permits even people of very ordinary ability to function at the edge of shadow, shining a little light here and there.

The goals of science are more modest than they used to be. People are learning to ask questions that have some hope of being answered, to go one step at a time and leave the next question on deposit for the future. When Leibniz read *Principia* he found that Newton had introduced a force of gravity without telling how it was supposed to be exerted, and so he concluded and told everybody that beyond some mathematics Newton had done nothing at all. Today we would say Newton was quite right to publish what he had found and leave further questions open. Einstein managed to trace gravity one step further back and explain that it occurs because the four-dimensional spacetime around a massive body is curved. But why spacetime? Why four dimensions? Why curved? One good answer generates many questions, and perhaps these will be answered one by one. Other questions for the future are not so easy: What do simple creatures think? How does evolution occur? Why are electrons so light? These too may be divided into simpler questions, and some will be answered. For the rest, perhaps not everything needs to be explained; perhaps it is not our duty to shine light into every shadow, for shadow shelters a life of its own.

In the evening, as light fades from the sky, birds clamp onto their twigs and go to sleep, but humanity does not; it enters a different phase. Minds that have spent the daytime hours attending to facts begin to share their recollections and order their thoughts in rooms where shadow is welcomed. It is a time for slowness.

The Japanese writer Junichiro Tanizaki has written an essay "In Praise of Shadows"; here is a scene from it.

> I think of an unforgettable vision of darkness I once had when I took a friend from Tokyo to the old Sumiya tea-house in Kyoto. It was in a wide room, the "Pine Gate," I think, since destroyed by fire, and the darkness, broken only by a few candles, was of a richness quite different from the darkness of a small room. As we came in the door an elderly waitress with shaven eyebrows and blackened teeth was kneeling by a candle behind which stood a large screen. On the far side of the screen, at the edge of the little circle of light, the darkness seemed to fall from the ceiling, lofty, intense, of a color, the fragile light of the candle unable to pierce its thickness, turned back as from a black wall. . . . It was a repletion, a fullness of tiny particles like fine ashes, each particle luminous as a rainbow. I blinked in spite of myself, as though to keep it out of my eyes.

Truth is finite; darkness is infinite and contains infinite possibilities. We come from darkness and end in darkness.

Michelangelo's Sonnet to Darkness

Any place that's closed, or any room
that solid walls shut in on every side
gives shelter to the night when daylight comes,
defends her from the sun's ferocious play.
  Fire and flame can drive her from her home
and every day she's hunted by the sun,
but even little lights can stain her beauty;
a firefly or two and it is gone.
  In any open field the burning sun
falls on a thousand seeds, a thousand leaves,
until a farmer turns them with his plow.
  But man is planted only in the dark
and therefore night is holier than day;
man's worth more than any farmer's crop.

TECHNICAL TERMS

Accident (*symbebekos*) — A quality that belongs to a thing, not necessarily or even usually, but here and now (Aristotle, *Metaphysics* 1025a14). There can be no scientific knowledge (*epistemē*) based on accidents (Aristotle, *Posterior Analytics* 75a, b; *Metaphysics* 1026b, in which the discussion of a house is amusing and illustrates Aristotle's marvelous use of analogy).

Angle of incidence — At the point where a ray strikes a surface, draw a line perpendicular to the surface. The ray's angle of incidence is the angle between the ray and the perpendicular. The same construction defines angles of reflection and refraction (see Figure 3.5).

Angular measurement — A circle is divided into 360 degrees. Each degree is 60 arc minutes; each arc minute is 60 arc seconds. A line about $1/3$ inch long seen at a distance of a mile defines an angle of 1 arc second.

Archetype — The single original from which copies or images are made.

Arc second — See ANGULAR MEASUREMENT.

Catoptrics — The theory of the reflection of light from flat and curved mirrors.

Chroma — The qualities of a visible surface, including its color and texture.

Diffraction — The bending of a ray of light as it passes close to the edge of an opaque object.

Doxa — See EPISTEMĒ.

Eidolon — In the theory of vision attributed to Democritus, an image that is like an object in very attenuated form, being continually emitted and somehow detected by the eye and transmitted to the soul.

Electromagnetic — Combining electricity and magnetism.

Epistemē — Plato contrasts the way we know Ideas with the way we know the world around us. The first is *epistemē*, absolute knowledge not derived from experience; the second is *doxa*, translatable as opinion. Aristotle uses *epistemē* for an understanding of things and events that we know must be as they are because we know their causes. *Doxa* refers to our knowledge of things that could be otherwise.

Form (*eidos*) — In Aristotle, the intelligible essence of an existing thing; the sum of a thing's properties.

Frequency — The frequency of a wave is the number of times per second that a crest moves past a given point of space. It can be calculated by dividing the wave's speed of travel by its wavelength.

Hertz — The scientific unit that measures the frequency of an oscillation. One hertz is one complete cycle per second.

Hexaëmeron — An account of the six days of creation that tries to fill in what is omitted from the biblical account.

Homogeneal color — Newton's term for the color at any particular point in the spectrum of white light, generally called pure color in this book. It corresponds to a single definite wavelength. To be compared with compounded color, which is the result of mixing several homogeneal colors.

Humor — In optics, one of several transparent and liquid or gelatinous substances inside the eyeball. In ancient and medieval medicine, one of the four fluids that are kept in balance in a healthy body.

Hyperboloid — The pair of curves in Figure 3.13 are hyperbolas. Points *a* and *d* are called their foci, and the hyperbolas are defined by the property that for all points on either of them, the difference between the distance to *a* and the distance to *d* is the same. A hyperboloid is the bowl-shaped figure made by rotating a hyperbola on its axis.

Image — An optical representation of a scene, formed by a lens or a mirror. There are two kinds, real and virtual. A real image is formed where rays draw together at the focus; it can be caught by putting a white surface there, as is done with a slide projector. A virtual image is formed by rays that diverge from the focus, and can be seen by putting the eye in the path of the rays, as one does with a magnifying glass.

Infrared light — Light invisible to the eye because its wavelength is greater than that of red light (see SPECTRUM).

Mathematician — In classical and medieval terminology, a scholar who is concerned with giving an accurate description of some aspect of Nature.

Mathematics — In classical and medieval terminology, the study of geometrical forms and of those aspects of Nature that can be analyzed by calculation.

Nanometer (nm) — A measure of length equal to a billionth of a meter, or a millionth of a millimeter. The wavelength of green light is about 500 nm; atomic sizes are about 0.1 nm.

Natural philosophy — The study of natural phenomena (see MATHEMATICS and PHYSICS).

Optics — The science of light, vision, and the scientific instruments involving them. Known in the middle ages as *perspectiva*.

Phase — If you draw a diagram of a sine wave, as in Figure 8.1, phase is the number that distinguishes points in the pattern: the crest, the trough, the places in between. If two waves of the same wavelength are brought together so that their phases are the same, they interfere constructively to make a larger wave. If their phases are opposite (crest to trough) they interfere destructively to make a wave that is weaker, perhaps zero.

Physics — In antiquity and the middle ages, the study of the philosophical principles explaining the causes of natural phenomena.

Physis — Roughly, nature (Aristotle, *Physics* 192b–93a). Whatever gives something an inherent tendency to change or not change, as when we say that the nature of heavy objects is to fall.

Pneuma — See SPIRIT.

Property (idion) — A quality essential to the thing that has it. For example, the ability to learn grammar belongs to all humans and only to humans (Aristotle, *Topics* 102a19). Democritus distinguishes between properties that are primary (residing in the atoms) and secondary (in the senses).

Quality (poion, poiotes) — A basic descriptor such as round, red, alive, made of iron, etc.

Ray — A straight line drawn to illustrate the path of light.

Reflection — What a mirror does.

Refraction — The bending of a light ray at the surface of a transparent medium (see Plate 3). Refraction also occurs when light enters the atmosphere, where the transition from empty space is gradual. In this case the ray from a star is curved.

Refrangibility — Newton's term for the property of a transparent medium that determines how much it will bend a ray of light that enters it at an angle. The equivalent modern term is index of refraction.

Science, in Aristotelian philosophy — The body of knowledge that explains the nature and causes of things. Because nature and cause are not perceived by the senses, science seeks its goal by arguing logically from general propositions, perhaps suggested by experience, which are assumed to be true.

Sine wave — A smoothly varying wave form like that shown in Figure 8.1.

Snel's law — When a light ray passes from one transparent medium to another, (sine of the angle of incidence) 5 (sine of the angle of refraction) = a number that depends on the two media and is called the index of refraction.

Soul — In Plato the idea develops through the dialogues until in later works it is the reasoning part of a human being, divine and immortal and reincarnated from time to time. In Aristotle, it is the actuality or organizing principle of a living body. In Thomas Aquinas: as in Aristotle, except that the soul is a form that exists independently of substance and therefore does not die with the body.

Species — At first, what flies through the air to produce the sensation of sight, an *eidolon*. Also, the visible manifestation of a genus, which is the form or idea that defines a class of things. Later, species was the power through which one thing acts on another. It transmitted a bundle of qualities but always had the general character of light.

Spectrum — The band of color seen when white light passes through a prism and is projected onto a white surface or (if not too bright) into the eye. The colors merge continuously but are conventionally divided into seven. Starting at the long-wavelength end they are red, orange, yellow, green, blue, indigo, and violet, but many people have trouble distinguishing indigo as a separate color.

Spirit — A delicate vapor which in Stoic philosophy (under the name *pneuma*) unites the disparate elements of the universe into a living and rational organism. In Galen's physiology, and into the eighteenth century, it flows through the nerves, reporting sensation to the brain and carrying commands from it.

Substance (ousia), in Aristotelian philosophy — The actual identity of a thing. Consider a stone, take away all its qualities of color, shape, weight, etc.; what remains is its substance.

Substantial form, in medieval philosophy — The property or properties that organize matter into a particular substance and determine what it is.

Ultraviolet light — Light invisible to the eye because its wavelength is less than that of violet light (see SPECTRUM).

Visual ray — A beam of light imagined by Empedocles, Plato, and others as issuing from the eye to unite with light from the environment at the surface of an object to bring its visual image back to the eye.

Wavelength — The length of a regularly periodic wave measured from crest to crest (see Figure 8.1).

REFERENCES AND FURTHER READING

PREFACE

"God is light": 1 John 1:5. Medicine and astronomy (we would say astrology): In Chaucer's *Canterbury Tales*, for example, the Prologue describes the Doctour of Physick as very skilled "To speke of physik and of surgerye / For he was grounded in astronomie."

CHAPTER ONE
THE POWER OF LIGHT

Epigraph. There is no room here, even if I were competent, for a discussion of Jewish mysticism. Suffice it to say that a passage like this is read symbolically, not literally. The *Zohar*, from which this passage comes, seems to be a late thirteenth-century work, written mostly by a Spanish Rabbi Moses and containing his version of the teachings called Cabala. *En-Sof*, here called the King, is the first hidden source of being and activity, from which will emanate the qualities that identify the God of Scripture. Its nature cannot be known, for it is beyond thought, but it makes scratches on the purity of nonexistence from which come a little spark, called black because it is still hidden. A few pages later the spark cracks open the air around it and becomes the light of Creation. See Lashower and Tishby 1989; the quoted passage is in vol. 1 on p. 309.

1.2. There is little robust information on the lives of the Presocratic philosophers. For a general reference see Diogenes Laertius 1925, but this is all secondhand and uncritically reported. The fragments of the Presocratic philosophers have been gathered together and numbered by Hermann Diels [1959]. They are translated in Freeman 1978. An excellent source of texts and judicious interpretation, containing almost all the known fragments, is Kirk, Raven, and Schofield 1983. See also Burnet 1892, Freeman 1978, and, for natural philosophy, Cohen and Drabkin 1948 and Sambursky 1959.

Empedocles: Diogenes Laertius 1925, VIII.76f.; see also Empedocles, Fragment 117, in Freeman 1978. On the elements: Diogenes Laertius VIII.73. Aristotle on the elements: *Metaphysics* 988a27; see also Kirk, Raven, and Schofield 1983. Aristotle vs. Empedocles on light: *On the Soul* 418b23.

1.3. Bailey 1928; Clay 1983; Engelert 1987; Epicurus 1964; Lucretius 1968; Strodach 1963.

Aristotle vs. vacuum: *Physics* IV.6, 7. Helmholtz on sensations as symbols: 1881, p. 47. Epicurus on the physical world: "Letter to Herodotus," in Epicurus 1964. Lucretius on the swerve: *The Way Things Are*, Book II, lines 218, 257ff.; Engelert 1987. Epictetus's opinion of Epicurus: *Discourses*, 1968, XX.2; see Engelert 1987.

1.4. Armstrong 1940; Bréhier 1958; Lindberg 1986; Plato 1961; 1962; Plotinus 1966.

Heavenly beds and tables: *Republic* X. "Pit of nonsense": *Parmenides* 130b–d. The Cave: *Republic*, Book VII. The soul is like an eye: *Republic* 508d. The Good is the cause of all things: *Republic* 509b. Light from the One: Plotinus 1966,

Book V; compare Dante, *Paradiso* 28; Lindberg 1986. God is perfectly simple: Aquinas 1964, Book I, Question 3, Article 7.

1.5. Aristotle 1983; Jaeger 1948.

Accidental qualities: *Posterior Analytics* I.4. Four elements: *On Generation and Corruption* II.3. Essence: *Metaphysics* VII.4. Potentiality is "every principle of movement or of rest": *Metaphysics* 1049b7. Matter desires form: *Physics* 192a23. Prime Mover: *Metaphysics* 1072a, b, 1074b. "The love that moves the sun . . .": last line of the *Divine Comedy*. "A certain continuity": *Meteorology* I.2.

1.6. Aristotle 1983.

Art and science: *Nicomachean Ethics* VI.3, 4, *Metaphysics* I.1. *Epistem FE*: *Metaphysics* 1064b30; *Nicomachean Ethics* 1139b20. *Doxa*: *Metaphysics* 1940a1. *Techn FE*: *Nicomachean Ethics* 1140a10; *Metaphysics* 981a6. *Theoria*: *Nicomachean Ethics* X.7. Object of desire causes motion: *Metaphysics* 1072a26, 1072b3.

1.7. Aristotle 1983; Neugebauer 1957.

Stars affect plants' growth: Aristotle 1983, *On Plants* 824b10. Babylonian planetary theory: Neugebauer 1957. Romans unfamiliar with Plato: Cicero 1949, *Topica*, I.3. Common opinion of the sun's powers: Vitruvius 1914, IX.1; Pliny 1938, II.13; Lucan 1984, X.201. Light and the planets: Vitruvius 1914, Pliny 1938, Lucan 1934. Archimedes's mirrors: Diodorus Siculus 1933, vol. 11, p. 193; for Buffon's experiment see Knowles Middleton 1961. Measurement of light pressure: Lebedew 1901.

1.8. Asmussen 1975; Burkitt 1925; Jonas 1963; Runciman 1947.

Mithraism: Cumont 1903, pp. 104ff. Earth of Darkness: Jonas 1963, p. 212. "Woe to the creator of my body": Burkitt 1925, p. 21. Manichaean hymns: Asmussen 1975. Augustine and the Manichaeans: *Confessions* (1979) III.6. Manichaeans in the middle ages: Runciman 1947. "Country full of barbarous customs": Grousset 1970, p. 122.

1.9. Augustine 1977, 1982; Basil of Caesarea in Schaff and Wace 1895; Gregory of Nyssa in Migne 1857, vol. 44; Milton: *Paradise Lost*; Robbins 1912.

Basil of Caesarea's *Hexaëmeron*: Schaff and Wace, vol. 8, Homily IX, 2.7ff.; the quotation is from Romans 1:16. Gregory of Nyssa's version: Armstrong 1967, pp. 447ff.; Duhem 1913, vol. 2, pp. 430f. Gregory's text: Migne 1857, vol. 44, pp. 62–125, especially 76–78. Augustine's version: formless matter existed before creation: *Confessions* 1977 Book III, pp. 3–6; Book XII, pp. 8, 12; Book XII, pp. 21–22. "Not absolutely nothing": XII.8. There was a deep: XII.3. The instant of creation: *Literal Meaning* (1982) I.2–3. Darkness and light created at the same time: *Confessions* (1977) XII.29. Satan's view of creation: end of Book II of *Paradise Lost*. Hymn to light: beginning of Book III; in the third line he hesitates: should he say that light was first created a few days ago or that God has always lived in brightness? The Creation: Book VII, lines 220ff.

CHAPTER TWO
THE IMAGE AND THE MIND

2.1 Barker 1760; Brecher 1977; Chapman-Rietschi 1995; see 1926.

Seneca on the Dog Star: Seneca 1971, vol. 1, p. 19. I have translated *lux pura* as "pure white." Ptolemy's star table: Ptolemy 1984, Books VII, VIII. Manilius 1977 says Sirius is blue: p. 37. Amateurs of English history may enjoy Gladstone (*the* Gladstone) 1858, vol. 3, pp. 457ff.; 1877; his hypothesis is that Greeks in the time of Homer did not clearly distinguish one color from another. For an attempt to explain Sirius through folklore see Ceragioli 1992.

2.2. Aristotle 1983; Freeman 1978; Kirk, Raven, and Schofield 1983.

The fragment of Empedocles: Aristotle, *Sense and Sensibilia* 437b27. His poem as reconstructed: Empedocles 1992.

2.3. Aristotle 1983; Burnet 1892; Epicurus 1964; Freeman 1978; Lucretius 1968; Plutarch 1927.

Leucippus's originality: Burnet 1892, Chapter 9. Democritus explains dreams: Aristotle, *On Divination in Sleep* 464a5, and Plutarch, *Symposium* (in *Moralia*, 1927, vol. 9), Book VIII, p. 735. Leonardo, c. 1510, explains that the eye is a magnet: Richter 1883 (1970 ed.), vol. 1, p. 37. Sweet by convention: Democritus, Fragment 9. Letter to Herodotus: Epicurus 1964, p. 14; original in Diogenes Laertius 1925, X.46. Eidola as membranes originate in Lucretius, *The Way Things Are*, Book IV, lines 217ff. Is the sun a bonfire? V.650ff. Lucretius on the source of the sun's light: V.592ff.

2.4. Augustine 1897, 1951, 1982; Plato 1961.

The visual stream of fire: Plato, *Timaeus* 45b, 58c. Plato on vision: *Meno* 76c. Theophrastus on vision: Burnet 1957, p. 246. Snow is not white: Cicero, *Academica*, 1933, 2.72, 2.100. Plato on white and black: Notopoulos 1944. On sensations of color: *Timaeus* 67–68, *Theaetetus* 153e. Vision is deceptive: *Phaedo* 65b. Augustine on the visual ray: *Literal Meaning of Genesis* 1982, I.16. On two classes of things that are known: *On the Trinity*, 1897, XV.12.21. On intelligible and sensible worlds: *Against the Academics* 1951, iii.17.37.

2.5. Aristotle 1983; also Hamlyn's critical translation of *On the Soul*, Books II and III, 1968.

Criticism of the visual ray: *Sense and Sensibilia* 437b11 and 438a25. A man with weak sight: *Meteorology* 373b1. Soul is actuality (*entelechia*): *On the Soul* 412a23; the best translation might be, "The soul is what completes a body." *Chroma*: *Sense and Sensibilia* 439a30. Definition of light: *Sense and Sensibilia* 439a26f, *On the Soul* 418b10 ("activity" is *energeia*, as opposed to *entelechia*, above). Imagination: *On the Soul*, III.3. Common sense defined: *On the Soul* 425a28, see also 426b14–427a15. The lens is the organ of vision: *Parts of Animals* 653b26; *Sense and Sensibilia*, Book II.

2.6. Diogenes Laertius 1925; Galen 1968; Lloyd 1973; Sambursky 1959; Siegel 1970; Zeller 1928.

"Artistically working fire": Diogenes Laertius, Book VII, pp. 142, 156. Stoics on vision: Diogenes Laertius, Book VII, p. 157. See also Cicero 1933, II.83. "The lens is the organ of vision": Siegel 1970, p. 38. Galen on the retina: Siegel, p. 56 (edited).

2.7. Aristotle 1983; Boyer 1987.

God's covenant: Genesis 9:12. Xenophanes on the rainbow: Fragment 32, in Freeman 1978. Aristotle on the rainbow: *Meteorology* III.4. The stars are very distant: Park 1988, p. 59.

CHAPTER THREE
RAYS

3.1. Aristotle 1941, 1983; Cumont 1912; Lucan 1988; Manilius 1977; Neugebauer 1957, 1975; Tester 1987; Wedel 1920.

Basil on the moon's influence: *Hexaëmeron*, Schaff and Wace 1895, Book IX. Vitruvius 1914 on Chaldean astrology: Book IX. Aristophanes's *Birds*: Frere

1874, vol. 3, p. 179. The stars are gods: Aristotle, *Metaphysics* 1074b. (I have used McKeon's translation, Aristotle 1941.) Causality originates in the heavenly bodies: Aristotle, *Meteorology* 339a29.

3.2. Diocles 1976; Euclid 1945, 1959; Fraser 1972.

Conjectures about the Museum: Fraser 1972, vol. 1, p. 325. Burning glasses available in the drugstore: Aristophanes, *Clouds*, line 766.

3.3. Heron 1899.

Catoptrics: in Heron 1899, vol. 2. Light makes no detours: Book II.

3.4. Ptolemy 1940, 1984, 1989.

Color affected by distance: *Optics* II.124. Ptolemy measures refraction: *Optics* V.8. Tacitus on astrologers: *Histories*, 1925, I.22. Intelligences guiding the planets: e.g., Robert Boyle, "A free inquiry into the received notion of nature," p. 244, in Boyle 1772, vol. 5. This long essay gives an excellent picture of the intellectual climate in which Isaac Newton and his contemporaries were born and educated.

3.5. Samic 1958; Sorabji 1987.

End of the Museum: Wissowa 1894, vol. 16 (1935), "Musaeon." For Aristotle the spheres are divine: *Metaphysics* 1074b1. Samburský 1958, p. 117, compare Sorabji 1987 for a somewhat different interpretation, which concludes that Philoponus and Aristotle both use "move" in a metaphorical way, since something that arrives instantaneously cannot be said to move.

3.6. Alkindi 1912; Lindberg 1976, 1986, 1988; Ronchi 1970.

Alkindi's views similar to Aristotle's: *De aspectibus*, On Vision, in Alkindi 1912, pp. 3–41, prop. 12; compare Aristotle, *On the Soul* 435a6–10. Later European scholars found that Alkindi gave the heavenly machinery far too much power over our destiny and God and our own efforts far too little. There is an interesting critique of Alkindi by Giles of Rome (1247–1316) in Shapiro 1964, pp. 384ff.

3.7. Alhazen in Latin 1572, in English 1989; Lindberg 1967, 1976, 1988; Omar 1974; Rashed 1993.

The Latin manuscript of *De aspectibus* was published by Risner as *Opticae thesaurus* in 1572. Life: A comprehensive article by A. I. Sabra in the *Dictionary of Scientific Biography* (hereafter abbreviated *DSB*); see also Alhazen 1989. Ibn Sahl's treatise is summarized in Rashed 1990; more details in Rashed 1993. Biographical note on Alhazen: 1989, vol. 2, p. xxii. Nothing issues from the eye: Alhazen 1989, vol. 1, p. 80, considerably edited. Alhazen on light and color: 1989, vol. 1, p. 109. Each point of the object is imaged separately: 1989, vol. 1, p. 75. After Galileo discussed the infinity of points imaged in the eye, Robert Hooke resumed the discussion: Hooke 1705, pp. 120ff. A point on the crystalline sphere becomes green: Alhazen 1989, p. 127. Aristotle on the shape of eclipse shadows: *Problems* XV.11; see also XV.6. Perception by judgment and inference: Alhazen 1989, p. 128.

CHAPTER FOUR
THE LIGHT THAT SHINETH IN DARKNESS

Epigraph. *City of God*, 1957, 10.1–2.

4.1. Plato 1961

How the world came into being: *Timaeus* 30b. "Moving image of eternity": *Timaeus* 37d. *Timaeus* prefigures the Gospel: Augustine 1977, VII.9. "The first is the true world": Augustine 1951, III.37, edited. "One does not say": Augustine 1982, IV.28. On the same passage see *City of God*, 1957, X.2, and *On the Trinity*, 1897, XIII.1.

4.2. Pseudo-Dionysius 1987.
Denis in the Bible: Acts 17:34. Present at the death of Mary: Pseudo-Dionysius 1987, p. 70. "Let us move on now": 1987, pp. 71, 74; Augustine, *On Free Will*, 1955b, XX.16. The conflation of Saint Denis with Dionysius the Areopagite was the work of Hilduin, Abbot of Saint Denis in the ninth century who was trying to provide a history for the founder of his church. The whole story is found in Jacobus de Voragine 1941, vol. 2, p. 616.

4.3. Von Simson 1962; Suger 1979.
"Thou hast ordered all things": Wisdom 11:20. The world is God's handiwork: Augustine, *On Free Will*, 1955b, II.20. Heaven's mathematical perfection: von Simson 1962, Part I. "Just proportions": Augustine, *Of True Religion*, 1955b, Books XXX, XXXII. "Even the old part shines": Suger 1979, p. 50. The Holy City: Revelations 21:2–5.

4.4. Crombie 1953; Dales 1961, 1967; Grant 1974; Grosseteste 1942; Lindberg 1976, 1988; McEvoy 1982; Southern 1992.
"Natural local motion": Claggett 1959, p. 445. Robert's life: McEvoy 1982, Southern 1992. The best translation of *On Light* is Grosseteste 1942, but it is also available in Shapiro 1964. Robert's theory of vision: Grant 1974, p. 389.

4.5. Crombie 1959; Grant 1974; Grosseteste 1982; Lindberg 1983a, 1988.
Earth is very small: Grosseteste, *Hexaëmeron*, pp. 119f. Skepticism concerning astrology: *Hexaëmeron*, pp. 165f. Robert's theory of sound: Crombie 1959, vol. 1, p. 103. "Species and perfection": shortened from Grosseteste 1942. "All causes of natural effects": Grant 1974, p. 385. "Action is stronger and better": Grant 1974, p. 386. Pyramids: Grant 1974, p. 388. "Visual species is a substance": Grant 1974, p. 389. Intelligence: compare Augustine, *City of God*, 1957, X.2. "Insofar as it lacked . . ." and "Everything that is manifest": *Hexaëmeron*, pp. 77f.; Ephesians 5:13. A more modern translation is "everything that becomes visible is light." "I rebel": McEvoy 1982, edited.

Interlude. Shapiro 1964; Thorndike 1944.
Paris curriculum in 1255: Thorndike 1944, pp. 64ff. Saint Jerome on lost virginity: Jerome 1933, Letter 22. Tempier's propositions: Hyman and Walsh 1967, pp. 542ff. See also Grant 1974, pp. 45ff. Ramée on logic: Waddington 1855, p. 23.

4.6. Bacon 1928; Easton 1952; Lindberg 1983a, b, 1986; Sharp 1930.
Two sparrows: Matthew 10:29. "Agent sends a species into matter": condensed from Lindberg 1983a, p. 7. "By its own force and species": 1928, vol. 1, p. 470. On reflection of light: 1928, vol. 2, p. 546. "Visible by its own force": 1928, vol. 2, p. 471. "Mathematicians do not presume": 1928, vol. 1, p. 270. Thomas Aquinas on stellar influences: *Summa theologica*, 1964, I.115.4 and 5. Speed of light is finite: Lindberg 1983a, p. 221.

4.7. Aristotle 1983; Grosseteste 1982; Lindberg 1976; Steneck 1976.
Augustine 1897 on species: *On the Trinity*, 1897, XI.9. Aristotle: definition of soul: *On the Soul* 412a28, *Metaphysics* 1070a26. On the location of the soul: *Movement of Animals* 703a2. On spirit: *Generation of Animals* II.3. On the soul: *Generation of Animals* 736b38. On faculties: *Parts of Animals* 647a25, 666a12; *On the Soul* 414a28. Galen 1956 on the faculties: pp. 235, 726. Internal senses: Avicenna 1968. Grosseteste 1982 says that spirit acts through light: *Hexaëmeron*,

p. 98. Willis 1664 on spirit: pp. 96f. Newton 1959 on spirit: vol. 1, pp. 368f.

4.8. Albertus Magnus 1890; Aquinas 1963, 1964; Aristotle 1983; Lindberg 1976; Weisheipl 1980.

Albert's theory of sensation: N. H. Steneck in Weisheipl 1980. Albert on vision: references in Lindberg 1976, p. 251, n. 7. Aristotelian source for eye and spirit: *Generation of Animals* 744a4; general discussion in *On the Soul*, Book II, p. 7. Summary in Lindberg 1976, Chapter 6. "Know the truth of things": LeGoff 1992, p. 117. Light is accidental: Aquinas 1964, *Summa*, Book I, Question 67, Articles 3 and 4. Natural and spiritual perception: 1964, I.78, 3. Aristotle on substances defined into existence: *Topics* 135a12. On knowing forms but not accidents: *Physics* 244b7, though the language is rather obscure. Ockham on species: Lindberg 1976, pp. 140–42.

CHAPTER FIVE
THE END OF CLASSICAL OPTICS

5.1. Lindberg 1983a; Witelo 1572, 1972, 1977, 1983.

Witelo's life: Lindberg in *DSB*. Lindberg, "Lines of influence in thirteenth-century optics," in Lindberg 1983a. See also his preface to Witelo 1972.

5.2. Ronchi 1970

The lens from Nimrud: Layard 1853, p. 197. Grosseteste on the rainbow: 1912. On his "telescope": Crombie 1953, p. 119. Bacon experiments with lenses: 1928, vol. 2, p. 574. Eyeglasses: Ronchi 1970, p. 69. *Roidi di ogli*: Singer 1957, vol. 3, p. 230.

5.3. Edgerton 1975; Kemp 1978, 1990.

Castiglione on painting as an art: 1928, p. 78. Perspective in Greek and Roman art: G.M.A. Richter [1970]; her figure 176 shows an attempt at perspective in a South Italian *krater* from Euclid's time. Vitruvius 1914 on Greek scene painting: 1914, vol. 1, Chapter 2, 2. "We understand nothing": Bacon 1928, vol. 1, p. 234. Brunelleschi and the mirror: need for caution, Kemp 1978; reconstruction, Edgerton 1975.

5.4. Alberti 1956; Edgerton 1975; Lindberg 1976.

Drawing the cross-lines: Alberti 1956, p. 57; Lindberg 1976, p. 151. "Concerned solely with what can be seen": 1956, p. 43.

5.5. Clubb 1965; Porta 1593, 1658; Ronchi 1970.

Life of Porta: M. H. Rienstra in *DSB*; Clubb 1965. To cure a wound: Porta 1658, p. 229. "This will please ingenious people": 1658, p. 367. Lenses: 1593, Chapter 8.

5.6. Biagioli 1993; Drake 1957, 1978; Galileo 1968; Langford 1971; Reston 1994; Ronchi 1958, 1970; Seeger 1966.

Modern study of Galileo's lenses: Greco, Molesini, and Quercioli 1992; Ringwood 1994. Galileo's microscope: Reston 1994; Galileo 1968, Book XIII, p. 208. *Sidereus nuncius* translated in Drake 1957. Changeless, unblemished ether: Aristotle, *On the Heavens*, Book II. Galileo intimidated: Galileo 1968, Book X, p. 68. Galileo's position at court: Biagioli 1993. Seven is the key to the universe: Cicero 1928, *De republica*, VI.18. Astrology of Jupiter's moons: Galileo 1968, Book XI, p. 112. "Sun, stand thou still": Joshua X:12 and Psalm 119. Porta's letter: Galileo 1968, Book XII, p. 101. Galileo's trial: Langford 1971. Mathematical understanding: Galileo 1953, p. 114; 1968, Book VII, p. 128. Niccolini

and the Pope: 1968, Book XV, p. 67.
5.7. Galileo 1968; Galileo et al. 1960.
"Philosophy is written": Galileo 1968, Book VI, p. 232 = Galileo et al. 1960, p. 183. "Nothing but words": 1968, Book VI, p. 336 = 1960, p. 311. "Light is created": 1968, Book VI, p. 351 = 1960, p. 313. Measurement of the speed of light: 1914, p. 43. Galileo in the dark: 1968, Book VIII, p. 208.

CHAPTER SIX
A NEW AGE BEGINS

6.1. Baumgardt 1951; Caspar 1959; Kepler 1609, 1937.
Galileo's early studies: Wallace 1977. Kepler's numerical relations: Park 1988, p. 153. Kepler describes himself: Baumgardt 1951, p. 52. Effects of the moon: Kepler 1601, p. 92; Ptolemy, *Tetrabiblos*, 1940, pp. 35, 45. Galileo casts a horoscope: Reston 1994, p. 220. Angular separations of planets: *Harmonices mundi*, 1937, vol. 6, Book IV, p. 5. "A vegetative, animate power": Kepler 1601, Proposition 43. The orbit of Mars: Park 1988, Chapter 9. "The seeds of great discoveries": 1609, Dedication.
6.2. Kepler 1604 = 1937, vol. 2; Lindberg 1976, 1988; Ronchi 1970.
Importance of the retina: *Ad Vitellionem paralipomena*, 1604, Book IV, Chapter 4. "I leave for others to decide.": 1604, Book V, Chapter 2.
6.3. Kepler 1611 = 1937, vol. 4, 1992; Stephenson 1987.
The sun's motive power is light: Kepler 1992, pp. 380–82, edited. Divinity in the sun: 1992, p. 385. The sun rotates: 1992, pp. 386f. A ball would move forever: Galileo 1914, pp. 215, 244. "Peaceful ground of eternity": 1937, vol. 11, Part I, p. 473.
6.4. Rashed 1990; Shirley 1951.
Harriot's life: J. A. Lohne in *DSB*. Harriot does not tell Kepler the law: Lohne in *DSB*.
6.5. Descartes 1964, 1980, 1983; Sabra 1967; Scott 1952; Shea 1991; Smith 1987.
Discourse on Method: Descartes 1964, vol. 6; trans. in Descartes 1980. Three dreams: e.g., Vrooman 1970, Chapter 2. A spectator of the world: 1980, p. 15. Why there can be no vacuum: 1983, p. 46. Light is pressure: 1964, vol. 11, p. 53. Spinning tennis ball: 1964, vol. 6, p. 90; see Sabra 1967, Chapters 1–4.
6.6. Fermat 1679, 1891; Ronchi 1970; Sabra 1967.
Fermat-Descartes correspondence: Fermat 1891, vol. 1, pp. 355, 464; Descartes 1964, vol. 1, p. 450; vol. 2, p. 15. Letter to M. de ***: 1891, vol. 2, pp. 485–96, trans. in Ronchi 1970, pp. 120–22. Fermat's objection to Descartes: 1891, vol. 2, p. 124; Sabra 1967, Chapter 4. Fermat's derivation: 1891, vol. 2, pp. 489–96. "Resistance" to motion: 1891, vol. 3, pp. 149ff.; see Sabra 1967, Chapter 5. Clerselier's objection: Fermat 1891, vol. 2, p. 465; Sabra 1967, Chapter 5.

CHAPTER SEVEN
THE RISE OF OPTICAL EXPERIMENT

Epigraph. Newton 1730, p. 124.
7.1. Boyer 1940; Cohen 1940; Roemer 1676.
Roemer's life: Z. Kopal in *DSB*. Light takes eleven minutes: Roemer 1676.

Huygens's calculation: Olmsted 1942, n. 86. Hooke 1765 on the speed of light: pp. 78, 108.

7.2. Grimaldi 1665; Ronchi 1970.

Life: B. S. Eastwood in *DSB*, Ronchi 1970, Chapter 4. "Only the blind": Grimaldi 1665, Preface. The "boundary ray" is not straight: Proposition I.i.17 (meaning Book I, Proposition 1, 17). "Light is a kind of fluid": Proposition I.ii.1. Mechanism of diffraction: Proposition I.ii.16. Reflection from a sheet of glass: Proposition I.viii.69. Light can be colored in three ways: Proposition I.xxxi. Species cannot explain vision: Proposition I.xl, esp. 84.

7.3. Birch 1756; Hooke 1667, 1765; Sabra 1967

The nature of light: Hooke 1667, p. 55. Lectures on light: Hooke 1705, pp 71–148. Studies of "Muscovie glasse:" 1667, p. 65. Hooke's "irises": 1667, p. 50; Birch 1756, vol. 3, p. 54.

7.4. Birch 1756; Brewster 1855; Hakfoort 1994; Newton 1958, 1959, 1984; Sabra 1967; Sepper 1994; Westfall 1980.

Newton's early reading notes: Westfall 1980, Chapter 3. The lectures on optics were published as Newton 1729 and 1984. "The oddest if not the most considerable detection": 1959, vol. 1, p. 82. "I procured me a triangular glass-Prisme": 1959, vol. 1, pp. 92–102. "To determine what light is . . . is not so easie.": 1959, vol. 1, p. 100. First letter on light and color: 1959, vol. 1, pp. 97ff. See Sabra 1967, Chapter 9.

7.5. Birch 1756; Buchwald 1989; Guerlac 1981; Newton 1687, 1704, 1730 (Dover reprint in 1952), 1959; Sabra 1967; Sepper 1994; Westfall 1980.

Hooke criticizes Newton's theory of white light: Newton 1959, vol. 1, p. 110. See also Sabra 1967, Chapter 10. Newton's reply: 1959, vol. 1, p. 171. Waves bend around corners; light doesn't: 1687, pp. 367ff. Relation between color and wavelength: 1959, vol. 1, p. 175. Definition of light: 1959, vol. 1, p. 164. Rays of light: *Opticks*, p. 1. See also Buchwald 1989, Chapter 1. "A sensation of light": 1959, vol. 1, p. 175. "An Hypothesis Explaining the Properties of Light": incomplete version in 1959, vol. 1, pp. 362–86; complete version in Birch 1756, vol. 3, pp. 248–93.

7.6. Newton 1704, 1730, 1959; Westfall 1980.

Simple convex lenses: *Opticks*, Dover reprint, p. 12, corresponding to p. 8 in the first edition. References will be written below in the form p. 12 (8). Newton explains refraction: *Opticks*, p. 80 (57). "Only one color": *Opticks*, p. 124 (90). Newton's rings: *Opticks*, p. 195 (p. 5, since the pagination of the first edition starts again at Book 2). An earlier account is in Newton 1959, vol. 1, pp. 378ff. Least parts are transparent: *Opticks*, pp. 248ff. (52ff.). "Of Natures Obvious Laws": Manuscript known as Burndy 16, summarized in Dobbs 1982. Von Guericke [1672] on the living Earth: p. 156. The many functions of spirit: *Principia*, p. 547; comets as its source: p. 530. Newton's alchemy: Westfall 1980, Chapter 8. Signs that the speed of light depends on color: Newton 1959, vol. 3, p. 164, and Flamsteed's answer on p. 202. (Newton is clearly anxious to see whether all the colors disappear at the same time when a satellite of Jupiter is eclipsed.) Einstein's admiration of Newton: Dover reprint of *Opticks*, p. lix. Newton and the Enlightenment: Berlin 1980, p. 144.

7.7. Huygens 1690, 1912; Sabra 1967.

Life: H.J.M. Bos in *DSB*. Modern justification of Huygens's principle: Born and Wolf 1970, 8.3.2. Light is not a beam of particles: 1912, p. 22. See Sabra 1967,

Chapters 7, 8.

7.8. Boyer 1987; Crombie 1953; Descartes 1964, vol. 6; Newton 1704, 1730; Wallace 1959.

General discussion: Boyer 1959. Dietrich of Freiberg: Boyer 1959, Chapter 5; Crombie 1953, Chapter 9. Extensive translation in Grant 1974, pp. 435ff. For the original text see Theodoric 1914. Analysis and commentary in Wallace 1959. Dietrich's experiment: Crombie 1953, p. 250, n. 4. Descartes's theory: 1964, vol. 6, pp. 325ff. Newton's theory: *Opticks* p. 168 (126).

7.9. Bradley 1729.

Life: A.F.O'D. Alexander in *DSB*.

CHAPTER EIGHT
ENLIGHTENMENT

Epigraph. From *Encyclopaedia Metropolitana* (1827), quoted in Whewell 1857, vol. 2, p. 349.

8.1. Jones 1966; Nicholson 1946; Voltaire 1738.

Opticks and literature: Nicolson 1946. "Order, proportion, and fitness": Smith 1956, p. 325. Lavoisier's list of elements: 1796, p. 246. "Mr. Newton is no physicist": Guerlac 1981, p. 50, n. 33. Voltaire on light: 1738, p. 127.

8.2. Euler 1711, 1768; Hakfoort 1994.

"A new theory of light and colors": Euler 1711, series 3, vol. 5, pp. 1–45. Sun composed of particles of different kinds: "New Theory," par. 88. "Particles like stretched strings": par. 113. *Letters of Euler . . .*: 1768, English trans., 1842, pp. 19ff. Detailed critique of the new theory: Hakfoort 1994, Chapter 4. "Not a single person of prominence": Cajori 1899, p. 102. Atoms of Newton and Boyle: Thackray 1970. Broken link between humanity and Nature: Nicolson 1946. "Reason and nature": Blake 1982, p. 501.

8.3. Goethe 1840, 1985.

Beiträge zur Optik: Goethe 1985, vol. 4.2, pp. 264ff. Experiments on black-edged surfaces: *Beiträge*, Part I, pp. 275ff.; experiments are continued in Part II with colored surfaces. Goethe's color wheel: *Beiträge*, Part II, pp. 419ff; also 1985, vol. 10, *Farbenlehre*, p. 23.

8.4. Goethe 1985, 1840; Sepper 1988,

Farbenlehre: Goethe 1985, vol. 10; didactic part translated in Goethe 1840. Cloudiness is essential: 1985, vol. 10 = 1840, pars. 143ff. First degree of cloudiness: par. 148. Negative capability: Keats 1958, vol. 1, p. 193. "A great mistake has been made": pars. 176, 177. Newton avoids "phantasms": 1959, vol. 1, p. 484. Goethe studies thin films: *Farbenlehre*, par. 452. "A consciousness of superiority": Eckermann 1930, p. 302. Helmholtz 1881 on Goethe: p. 39.

8.5. Buchwald 1989; Cantor 1975; Crew 1900; Fresnel 1866; Kipnis 1991; Young 1802, 1807, 1855.

"On the theory of light and colors": Young 1855, pp. 140–69. "Experiments and calculations": Young 1855, pp. 179–91. Young's calculation reconstructed: Kipnis 1991, Chapter 5. Young's lectures: Young 1807. Fresnel on economy in natural philosophy: 1966, vol. 1, p. 259. His theory of diffraction: any modern text: e.g., Meyer-Arendt 1984. Planck 1949 on new scientific truths: p. 33. Helmholtz 1881 on Young: p. 220. Light waves may be partly transverse: Young 1855, vol. 1, pp. 279–342, especially pp. 332f. Light waves are entirely transverse: Fresnel 1866, vol. 1, p. 630.

8.6. Bedini 1994; Mrs. J. Herschel 1876; Hevelius 1673; Hoskin 1963; King 1955; Lubbock 1933; McCann 1994; Ronchi 1964.

Newton's error: 1730, Book I, Part II, Proposition 3. Herschel learns English and reads Locke: Lubbock 1933, p. 8. "A mighty bewilderment": Holmes 1900, vol. 3, p. 219. Distance to the Andromeda nebula: Hubble 1927.

8.7. Bedini 1994; Bradbury 1967; Dobell 1932; Ford 1985.

Italian compound microscopes contemporary with Hooke: Bedini 1994, Chapter 7. Figure 9 shows the use of one in the clinic and also how, as Galileo had said, the same instrument could be readjusted for use as a telescope. Single lens is offensive to the eye: Hooke 1667, Preface. Leeuwenhoek's life: J. Henigerin in *DSB*, Dobell 1932. History of the single-lens microscope: Ford 1985.

8.8. Fraunhofer 1823; Hearnshaw 1986; Herschel 1800; Kayser 1900, vol. 1; Wollaston 1802.

Early studies of spectra: Melvil 1752, Morgan 1785, Wollaston 1802. Detection of infrared light: Herschel 1800. Glassworks at Benediktbeuern: Chance 1937. Dark lines in solar spectrum: Fraunhofer 1823, p. 296. Dark spectral lines in the laboratory: Brewster 1836. "We will never know anything . . .": Comte 1835, vol. 2, Lesson 19.

CHAPTER NINE
UNIFICATION

Epigraph. Poincaré 1904.

9.1. Dijksterhuis 1961; Swenson 1972; Lodge 1909, 1925; Whittaker 1953.

Stars and planets are made of ether: Aristotle, *On the Heavens* 270b. Action at a distance is absurd: Newton 1958, p. 302. Ether is absolutely solid: Young 1855, vol. 1, pp. 412–17.

9.2. Agassi 1971; Bence Jones 1870; Faraday 1839; Gilbert 1600; Hesse 1961; Tyndall 1898.

Earth as a magnet: Gilbert 1600; Whittaker 1952, vol. 1, pp. 34ff. Magnetism different from electricity: 1600, pp. 52f.; Mottelay trans., p. 85. Earth has a soul: 1600, p. 209; Mottelay trans., p. 310. Grand Old Man: Agassi 1971. "Atmosphere of force": Faraday 1839, vol. 2, p. 290. Faraday's drawings of lines of force: 1839, vol. 3. Physical nature of magnetic lines of force: 1839, vol. 3, nos. 3263, 3297; lateral repulsion, pp. 3267f.; tension, 3298; connection with ether, 3263; Maxwell's summary, 1890, vol. 2, p. 320. "Faraday, in his mind's eye . . ."; Maxwell 1873, p. x. "Magnetizing and electrifying a ray of light": 1839, vol. 3, nos. 2146–48. "Thoughts on Ray-Vibrations": 1839, vol. 3, pp. 447–52.

9.3. Campbell and Garnett 1882; Foucault 1850; Maxwell 1873; 1890.

Maxwell's life: Campbell and Garnett 1882. Foucault's measurement: 1850. Magnetism analogous to gas flowing through steel wool: Maxwell 1890, vol. 1, pp. 155–229 Maxwell's model of a magnetic field: 1890, vol. 1, pp. 451–513; Lodge 1889, pp. 186, 472. It worked for him: Siegel 1991. "We find ourselves in a factory": Duhem 1962, p. 70 "A dynamical theory of the electromagnetic field": 1890, vol. 1, pp. 526–97. "Ether" in Encyclopedia Britannica: 1890, vol. 1, pp. 763–75.

9.4. Henry 1886; Hertz 1893, 1896.

Henry detects electromagnetic waves: Henry 1886, vol. 1, p. 204. "Maxwell's theory is Maxwell's equations": Hertz 1893, p. 21. Hertz on ether: 1896, p. 312. What he was really looking for when he discovered electric waves: Buchwald 1994. Discovery of the photoelectric effect: 1893, Chapter 4. "Icy summits": 1896, p. 326.

9.5. Röntgen 1898.

Life: G. L'E. Turner in *DSB*. Discovery of X rays: Röntgen 1895 (trans. in Magie 1935, p. 600); 1898. Reportage: *New York Times*, January 16, 1896.

9.6. Livingston 1973; Michelson and Morley 1887 (in Swenson 1972); Swenson 1972.

Michelson's life: Livingston 1973. The Michelson-Morley experiment: Michelson and Morley 1887. Lorentz-FitzGerald contraction: Lorentz, in Lorentz et al. 1923; FitzGerald 1889. "With our mothers' milk": Boltzmann 1891, p. 2.

9.7. Bergman 1968; Lorentz et al. 1923; Einstein 1922; Pais 1982; Poincaré 1905, 1907.

Einstein's life: Whittaker 1948, Pais 1982. "Engrossed us and held us spellbound": Solovine 1956, p. 9. Alhazen and Witelo had made the same point: Alhazen, *Perspectiva*, Book II, 49; Witelo, *Optica* IV, 110, p. 167. Poincaré's loyalty to ether: Poincaré 1905; see also 1907, p. 94. "Unconscious opportunism": Poincaré 1907, p. 36. Invited to address a Congress of Arts and Sciences at the Saint Louis Exposition of 1904, Poincaré surveyed the ruined landscape of physics and predicted in general terms several features of the relativity theory that Einstein published in the following year: Rogers 1905, vol. 1, pp. 604–22. The principle of Relativity in classical literature: Alhazen 1572, Book II, 49; Witelo 1572, Book IV, 110. "On the electrodynamics of moving bodies": in Lorentz et al. 1923. "Must be read about elsewhere": a classic, but not easy, is Einstein 1920. The principle of equivalence: Einstein 1911, in Lorentz et al. 1923. "Ether and Relativity": Einstein 1922.

9.8. Gregory 1990; Helmholtz 1924; Kandel, Schwartz, and Jessell 1991; Wasserman 1978; Young 1802.

Young's three-color theory: Young 1802, p. 147. Maxwell's comment on Young: 1890, vol. 2, "On colour vision," pp. 167–79. Three kinds of cones: Brown and Wald 1964. On the brain's construction of sensations of color see Sacks 1995, "The Case of the Colorblind Painter." Shadow in candlelight at dawn: von Guericke [1672], p. 142.

CHAPTER TEN
WHAT IS LIGHT?

10.1. Einstein 1905a; Hermann 1971; Kangro 1972; Kuhn 1978, 1984; Mehra and Rechenberg 1982, vol. 1; Planck 1906, 1913, 1914; Weiner 1977.

Planck's life: *DSB*; Born 1948, reprinted in Boorse and Motz 1966, p. 462. Planck develops his radiation theory: a clear account in Planck 1906, 1913; Nobel address (1919), in Boorse and Motz 1966, p. 491; Klein and Heilbron in Weiner 1977; Kuhn 1978. "A purely formal assumption": Hermann 1971, p. 24. Einstein's light quantum hypothesis, 1905a, trans. in Boorse and Motz 1966, p. 544. Einstein on emission theory: 1987, vol. 2, p. 564.

10.2. Bohr 1913; Compton 1923; Heilbron and Kuhn 1969; Mehra and Rechenberg 1982, vol. 1; Pais 1991; Rutherford 1911; Weiner 1977.

Rutherford invents the nuclear atom: Rutherford 1911, in Boorse and Motz

1966, p. 701. Bohr, life: Cockcroft 1963, Pais 1991. Theory of the hydrogen atom: 1913, reprinted in Boorse and Motz 1966, p. 751; Heilbron in Weiner 1977; Mehra 1982, vol. 1. Compton effect: Compton 1923, in Boorse and Motz 1966, p. 911.

10.3. Bohr 1934, 1938, 1963; de Broglie 1924; Heisenberg 1971.

Images and parables: Heisenberg 1971, pp. 229f. (I have changed a few words). De Broglie's electron waves: de Broglie 1924, also in Boorse and Motz 1966, p. 1048; Mehra 1982, vol. 1, V.4. Experimental verification: Thomson and Reid 1927, reprinted in Boorse and Motz, pp. 1143f. "The great veil": Mehra 1982, vol. 1, p. 604. A calculus of individual cases: Omnès 1994.

10.4. Bell 1964, 1987; Polkinghorne 1984; Squires 1986.

Hidden variables cannot explain Aspect's experiment: Bell 1964. Einstein on spooky actions: Born 1971, p. 158.

10.5. Optics of modern telescopes: Schroeder 1987. Artificial bright star: Fugate and Wild 1994; Wild and Fugate 1994. Scanning interferometric apertureless microscope (SIAM): Zenhausern, Martin, and Wickramasinghe 1995. Invention of the laser: Maiman 1960.

10.6. Feynman 1985.

10.7. Tanizaki 1955.

BIBLIOGRAPHY

Abney, W. de W. 1895. *Colour Vision*. London: Sampson Low, Marston.

Agassi, J. 1971. *Faraday as a Natural Philosopher*. Chicago: University of Chicago Press.

Alberti, L. B. 1956. *On Painting*. Trans. J. R. Spencer. New Haven: Yale University Press.

Albertus Magnus. 1890–99. *Opera omnia*. 38 vols. Ed. A. Borgnet. Paris: Vivès.

Alexander, P. 1985. *Ideas, Qualities, and Corpuscles*. Cambridge: Cambridge University Press.

Alhazen (al-Haytham). 1572. In *Opticae thesaurus*. Basel: Risner. Reprinted with an intro. by D. C. Lindberg, New York: Johnson, 1972.

———. 1989. *The Optics of Ibn al-Haytham, Books I–III*. Trans. with commentary by A. I. Sabra. 2 vols. London: Warburg Institute.

Alkindi (al-Kindi) 1912. *De aspectibus*. Ed. A. A. Björnbo and S. Vogl. In "Al-Kindi, Tidaeus, und Pseudo-Euclid. Drei optische Werke," *Abhandlungen zur Geschichte der mathematischen Wissenschaften* 26.3, pp. 1–176. On Alkindi, pp. 3–41.

———. 1974. *De radiis*. Ed. M.-Th. d'Alverny and F. Hudry. In *Archives d'histoire doctrinaire et littéraire du moyen âge* 41:139–260.

Aquinas, Thomas. 1964–. *Summa theologica*. Latin and English. 56 vols. London: Blackfriars.

———. 1963. *On the Physics*. Trans. R. J. Blackwell, R. J. Spaeth, and W. E. Thirlkel. New Haven: Yale University Press.

Aristotle. 1941. *The Basic Works of Aristotle*. Ed. and trans. R. McKeon. New York: Random House.

———. 1969. *On the Soul, Books II and III*. Trans. D. Hamlyn. Oxford: Clarendon Press.

———. 1983. *Complete Works*. Ed. J. Barnes. 2 vols. Princeton: Princeton University Press.

Armstrong, A. H. 1940. *The Architecture of the Intelligible Universe According to Plotinus*. Cambridge: Cambridge University Press.

———, ed. 1967. *Cambridge History of Later Greek and Early Medieval Philosophy*. Cambridge: Cambridge University Press.

Asmussen, J. P. 1975. *Manichaean Literature*. Delmar, N.Y.: Scholars' Facsimiles and Reprints.

Aspect, A., P. Grangier, and G. Roger. 1981. "Experimental Tests of Realistic Local Theories via Bell's Theorem." *Physical Review Letters* 47:460–63.

Aspect, A., J. Dalibard, and G. Roger. 1982. "Experimental Test of Bell's Inequalities Using Time-Varying Analyzers." *Physical Review Letters* 49:1804–7.

Auerbach, E. 1953. *Mimesis*. Trans. W. R. Trask. Princeton: Princeton University Press.

Augustine of Hippo. 1897. "On the Trinity." In *A Select Library of Nicene and Post-Nicene Fathers of the Christian Church*. First ser., vol. 3. Ed. P. Schaff. Buffalo: Christian Literature Co.

———. 1948. *The Writings of Saint Augustine*. 3 vols. New York: Cima.

———. 1951. *Against the Academics*. Trans. J.J.O. Meara. Westminster, Md.: Newman Press.

———. 1953. *Earlier Writings*. Trans. J.H.S. Burleigh. Philadelphia: Westminster Press.

———. 1955a. *Later Works*. Trans. J. Burnaby. Philadelphia: Westminster Press.

———. 1955b. *On Free Will*. Trans. M. Pontifex. Westminster, Md.: Newman Press.

———. 1957–72. *The City of God*. 5 vols. Several trans. Cambridge, Mass.: Loeb Classical Library, Harvard University Press.

———. 1962. *De doctrina Christiana* and *De vera religione*. Turnhout: Brepols.

———. 1977. *St. Augustine's Confessions*. 2 vols. Trans. W. Watts. Cambridge, Mass.: Loeb Classical Library, Harvard University Press.

———. 1982. *The Literal Meaning of Genesis*. 2 vols. Trans. J. H. Taylor. New York: Newman Press.

Avicenna. 1968–72. *Liber de anima*. 2 vols. Ed. S. van Riet. Louvain: Editions orientalistes; and Leiden: Brill.

Bacon, R. 1928. *The Opus Majus of Roger Bacon*. 2 vols. Trans. R. B. Burke. Philadelphia: University of Pennsylvania Press.

Bailey, C. 1926. *Epicurus, the Extant Remains*. Oxford: Clarendon Press.

———. 1928. *The Greek Atomists and Epicurus*. Oxford: Oxford University Press.

Barker, T. 1760. "Remarks on the Mutations of the Stars" *Philosophical Transactions of the Royal Society* 51:498–99.

Baumgardt, C. 1951. *Johannes Kepler: Life and Letters*. New York: Philosophical Library.

Bedini, S. A. 1994. *Science and Instruments in Seventeenth-Century Italy*. Aldershot: Variorum.

Bell, J. S. 1964. "On the Einstein Podolsky Rosen Paradox." *Physics (N.Y.)* 1:195–200.

———. 1987. *Speakable and Unspeakable in Quantum Mechanics*. Cambridge: Cambridge University Press.

Bergmann, P. G. 1968. *The Riddle of Gravitation*. New York: Scribner.

Berlin, I. 1980. *Personal Impressions*. New York: Viking.

Biagioli, M. 1993. *Galileo Courtier*. Chicago: University of Chicago Press.

Birch, T. 1756–57. *History of the Royal Society of London*. 4 vols. Reprint. New York: Johnson, 1968.

Blake, W. 1982. *The Complete Poetry and Prose of William Blake*. Ed. W. B. Erdman and H. Bloom. Garden City, N.Y.: Anchor Books.

Bohr, N. 1913. "On the Constitution of Atoms and Molecules." *Philosophical Magazine* 26: 1–19.

———. 1934. *Atomic Theory and the Description of Nature*. Cambridge: Cambridge University Press.

———. 1958. *Atomic Theory and Human Knowledge*. New York: Wiley.

———. 1963. *Essays, 1958–1962, on Atomic Physics and Human Knowledge*. New York: Wiley.

———. 1972–. *Collected Works*. Ed. L. Rosenfeld et al. Amsterdam: North-Holland.

Boltzmann, L. 1891. *Vorlesungen Über Maxwells Theorie der Elektricität und des Lichtes*. Leipzig: Barth.

Boorse, H. A., and L. Motz, eds. 1966. *The World of the Atom*. 2 vols. New York:

Basic Books.

Born, M., and W. Heisenberg. 1923. "Die Elektronenbahnen im angeregten Heliumatom." *Zeitschrift für Physik* 16:229–43.

Born, M. 1948. "Max Karl Ernst Ludwig Planck." *Obituary Notices of Fellows of the Royal Society* 6:161–88.

———. 1971. *The Born-Einstein Letters.* Trans. I. Newton-John. New York: Walker.

Boyer, C. B. 1940. "Early Estimates of the Velocity of Light." *Isis* 33:24–40.

———. 1987. *The Rainbow.* Princeton: Princeton University Press.

Bradbury, S. 1967. *The Evolution of the Microscope.* Oxford: Pergamon.

Bradley, J. 1729. "A New Apparent Motion Discovered in the Fixed Stars." *Philosophical Transactions of the Royal Society (1809)* 35:308–21.

Brecher, K. 1977. "Sirius Enigmas." *Technology Review* 80, no. 2:53–63.

Bréhier, E. 1958. *The Philosophy of Plotinus.* Trans. J. Thomas. Chicago: University of Chicago Press.

Brewster, D. 1836. "Observations on the Lines of the Solar Spectrum . . . ." *Philosophical Magazine* 8:384–92.

———. 1855. *Memoirs of the Life, Writings, and Discoveries of Isaac Newton.* 2 vols. Edinburgh: Constable.

Broglie, L. de. 1924. "A Tentative Theory of Light Quanta." *Philosophical Magazine* 47:362–72.

———. 1925. "Recherches sur la théorie des quanta." *Annales de physique* 3:22–128.

Brown, P. K., and G. Wald. 1964. "Visual Pigments in Single Rods and Cones of the Human Retina." *Science* 144:49–52.

Buchwald, J. Z. 1989. *The Rise of the Wave Theory of Light.* Chicago: University of Chicago Press.

———. 1994. *The Creation of Scientific Effects: Heinrich Hertz and Electric Waves.* Chicago: University of Chicago Press.

Burkhardt, J. 1944. *The Civilization of the Renaissance in Italy.* Trans. J. Middlemore. New York: Phaidon.

Burkitt, F. C. 1925. *The Religion of the Manichees.* Cambridge: Cambridge University Press.

Burnet, J. 1892. *Early Greek Philosophy.* London: Macmillan. 4th ed. Reprint, New York: Meridian Books, 1957.

Cajori, F. 1899. *A History of Physics.* New York: Macmillan.

Campbell, L., and W. Garnett. 1882. *The Life of James Clerk Maxwell.* London: Macmillan.

Cantor, J. 1975. "The Reception of the Wave Theory of Light in Britain." *Historical Studies in the Physical Sciences* 6:109–32.

Caspar, M. 1959. *Kepler.* Trans. C. D. Hellmann. London: Abelard-Schuman. Reprint, New York: Dover, 1993.

Castiglione, B. 1928. *The Book of the Courtier.* Trans. T. Hoby. London: Dent.

Ceragioli, R. 1992. "Behind the 'Red Sirius' Myth." *Sky and Telescope* 83 (June): 613–15.

Chance, W.H.S. 1937. "The Optical Glassworks at Benediktbeuern." *Proceedings of the Physical Society* 49:433–43.

Chapman-Rietschi, P.A.L. 1995. "The Colour of Sirius in Ancient Times." *Quarterly Journal of the Royal Astronomical Society* 36:337–50.

Cicero, M. T. 1928. *De re publica and De legibus*. Ed. and trans. C. W. Keyes. Cambridge, Mass: Loeb Classical Library, Harvard University Press.

———. 1933. *De natura deorum* and *Academica*. Trans. H. Rackham. London: Loeb Classical Library, Heinemann.

———., 1949. *De inventione, De optimo genere oratorum, Topica*. Trans. H. M. Hubbell. Cambridge, Mass.: Loeb Classical Library, Harvard University Press.

Clagett, M. 1959. *The Science of Mechanics in the Middle Ages*. Madison: University of Wisconsin Press.

Clauser, J. F., and A. Shimony. 1978. "Bell's Theorem: Experimental Tests and Implications." *Reports on Progress in Physics* 41:1881–1927.

Clay, D. 1983. *Lucretius and Epicurus*. Ithaca: Cornell University Press.

Clubb, L. G. 1965. *Giambattista della Porta Dramatist*. Princeton: Princeton University Press.

Cockcroft, J. 1963. "Niels Henrik David Bohr." *Biographical Memoirs of Fellows of the Royal Society* 9:37–53.

Cohen, I. B. 1940. "Roemer and the First Determination of the Velocity of Light." *Isis* 31:327–79.

Cohen, M. R., and I. E. Drabkin. 1948. *A Source Book in Greek Science*. Cambridge, Mass.: Harvard University Press.

Comte, A. 1830–42. *Cours de philosophie positive*. 6 vols. Paris: Bachelier.

Crew, H., ed. 1900. *The Wave Theory of Light*. New York: American Book Co.

Crombie, A. C. 1953. *Robert Grosseteste and the Origins of Experimental Science, 1100–1170*. Oxford: Oxford University Press. Reprint, 1962.

———. 1959. *Medieval and Early Modern Science*. New York: Doubleday. First published as *Augustine to Galileo: The History of Science* A.D. 400–1650. Cambridge: Harvard University Press, 1953.

Cumont, F. 1912. *Astrology and Religion among the Greeks and Romans*. London: Constable. Reprint, New York: Dover, 1960.

Dales, R. C. 1961. "Robert Grosseteste's Scientific Works." *Isis* 52:381–402.

Descartes, R. 1964–74. *Oeuvres*. Ed. C. Adam and C. Tannery. New ed., 11 vols. Paris: Vrin.

———. 1980. *Discourse on the Method for Rightly Conducting One's Reason and for Seeking Truth in the Sciences*. Trans. D. A. Cress. Indianapolis: Hackett.

———. 1983. *Principles of Philosophy*. Trans. V. R. Miller and R. P. Miller. Dordrecht: Reidel.

———. 1984. *The Philosophical Writings of Descartes*. 2 vols. Trans. J. Cottingham, R. Stoothoff, and D. Murdoch. Cambridge: Cambridge University Press.

d'Espouy, H. C. 1895. *Fragments d'architecture antique*. Paris: Schmid.

Dictionary of Scientific Biography. 1970–80. 18 vols. Ed. C. C. Gillispie. New York: Scribner. (Abbreviated *DSB* in "References and Further Reading" section.)

Diderot, D., and J. le R. d'Alembert, eds. 1751–72. *Encyclopédie ou Dictionnaire raisonné des sciences, des arts et des métiers*. Paris: Briasson et al.

Diels, H. 1959–60. *Die Fragmente der Vorsokratiker*. Ed. W. Kranz. 9th ed. Berlin: Weidmann.

Dietrich von Freiberg. 1914. "Über den Regenbogen und die durch Strahlen erzeugten Eindrücke," ed. J. Würschmidt. *Beiträge zur Geschichte der Philosophie und Theologie des Mittelalters* 12, parts 5–6. Münster in Westfalen.

Dijksterhuis, E. J. 1961. *The Mechanization of the World Picture*. Oxford: Oxford University Press. Reprint, Princeton University Press, 1986.

Diocles. 1976. *On Burning Mirrors.* Trans. G. J. Toomer. New York: Springer-Verlag.

Diodorus Siculus. 1933–67. *Diodorus of Sicily.* 12 vols. Trans. C. H. Oldfather et al. London: Loeb Classical Library, Heinemann.

Diogenes Laertius. 1925. *Lives of Eminent Philosophers.* 2 vols. Trans. R. D. Hicks. London: Loeb Classical Library, Heinemann.

Dobbs, B.J.T. 1982. "Newton's Alchemy and His Theory of Matter." *Isis* 73:511–28.

Dobell, C. 1932. *Antony van Leewenhoek and His "Little Animals."* London: Bale and Danielsson. Reprint, New York: Dover, 1960.

Drake, S. 1957. *Discoveries and Opinions of Galileo.* New York: Doubleday.

———. 1978. *Galileo at Work.* Chicago: University of Chicago Press.

Dreyer, J.L.E. 1906. *History of the Planetary Systems from Thales to Kepler.* Cambridge: Cambridge University Press. Reprinted as *A History of Astronomy from Thales to Kepler.* New York: Dover, 1953.

Duhem, P. 1913–59. *Le système du monde.* 10 vols. Paris: Hermann.

———. 1954. *The Aim and Structure of Physical Theory.* Trans. P. P. Wiener. Princeton: Princeton University Press. Reprint, New York: Athenaeum, 1962.

Easton, S. C. 1952. *Roger Bacon and his Search for a Universal Science.* Oxford: Oxford University Press.

Edgerton, S. 1975. *The Renaissance Rediscovery of Linear Perspective.* New York: Basic Books.

Einstein, A. 1905a. "Über einen die Erzeugung und Verwandlung des Lichtes betreffenden heuristische Gesichtspunkt." *Annalen der Physik* 17:132–48. Trans. in *The World of the Atom,* ed. H. A. Boorse and L. Motz, vol. 1, p. 544. New York: Basic Books, 1966.

———. 1905b. "Zur Elektrodynamik bewegter Körper." *Annalen der Physik* 17:891–921. Trans. in H. A. Lorentz, A. Einstein, H. Minkowski, and H. Weyl, *The Principle of Relativity,* ed. A. Sommerfeld, trans. W. Perrett and G. H. Jeffery. London: Methuen, 1923. Reprint, New York: Dover, 1952.

———. 1911. "Einfluss der Schwerkraft auf die Ausbreitung des Lichtes." *Annalen der Physik* 35:898–908. Trans. in H. A. Lorentz, A. Einstein, H. Minkowski, and H. Weyl, *The Principle of Relativity,* ed. A. Sommerfeld, trans. W. Perrett and G. H. Jeffery. London: Methuen, 1923. Reprint, New York: Dover, 1952.

———. 1912. "Lichtgeschwindigkeit und Statik des Gravitationfeldes." *Annalen der Physik* 38:355–69. Trans. in H. A. Lorentz, A. Einstein, H. Minkowski, and H. Weyl, *The Principle of Relativity,* ed. A. Sommerfeld, trans. W. Perrett and G. H. Jeffery. London: Methuen, 1923. Reprint, New York: Dover, 1952.

———. 1916. "Die Grundlage der allgemeinen Relativitätstheorie." *Annalen der Physik* 49:769–822. Trans. in H. A. Lorentz, A. Einstein, H. Minkowski, and H. Weyl, *The Principle of Relativity,* ed. A. Sommerfeld, trans. W. Perrett and G. H. Jeffery. London: Methuen, 1923. Reprint, New York: Dover, 1952.

———. 1922. *Sidelights on Relativity.* Trans. G. B. Jeffery and W. Perrett. London: Methuen. Reprint, New York: Dover, 1982.

———. 1987–. *The Collected Papers of Albert Einstein.* Ed. J. Stachel. Princeton: Princeton University Press.

Empedocles. 1992. *The Poem of Empedocles.* Ed. and trans. B. Inwood. Toronto: University of Toronto Press.

Engelert, W. G. 1987. *Epicurus on the Swerve and Voluntary Action.* Atlanta: Scholars Press.

Epictetus. 1916. *The Discourses and Manual.* Trans. P. E. Matheson. Oxford: Oxford University Press.

Epicurus. 1964. *Letters, Principal Doctrines, and Vatican Sayings.* Trans. R. M. Geer. Indianapolis: Bobbs-Merrill.

Euclid. 1945. "The Optics of Euclid." Trans. H. E. Burton. *Journal of the Optical Society of America* 35:357–72.

———. 1959. *L'Optique et la catoptrique.* Trans. P. Ver Ecke. Paris: Blanchard.

Euler, L. 1911–. *Opera omnia.* Ed. F. Rudio, A. Krazer, P. Stäckel, et al. Leipzig: Teubner, and elsewhere.

———. 1768–74. *Lettres à une princesse d'Allemagne sur diverses sujets de physique et de philosophie.* 3 vols. Saint Petersburg: Académie impériale des sciences. Trans. as *Letters of Euler on Different Subjects in Natural Philosophy Addressed to a German Princess.* 2d ed. 2 vols. New York: Harper, 1842.

Faraday, M. 1839–55. *Experimental Researches in Electricity.* 3 vols. London: Taylor and Francis.

Fermat, P. 1679. *Varia opera mathematica D. Petri De Fermat.* Toulouse: Pech.

———. 1891–1912. *Oeuvres de Fermat.* 4 vols. Ed. C. Henry and P. Tannery. Paris: Gauthier-Villars.

Feynman, R. 1982. "Simulating Physics with Computers." *International Journal of Theoretical Physics* 21:467–88.

———. 1985. *QED.* Princeton: Princeton University Press.

FitzGerald, G. F. 1889. "The Ether and the Earth's Atmosphere." *Science* 13:390.

Ford, B. J. 1985. *Single Lens.* New York: Harper and Row.

Foucault, L. 1850. "Méthode générale pour mesurer la vitesse de la lumière dans l'air et les milieux transparents." *Comptes rendus de l'Académie des Sciences* 30:551–60.

Fraser, P. M. 1972. *Ptolemaic Alexandria.* 3 vols. Oxford: Clarendon Press.

Fraunhofer, J. 1823–24. "On the Refractive and Dispersive Power of Different Species of Glass." *Edinburgh Philosophical Journal* 9:288–99, and 10:26–40.

Freeman, K. 1978. *Ancilla to the Presocratic Philosophers.* Cambridge, Mass.: Harvard University Press.

Frere, J. H. 1874. *The Works of the Right Honourable John Hookham Frere.* 2d ed. Ed. W. E. Frere. 3 vols. New York: Denham.

Fresnel, A. J. 1819. "Mémoire sur la diffraction de la lumière." *Mémoires de l'Académie des Sciences* 5 (1821–22, printed in 1826): 339–475; and *Oeuvres* 1:247–363. There is an English translation of the greater part of this memoir in *The Wave Theory of Light,* ed. H. Crew. New York: American Book Co., 1900.

———. 1866–70. *Oeuvres complètes.* Ed. H. de Senarmont, E. Verdet, and L. Fresnel. Paris: Imprimerie Royale. Reprint, New York: Johnson, 1965.

Fugate, R. Q., and W. J. Wild. 1994. "Untwinkling the Stars, Part I." *Sky and Telescope* 87 (May): 24–31.

Galen. 1968. *On the Usefulness of the Parts of the Body.* Trans. M. T. May. Ithaca: Cornell University Press.

Galileo, G. 1914. *Dialogues Concerning Two New Sciences.* Trans. H. Crew and A. de Salvio. New York: Macmillan. Reprint, New York: Dover, n.d.

———. 1953. *Dialogue on the Great World Systems.* Trans. T. Salusbury. Chicago: University of Chicago Press.

———. 1968. *Edizione Nazionale delle Opere di Galileo.* 20 vols. Ed. A. Favaro.

Florence: Barbèra.

Galileo Galilei, H. Grassi, M. Guiducci, and J. Kepler. 1960. *The Controversy on the Comet of 1618*. Trans. S. Drake and C. D. O'Malley. Philadelphia: University of Pennsylvania Press.

Gilbert, W. 1600. *De magnete*. London: Short. Reprint, Brussels: Culture et Civilisation, 1967. Trans. P. F. Mottelay as *On the Loadstone and Magnetic Bodies, and on the Great Magnet the Earth*. New York: Wiley, 1893.

Gladstone, W. E. 1857. *Studies on Homer and the Homeric Age*. 3 vols. Oxford: Oxford University Press.

———. 1877. "The Colour-Sense." *Nineteenth Century* 2:366–88.

Goethe, J. W. 1985–. *Sämtliche Werke*. ed. K. Richter et al. Munich: Hanser.

———. 1840. *Goethe's Theory of Colours*. Trans. C. L. Eastlake. London: Murray. Reprint, London: Cass, 1967.

Grant, E. 1974. *A Source Book in Medieval Science*. Cambridge, Mass.: Harvard University Press.

Greco, V., G. Molesini, and F. Quercioli. 1992. "Optical Tests of Galileo's Lenses." *Nature* 358:101.

Gregory of Nyssa. *In Hexaëmeron liber*. In J. P. Migne, *Patrologia Graeca*, vol. 44, pp. 71–78. Paris: Migne, 1857.

Gregory, R. 1990. *Eye and Brain*. 4th ed. Princeton: Princeton University Press.

Grimaldi, F. M. 1665. *Physico-mathesis de lumine, coloribus, et iride*. Bologna: Bernia.

Grosseteste, Robert. 1912. *Die Philosophischen Werke des Robert Grosseteste, Bischofs von Lincoln*. Ed. L. Baur. Vol. 9 of *Beiträge zur Geschichte der Philosophie des Mittelalters*. Münster in Westfalen: Aschendorff.

———. 1942. *On Light*. Trans. C. Riedl. Milwaukee: Marquette University.

———. 1982. *Hexaëmeron*. Ed. R. C. Dales and S. Gieben. London: Oxford University Press.

Grousset, René. 1970. *The Empire of the Steppes*. Trans. N. Walford. New Brunswick, N.J.: Rutgers University Press.

Guericke, O. von. 1672. *Experimenta nova Magdeburgica de vacuo spatio*. Amsterdam: Jansson.

Guerlac, H. 1981. *Newton on the Continent*. Ithaca: Cornell University Press.

Hakfoort, C. 1995. *Optics in the Age of Euler*. Cambridge: Cambridge University Press.

Hearnshaw, J. B. 1986. *The Analysis of Starlight*. Cambridge: Cambridge University Press.

Heilbron, J. L., and T. S. Kuhn. 1969. "The Genesis of the Bohr Atom." *Historical Studies in the Physical Sciences* 1:211–90.

Heisenberg, W. 1971. *Physics and Beyond*. Trans. A. J. Pomerans. New York: Harper and Row.

Helmholtz, Hermann von. 1881. *Popular Lectures on Scientific Subjects*. 2nd ed. Trans. E. Atkinson. London: Longmans, Green.

———. 1924. *Helmholtz's Treatise on Physiological Optics*. Trans. of the 3d ed., ed. P. C. Southall. 3 vols. New York: Optical Society of America. Reprint (in 2 vols.), New York: Dover.

Henry, J. 1832. "On the Production of Currents and Sparks of Electricity from Magnetism." *American Journal of Science* (Silliman) 22:403–8.

———. 1886. *Scientific Writings of Joseph Henry*. 2 vols. Washington, D.C.: Smithsonian Institution.

Hermann, A. 1971. *The Genesis of the Quantum Theory (1899–1913)*. Trans. C. W. Nash. Cambridge, Mass.: MIT Press.

Hero of Alexandria. 1899–1914. *Heronis Alexandrini opera quae supersunt omnia*. 5 vols. Ed. and trans. W. Schmidt et al. Stuttgart: Teubner. Reprint, 1976.

Herschel, Mrs. John. 1876. *Memoir and Correspondence of Caroline Herschel*. New York: Appleton.

Herschel, W. 1800. "Experiments on the Refrangibility of the Invisible Rays of the Sun." *Philosophical Transactions of the Royal Society* 90:284–93.

Hertz, H. 1893. *Electric Waves*. Trans. D. E. Jones. London: Macmillan.

———. 1896. *Miscellaneous Papers*. Trans. D. E. Jones and G. H. Schott. London: Macmillan.

Hesse, M. B. 1961. *Forces and Fields*. London: Nelson. Reprint, Totowa, N.J.: Littlefield, Adams, 1965.

Hevelius, J. 1673–79. *Machina cœlestis*. 3 vols. Gdánsk: Reiniger. Reprint, Osnabrück: Zeller, 1969.

Holmes, O. W. 1900. "The Poet at the Breakfast Table." In *The Complete Works of Oliver Wendell Holmes*, vol. 3. New York: Sully and Kleinteich.

Hooke, R. 1665. *Micrographia*. London: Martin. Reprint, New York: Dover, 1961.

———. 1705. *The Posthumous Works of Robert Hooke*. London: Waller. Reprint, New York: Johnson, 1969.

Hoskin, M. A. 1963. *William Herschel and the Construction of the Heavens*. London: Oldbourne.

Hubble, E. P. 1927. "Cepheids in Spiral Nebulae." *Publications of the American Astronomical Society* 5:261–64.

———. 1929. "A Relation between Distance and Radial Velocity among Extra-Galactic Nebulae." *Proceedings of the National Academy of Sciences* 15:168–73.

Hubel, D. H. 1988, 1995. *Eye, Brain, and Vision*. New York: Scientific American Library.

———. and Wiesel, T. N. 1959. "Receptive Fields of Single Neurones in the Cat's Striate Cortex." *Journal of Physiology* 148:574–91.

Huygens, C. 1690. *Traité de la lumière*. Leiden: Vander Aa. Trans. S. P. Thompson as *Treatise on Light*. London: Macmillan, 1912. Reprint, New York: Dover, 1962.

Isidore of Seville. 1911. *Etymologiae*. 2 vols. Ed. W. M. Lindsay. Oxford: Clarendon Press.

Jacobus de Voragine. 1941. *The Golden Legend of Jacobus de Voragine*. 2 vols. Ed. and trans. G. Ryan and H. Ripperger. New York: Longmans, Green.

Jerome, Saint. 1933. *Select Letters of Saint Jerome*. Trans. F. A. Wright. London: Loeb Classical Library, Heinemann.

Jonas, H. 1963. *The Gnostic Religion*. 2d ed. Boston: Beacon.

Jones, H. Bence. 1870. *The Life and Letters of Faraday*. 2 vols. Philadelphia: Lippincott.

Jones, W. P. 1966. *The Rhetoric of Science: A Study of Scientific Ideas and Imagery in Eighteenth-Century English Poetry*. Berkeley and Los Angeles: University of California Press.

Kangro, H. 1972. *Planck's Original Papers in Quantum Physics*. Trans. D. ter Haar and S. G. Brush. London: Taylor and Francis.

Kandel, E. R., J. H. Schwartz, and T. M. Jessell, eds. 1991. *Principles of Neural Science*. 3d ed. New York: Elsevier.

Kayser, H. 1900. *Handbuch der Spektroscopie.* 8 vols. Leipzig: Hirzel.

Keats, J. 1958. *The Letters of John Keats.* 2 vols. Ed. H. E. Rollins. Cambridge: Harvard University Press.

Kemp, M. 1978. "Science, Non-Science, and Nonsense." *Art History* 1:134–61.

———. 1990. *The Science of Art.* New Haven: Yale University Press.

Kepler, J. 1601. *De fundamentis astrologiae certerioribus.* Prague: Schuman. In *Gesammelte Werke*, vol. 4. Trans. as "On the More Certain Fundamentals of Astrology" by M. A. Rossi. Ed. J. B. Brackenridge. *Proceedings of the American Philosophical Society* 123, no. 2 (1979): 85–116.

———. 1604. *Ad Vitellionem paralipomena.* Frankfurt: Marinus. Reprint, Brussels: Culture et Civilization, 1968. Trans. as *Paralipomènes à Vitellion* by C. Chevalley. Paris: Vrin, 1980.

———. 1609. *Astronomia nova.* Heidelberg: [De Thou]. In *Gesammelte Werke*, vol. 3. Trans. as *New Astronomy* by W. H. Donahue. Cambridge: Cambridge University Press, 1992.

———. 1610. *Narratio de observatis a se quatuor Jovis satellitibus erronibus.* In *Gesammelte Werke*, vol. 4.

———. 1611. *Dioptrice.* Augsburg: Franck. In *Gesammelte Werke*, vol. 4.

———. 1937–75. *Gesammelte Werke.* 22 vols. Ed. M. Caspar et al. Munich: Beck.

King, H. C. 1955. *The History of the Telescope.* High Wycombe: Griffin. Reprint, New York: Dover, 1979.

Kipnis, N. 1991. *History of the Principle of Interference of Light.* Basel: Birkhäuser.

Kirchhoff, G., and R. Bunsen. 1860. "Chemische Analyse durch Spektralbeobachtungen." *Annalen der Physik* (Poggendorff) 110:160–89. See also *Philosophical Magazine* 22 (1861): 320–49.

Kirk, G. S., J. E. Raven, and M. Schofield. 1983. *The Presocratic Philosophers.* 2d ed. Cambridge: Cambridge University Press.

Kuhn, T. S. 1978. *Black-Body Theory and the Quantum Discontinuity, 1894–1912.* Oxford: Clarendon Press.

———. 1984. "Revisiting Planck." *Historical Studies in the Physical Sciences* 14, no. 2:231–52.

Land, E. H. 1977. "The Retinex Theory of Color Vision." *Scientific American* 237 (December): 108–28.

Langford, J. J. 1971. *Galileo, Science, and the Church.* Ann Arbor: University of Michigan Press.

Larmor, J. 1900. *Aether and Matter.* Cambridge: Cambridge University Press.

Lashower, E., and I. Tishby, eds. 1989. 5 vols. *The Wisdom of the Zohar.* Trans. D. Goldstein. Oxford: Oxford University Press.

Layard, A. H. 1853. *Discoveries in the Ruins of Nineveh and Babylon.* New York: Putnam.

Lebedew, P. 1901. "Untersuchungen Über die Druckkräfte des Lichtes." *Annalen der Physik* 6:422–58.

Le Goff, J. 1992. *Intellectuals in the Middle Ages.* Trans. T. L. Fagan. Cambridge, Mass.: Blackwell.

Lejeune, A. 1948. *Euclide et Ptolemée. Deux stades de l'optique géométrique grecque.* Université de Louvain, Receuil de travaux d'histoire et de philologie, series 3, nos. 31–, Louvain.

Leonardo da Vinci. 1938. *Notebooks of Leonardo da Vinci.* 2 vols. Ed. and trans. E. MacCurdy. New York: Reynal and Hitchcock.

Lindberg, D. 1967. "Alhazen's Theory of Vision and Its Reception in the West."

Isis 62:469–89. In D. Lindberg, *Studies in the History of Medieval Optics.* London: Variorum Reprints.

———. 1971. "Lines of Influence in Thirteenth-Century Optics." *Speculum* 46:66–83. In D. Lindberg, *Studies in the History of Medieval Optics.* London: Variorum Reprints.

———. 1976. *Theories of Vision from al-Kindi to Kepler.* Chicago: University of Chicago Press.

———. 1978. "Medieval Latin Theories of the Speed of Light." In R. Taton, ed. (preface), *Roemer et la vitesse de la lumière.* Paris: J. Vrin. In D. Lindberg, *Studies in the History of Medieval Optics.* London: Variorum Reprints.

———. 1983a. *Studies in the History of Medieval Optics.* London: Variorum Reprints.

———. 1983b. *Roger Bacon's Philosophy of Nature.* Oxford: Oxford University Press.

———. c. 1984. "On Laying the Foundations of Geometrical Optics: Maurolico, Kepler, and the Medieval Tradition." Los Angeles: William Andrews Clark Memorial Library.

———. 1988. "The Genesis of Kepler's Theory of Light: Light Metaphysics from Plotinus to Kepler." *Osiris,* 2d ser., 2:5–42.

Livingston, D. M. 1973. *The Master of Light.* New York: Scribner.

Lloyd, G.E.R. 1973. *Greek Science after Aristotle.* New York: Norton.

———. 1979. *Magic Reason and Experience.* Cambridge: Cambridge University Press.

Lodge, O. 1889. *Modern Views of Electricity.* London: Macmillan.

———. 1909. *The Ether of Space.* New York: Harper.

———. 1925. *Ether and Reality.* New York: Doran.

Lorentz, H. A., A. Einstein, H. Minkowski, and H. Weyl. 1923. *The Principle of Relativity.* Ed. A. Sommerfeld. Trans. W. Perrett and G. H. Jeffery. London: Methuen. Reprint, New York: Dover, 1952.

Lubbock, C. A. 1933. *The Herschel Chronicle.* New York: Macmillan.

Lucan. 1988. *Lucan's Civil War.* Trans. P. F. Widdows. Bloomington: Indiana University Press. Prose version: *The Pharsalia.* Trans. H. T. Riley. London: Bell.

Lucretius. 1968. *The Way Things Are.* Trans. R. Humphries. Bloomington: Indiana University Press.

Magie, W. F. 1935. *A Source Book in Physics.* New York: McGraw-Hill.

Maiman, T. H. 1960. "Stimulated Optical Radiation in Ruby." *Nature* 187:493–94.

Manilius, M. 1977. *Astronomica.* Ed. and trans. G. P. Goold. Cambridge, Mass.: Loeb Classical Library, Harvard University Press.

Mariotte, E. 1681. *De la nature des couleurs.* Paris: Michallet.

Marks, W. B., W. H. Dobelle, and E. F. MacNichol. 1964. "Visual Pigments of Single Primate Cones." *Science* 143:1181–83.

Maxwell, J. C. 1873. *A Treatise on Electricity and Magnetism.* 2 vols. Oxford: Clarendon Press.

———. 1890. *Scientific Papers.* 2 vols. Ed. W. D. Niven. Cambridge: Cambridge University Press.

McCann, G. W. 1994. "Sir John Herschel and the Birth of Celestial Photography." *Griffith Observer* 58, no. 9:2–11.

McEvoy, J. J. 1982. *The Philosophy of Robert Grosseteste.* New York: Oxford University Press.

Mehra, J., and H. Rechenberg. 1982–. *The Historical Development of Quantum Theory*. New York: Springer-Verlag.

Melvil, Thomas. 1752. *Edinburgh Physical and Literary Essays* 2:34.

Meyer-Arendt, J. R. 1984. *Introduction to Classical and Modern Optics*. New York: Prentice-Hall.

Michelson, A. A. 1881. "The Relative Motion of the Earth and the Luminiferous Ether." *American Journal of Science*, n.s. 3, 22:120–29. Reprinted in L. S. Swenson, Jr., *The Ethereal Aether*. Austin: University of Texas Press, 1972.

———, and E. W. Morley. 1887. "On the Relative Motion of the Earth and the Luminiferous Ether." *American Journal of Science*, n.s. 3, 34:333–45. Also published in *Philosophical Magazine* 24 (1887): 449–63; and reproduced in L. S. Swenson, Jr., *The Ethereal Aether*. Austin: University of Texas Press, 1972.

Middleton, W. E. Knowles. 1961. "Archimedes, Kircher, Buffon and the Burning Mirrors." *Isis* 52:533–43.

Migne, J. P. 1857. *Patrologia Graeca*. Vol. 44. Paris: Migne.

Miller, A. L. 1981. *Albert Einstein's Special Theory of Relativity*. Reading, Mass.: Addison-Wesley.

Millikan, R. A. 1914. "Direct Determination of 'h.'" *Physical Review* 4:73–75.

Morgan, G. C. 1785. "Observations and Experiments on the Light of Bodies in a State of Combustion." *Philosophical Transactions of the Royal Society* 75:665–80.

Needham, J. 1954–. *Science and Civilisation in China*. Cambridge: Cambridge University Press.

Neugebauer, O. 1957. *The Exact Sciences in Antiquity*. 2d ed. Providence, R.I.: Brown University Press. Reprint, New York: Harper, 1962.

———. 1975. *A History of Ancient Mathematical Astronomy*. 3 vols. New York: Springer-Verlag.

Newton, I. 1687. *Principia Mathematica*. London: Streater. Reprint, Glasgow: Maclehose, 1871. 4th ed., ed. and trans. F. Cajori, Berkeley and Los Angeles: University of California Press, 1934. Contains *The System of the World*.

———. 1704. *Opticks*. London: Smith and Walford. Reprint, Brussels: Culture et civilisation, 1966.

———. 1728a. *Optical Lectures*. London: Fayram. Also in *The Optical Papers of Isaac Newton*, ed. and trans. A. E. Shapiro. Cambridge: Cambridge University Press, 1984.

———. 1728b. *The Chronology of Ancient Kingdoms Amended*. London: Tonson.

———. 1728c. *Treatise of the System of the World*. Trans. A. Motte. London: Fayram.

———. 1733. *Observations upon the Prophesies of Daniel and the Apocalypse of St. John*. Dublin: Powell. Reprint, London: Nisbet, 1831.

———. 1730. *Opticks*. 4th ed. London: Innys. Reprint, New York: Dover, 1952.

———. 1958. *Isaac Newton's Papers and Letters on Natural Philosophy*. Ed. I. B. Cohen. Cambridge: Harvard University Press.

———. 1959–77. *The Correspondence of Isaac Newton*. 7 vols. Ed. H. W. Turnbull et al. Cambridge: Cambridge University Press.

———. 1984. *The Optical Papers of Isaac Newton*. Ed. and trans. A. E. Shapiro. Cambridge: Cambridge University Press.

Nicholson, M. 1946. *Newton Demands the Muse: Newton's Opticks and the Eighteenth-Century Poets*. Princeton: Princeton University Press.

Notopoulos, J. A. 1944. "The Symbolism of the Sun and Light in the *Republic* of

Plato." *Classical Philology* 39:163–72; 223–40.

Olmsted, J. W. 1942–43. "The Scientific Expedition of Jean Richer to Cayenne" *Isis* 34:117–28.

Omar, S. B. 1977. *Ibn al-Haytham's Optics*. Minneapolis: Bibliotheca Islamica.

Omnès, R. 1994. *The Interpretation of Quantum Mechanics*. Princeton: Princeton University Press.

Pais, A. 1982. *"Subtle Is the Lord . . . ."* New York: Oxford University Press.

———. 1991. *Niels Bohr's Times*. Oxford: Oxford University Press.

Park, D. 1988. *The How and the Why*. Princeton: Princeton University Press.

———. 1992. *Introduction to the Quantum Theory*. 3d ed. New York: McGraw-Hill.

Pico della Mirandola, G. 1977. *Heptaplus*. Trans. J. B. McGaw. New York: Philosophical Library.

Planck, M. 1900. "Zur Theorie des Gesetzes der Energieverteilung im Normalspektrum." *Verhandlungen der Deutschen Physikalischen Gesellschaft* 2:237–47. Reprinted and trans. in H. Kangro, *Planck's Original Papers in Quantum Physics*, trans. D. ter Haar and S. G. Brush. London: Taylor and Francis, 1972.

———. 1913. *The Theory of Heat Radiation*. 2d. ed. Trans. M. Masius. Philadelphia: Blakiston. Reprint, New York: Dover, 1959.

———. 1949. *Scientific Autobiography*. Trans. F. Gaynor. New York: Philosophical Library.

Plato. 1961. *Collected Dialogues*. Ed. E. Hamilton and H. Cairns. Princeton: Princeton University Press.

———. 1962. *Timaeus a Calcidio translatus*. Ed. J. H. Waszink. Leiden: Brill.

Pliny. 1938. *Natural History*. 10 vols. Trans. H. Rackham et al. Cambridge, Mass.: Harvard University Press.

Plotinus. 1966–88. *Enneads*. 7 vols. Trans. A. H. Armstrong. Cambridge, Mass.: Loeb Classical Library, Harvard University Press.

Plutarch. 1927–76. *Moralia*. 16 vols. Several trans. Cambridge, Mass.: Loeb Classical Library, Harvard University Press.

Poincaré, H. 1898. "La mesure du temps." *Revue de métaphysique et de morale* 6:1–13.

———. [1904]. *La théorie de Maxwell et les oscillations Hertziennes*. Paris: Carré et Naud.

———. 1905. *Science and Hypothesis*. Trans. G. B. Halsted. London: Scott. Reprint, New York: Dover, 1952.

———. 1907. *The Value of Science*. Trans. G. B. Halsted. New York: Science Press. Reprint, New York: Dover, 1958.

Polkinghorne, J. 1984. *The Quantum World*. Princeton: Princeton University Press.

Porta, G. B. della. 1558. *Magiae naturalis libri IV*. Naples: Cancer.

———. 1589. *Magiae naturalis libri XX*. Naples: Salviani.

———. 1658. *Natural Magick*. London: Wright. Reprint, New York: Basic Books, 1957.

———. 1593. *De refractione optices parte libri novem*. Naples: Salviani.

———. 1962. *De telescopio*. Florence: Olschki.

Porter, P. B. 1954. "Another Puzzle-Picture." *American Journal of Psychology* 67:550–51.

Priestley, J. 1772. *History and Present State of Discoveries Relating to Vision, Light, and Colours.* London: Johnson.

Pseudo-Dionysius. 1987. *The Complete Works.* Trans. C. Luibheid. New York: Paulist Press.

Ptolemy, C. 1898–1954. *Claudii Ptolemei Opera quae existant omnia.* 5 vols. Ed. J. L. Heiberg et al. Leipzig: Teubner.

———. 1940. *Tetrabiblos.* Ed. and trans. F. E. Robbins. Cambridge, Mass.: Loeb Classical Library, Harvard University Press.

———. 1984. *Ptolemy's Almagest.* Trans. G. J. Toomer. New York: Springer-Verlag.

———. 1989. *L'Optique de Claude Ptolemée dans la version latine d'après l'arabe de l'émir Eugène de Sicilie.* Ed. and trans. A. Lejeune. Leiden: Brill.

Rashed, R. 1968. "Le 'Discours de la lumière' d'Ibn al-Haytham (Alhazen)." *Revue d'histoire des sciences et de leurs applications* 21:197–224.

———. 1970. "Optique géometrique et doctrine optique chez Ibn Al-Haytham." *Archive for the History of the Exact Sciences* 6:271–98.

———. 1990. "A Pioneer in Anaclastics: Ibn Sahl on Burning Mirrors and Lenses." *Isis* 81:464–91.

———. 1993. *Géométrie et dioptrique au Xe siècle.* Paris: Les Belles Lettres.

Reisch, Gregor. 1503. *Margarita Philosophica.* Strasbourg and Fribourg: Schott. Reprint, Düsseldorf: Stern, 1973.

Reston, James, Jr. 1994. *Galileo.* New York: Harper-Collins.

Ringwood, S. D. 1994. "A Galilean Telescope." *Quarterly Journal of the Royal Astronomical Society* 35:43–50.

Richter, G.M.A. [1970?]. *Perspective in Greek and Roman Art.* London: Phaidon.

Richter, J. P., ed. 1883. *The Literary Works of Leonardo da Vinci.* 2 vols. London: Sampson Low, Marston, Searle, and Rivington. Reprinted as *The Notebooks of Leonardo da Vinci.* New York: Dover, 1970.

Robbins, F. 1912. *The Hexaëmeral Literature: A Study of the Greek and Latin Commentaries on Genesis.* Chicago: University of Chicago Press.

Roemer, O. 1676. "Démonstration touchant le mouvement de la lumière." *Journal des sçavans* (1676): 233–36. Trans. in *Philosophical Transactions of the Royal Society of London* 12 (1677): 893–94.

Rogers, H. J., ed. 1905–7. *Congress of Arts and Science.* 8 vols. Boston: Houghton, Mifflin.

Ronchi, V. 1958. *Il cannocchiale di Galileo e la scienza del Seicento.* Torino: Einaudi.

———. 1970. *The Nature of Light.* Trans. V. Barocas. London: Heinemann.

Röntgen, W. C. 1898. "Über eine neue Art von Strahlen." *Annalen der Physik und Chemie* 64:18–37.

Rosen, E. 1965. *Kepler's Conversation with Galileo's Sidereal Messenger.* New York: Johnson.

Runciman, S. 1947. *The Medieval Manichee.* Cambridge: Cambridge University Press.

Rutherford, E. 1911. "The Scattering of α- and β-Particles and the Structure of Atoms." *Philosophical Magazine* 6:238–54.

Sabra, A. I. 1967. *Theories of Light from Descartes to Newton.* London: Oldbourne.

Sacks, O. 1995. *An Anthropologist on Mars.* New York: Knopf.

Sambursky, S. 1959. *The Physical World of the Greeks.* London: Routledge and Kegan Paul.

———. 1958. "Philoponus' Interpretation of Aristotle's Theory of Light." *Osiris* 13:114–26.

Schaff, P., and H. Wace, eds. 1895. *A Select Library of Nicene and Post-Nicene Fathers of the Christian Church.* 2d ser. 14 vols. New York: Christian Literature Co.

Scheiner, C. 1630. *Rosa ursina, sive sol.* Bracciano: Fei.

Schmitt, C. B., ed. 1988. *The Cambridge History of Renaissance Philosophy.* Cambridge: Cambridge University Press.

Schroeder, D. J. 1986. *Astronomical Optics.* New York: Academic Press.

Scott, J. F. [1952]. *The Scientific Work of René Descartes.* London: Taylor and Francis.

See, T.J.J. 1926. "Historical Researches Indicating a Change in the Color of Sirius, between the Epochs of Ptolemy, 138, and AlSûfi, 980, A.D." *Astronomische Nachrichten* 229:246–71.

Seeger, R. J. 1966. *Galileo Galilei, His Life and His Works.* Oxford: Pergamon.

Seneca, L. A. 1971. *Naturales Questiones.* 2 vols. Trans. T. H. Corcoran. London: Loeb Classical Library, Heinemann.

Sepper, D. L. 1988. *Goethe contra Newton.* Cambridge: Cambridge University Press.

———. 1994. *Newton's Optical Writings.* New Brunswick, N.J.: Rutgers University Press.

Shapiro, H. 1964. *Medieval Philosophy.* New York: Modern Library, Random House.

Sharp, D. E. 1930. *Franciscan Philosophy at Oxford.* Oxford: Oxford University Press.

Shea, W. R. 1991. *The Magic of Numbers and Motion.* Canton, Mass.: Science History Publications.

Shirley, J. W. 1951. "An Early Experimental Determination of Snell's Law." *American Journal of Physics* 19:507–8.

Siegel, D. M. 1991. *Innovation in Maxwell's Electromagnetic Theory.* New York: Cambridge University Press.

Siegel, R. E. 1970. *Galen on Sense Perception.* Basel: Karger.

Simson, O. von. 1962. *The Gothic Cathedral.* Princeton: Princeton University Press.

Smith, A. M. 1987. "Descartes's Theory of Light and Refraction." *Transactions of the American Philosophical Society* 77, part 3.

Smith, C. P. 1956. *James Wilson.* Chapel Hill: University of North Carolina Press.

Sorabji, R., ed. 1987. *Philoponus and the Rejection of Aristotelian Science.* Ithaca: Cornell University Press.

Squires, E. 1986. *The Mystery of the Quantum World.* Bristol: Hilger.

Steffens, H. J. 1977. *The Development of Newtonian Optics in England.* New York: Science History Publications.

Steneck, N. H. 1976. *Science and Creation in the Middle Ages.* Notre Dame, Ind.: University of Notre Dame Press.

Stephenson, B. 1987. *Kepler's Physical Astronomy.* Princeton: Princeton University Press.

Strodach, G. K. 1963. *The Philosophy of Epicurus.* Evanston, Ill.: Northwestern University Press.

Suger, Abbé. 1979. *On the Abbey Church of St. Denis and Its Art Treasures*. Ed. and trans. W. Panofsky. Princeton: Princeton University Press.

Swenson, L. S., Jr. 1972. *The Ethereal Aether*. Austin: University of Texas Press.

Tacitus, P. C. 1925–37. *The Histories*. 4 vols. Trans. C. H. Moore. London: Loeb Classical Library, Heinemann.

Tanizaki, J. 1955. "In Praise of Shadows." Trans. E. Seidensticker. *Atlantic Monthly* 195 (January): 141–44.

Tester, S. J. 1987. *A History of Western Astrology*. Woodbridge, Suffolk: Boydell Press.

Thackray, A. 1970. *Atoms and Powers*. Cambridge, Mass.: Harvard University Press.

Theodoric of Freiberg. 1914. *De iride et radialibus impressionibus*. Published as Dietrich von Freiberg, "Über den Regenbogen und die durch Strahlen erzeugenen Eindrücke," ed. J. Würschmidt, in *Beiträge zur Geschichte der Philosophie des Mittelalters* 12 (1914), nos. 5 and 6.

Thompson, E. F. 1907. *Fitzgerald's Rubayat of Omar Khayyam with a Persian Text, a Transliteration, and a Close Prose and Verse Translation*. Privately printed.

Thompson, S. P. 1894. *Elementary Lessons in Electricity and Magnetism*. London: Macmillan. 2d ed., 1905.

Thomson, W. (Lord Kelvin). 1891. *Popular Lecturers and Addresses*. 2nd ed. 3 vols. London: Macmillan.

Thorndike, L. 1944. *University Records and Life in the Middle Ages*. New York: Columbia University Press.

"Veritas." 1790. *Gentleman's Magazine* 60:890–91.

Vitruvius, M. P. 1914. *The Ten Books on Architecture*. Trans. M. H. Morgan. Cambridge, Mass.: Harvard University Press. Reprint, New York: Dover, 1960.

Voltaire. 1738. *Elémens de la philosophie de Neuton*. Amsterdam: Ledet.

Waddington, C. P. 1855. *Ramus*. Paris: Meyrueis. Reprint, Geneva: Slatkine, 1969.

Wallace, W. A. 1959. *The Scientific Methodology of Theodoric of Freiberg*. Fribourg: University Press.

———. 1974. *Galileo's Early Notebooks: The Physical Questions*. Notre Dame, Ind.: University of Notre Dame Press.

Wasserman, G. S. 1978. *Color Vision: An Historical Introduction*. New York: Wiley.

Wedel, T. O. 1920. *The Medieval Attitude toward Astrology*. New Haven: Yale University Press.

Weiner, C., ed. 1977. *History of Twentieth-Century Physics*. New York: Academic Press, New York.

Weisheipl, J. A. 1980. *Albertus Magnus and the Sciences*. Toronto: Pontifical Institute of Medieval Studies.

Westfall, R. S. 1980. *Never at Rest: A Biography of Isaac Newton*. Cambridge: Cambridge University Press.

Whewell, W. 1857. *History of the Inductive Sciences*. 3d ed. 2 vols. London: Parker.

Whittaker, E. T. 1953. *A History of the Theories of Aether and Electricity*. 3d ed. 2 vols. London: Nelson.

———. 1955. *Biographical Memoirs of Fellows of the Royal Society* 1: 37–67.

Wild, W. J., and R. Q. Fugate. 1994. "Untwinkling the Stars, Part II." *Sky and Telescope* 87 (June): 20–27.

Willis, Thomas. 1664. *Anatomy of Brain and Nerves.* London: Martyn and Allefry. Reprint, Birmingham, Ala.: Classics of Medicine Library, 1978.

Wissowa, G., et al., eds. 1894–1972. *Paulys Real-Encyclopädie der classischen Altertumswissenschaft.* Stuttgart: Metzler.

Witelo. 1572. *Perspectiva.* In *Opticae thesaurus Alhazeni libri septem,* ed. F. Risner. Basel. Reprinted with intro. by D. C. Lindberg, New York: Jonson, 1972.

―――. 1977. *Perspectivae liber primus.* Ed. and trans. S. Unguru. Wrocław: Ossolineum.

―――. 1983. *Perspectivae liber quintus.* Ed. and trans. A. M. Smith. Wrocław: Ossolineum.

Wollaston, W. H. 1802. "A Method of Examining Refractive and Dispersive Powers by Prismatic Reflection." *Philosophical Transactions of the Royal Society* 20:365–80.

Young, T. 1802. "On the Theory of Light and Colours." *Philosophical Transactions of the Royal Society* 20:12–48.

―――. 1807. *A Course of Lectures on Natural Philosophy and the Mechanical Arts.* 2 vols. London: Johnson. Reprint, New York: Johnson, 1971.

―――. 1855. *Miscellaneous Works of the Late Thomas Young.* 3 vols. Ed. G. Peacock and J. Leitch. London: Murray. Reprint, New York: Johnson, 1972.

Zeller, E. 1931. *Outlines of the History of Greek Philosophy.* 13th ed. London: Routledge and Kegan Paul. Reprint, New York: Dover, 1980.

Zenhausern, F., Y. Martin, and H. K. Wickramasinghe. 1995. "Scanning Interferometric Apertureless Microscopy." *Science* 269:1083–85.

Zihl, J., D. von Cramon, and N. Mai. 1983. "Selective Disturbance of Movement Vision after Bilateral Brain Damage." *Brain* 106:313–40.

INDEX